VOLUME SIX HUNDRED AND FORTY SEVEN

METHODS IN ENZYMOLOGY

Linkers in Biomacromolecules

METHODS IN ENZYMOLOGY

Editors-in-Chief

ANNA MARIE PYLE
Departments of Molecular, Cellular and Developmental Biology and Department of Chemistry
Investigator, Howard Hughes Medical Institute
Yale University

DAVID W. CHRISTIANSON
Roy and Diana Vagelos Laboratories
Department of Chemistry
University of Pennsylvania
Philadelphia, PA

Founding Editors

SIDNEY P. COLOWICK and NATHAN O. KAPLAN

VOLUME SIX HUNDRED AND FORTY SEVEN

METHODS IN ENZYMOLOGY

Linkers in Biomacromolecules

Edited by

MAARTEN MERKX
Laboratory of Chemical Biology and Institute for Complex Molecular Systems, Eindhoven University of Technology, Eindhoven, The Netherlands

ELSEVIER

ACADEMIC PRESS
An imprint of Elsevier

Academic Press is an imprint of Elsevier
50 Hampshire Street, 5th Floor, Cambridge, MA 02139, United States
525 B Street, Suite 1650, San Diego, CA 92101, United States
The Boulevard, Langford Lane, Kidlington, Oxford OX5 1GB, United Kingdom
125 London Wall, London, EC2Y 5AS, United Kingdom

First edition 2021

ISBN: 978-0-12-820818-2
ISSN: 0076-6879

For information on all Academic Press publications
visit our website at https://www.elsevier.com/books-and-journals

Publisher: Zoe Kruze
Editorial Project Manager: Shellie Bryant
Production Project Manager: James Selvam
Cover Designer: Miles Hitchen

Typeset by SPi Global, India

Contents

Contributors

Friso Aalbers
Molecular Enzymology Group, University of Groningen, Groningen, The Netherlands; Department of Biology and Biotechnology "L. Spallanzani", University of Pavia, Pavia, Italy

Ryoichi Arai
Department of Biomolecular Innovation, Institute for Biomedical Sciences; Department of Applied Biology, Faculty of Textile Science and Technology, Shinshu University, Ueda, Nagano, Japan

Arnold J. Boersma
DWI-Leibniz Institute for Interactive Materials; Institute of Technical and Macromolecular Chemistry, RWTH Aachen University, Aachen, Germany

Eric Ceballos-Alcantarilla
Laboratory of Chemical Biology and Institute for Complex Molecular Systems, Eindhoven University of Technology, Eindhoven, The Netherlands

Lucia Beatriz Chemes
Instituto de Investigaciones Biotecnológicas "Dr. Rodolfo A. Ugalde", IIB-UNSAM, IIBIO-CONICET, Universidad Nacional de San Martín, San Martín, Argentina

Marco W. Fraaije
Molecular Enzymology Group, University of Groningen, Groningen, The Netherlands

Juliana Glavina
Instituto de Investigaciones Biotecnológicas "Dr. Rodolfo A. Ugalde", IIB-UNSAM, IIBIO-CONICET, Universidad Nacional de San Martín, San Martín, Argentina

Alejandro Gran-Scheuch
Molecular Enzymology Group, University of Groningen, Groningen, The Netherlands; Department of Chemical and Bioprocesses Engineering, Pontificia Universidad Católica de Chile, Santiago, Chile

Alexander Gräwe
Fachbereich Biologie; Centre for Synthetic Biology, Technische Universität Darmstadt, Darmstadt, Germany

Tejas M. Gupte
Department of Genetics, Cell Biology, and Development, College of Biological Sciences, University of Minnesota, Minneapolis, MN, United States

Ziliang Huang
Department of Bioengineering, University of California San Diego, La Jolla, CA, United States; Department of Chemical Engineering, Institute of Biochemical Engineering; Key Laboratory for Industrial Biocatalysis, Ministry of Education, Beijing, China

Robert M. Hughes
Department of Chemistry, East Carolina University, Greenville, NC, United States

Magnus Kjaergaard
Department of Molecular Biology and Genetics, Aarhus University; The Danish Research Institute for Translational Neuroscience (DANDRITE); Center for Proteins in Memory (PROMEMO), Aarhus, Denmark

Maarten Merkx
Laboratory of Chemical Biology and Institute for Complex Molecular Systems, Eindhoven University of Technology, Eindhoven, The Netherlands

Collin T. O'Bryant
Department of Chemistry, East Carolina University, Greenville, NC, United States

Loreto Parra
Institute for Biological and Medical Engineering, Pontificia Universidad Católica de Chile, Santiago, Chile

Theodoros Pittas
DWI-Leibniz Institute for Interactive Materials; Institute of Technical and Macromolecular Chemistry, RWTH Aachen University, Aachen, Germany

Jan Ranglack
Fachbereich Biologie; Centre for Synthetic Biology, Technische Universität Darmstadt, Darmstadt, Germany

Michael Ritt
Department of Genetics, Cell Biology, and Development, College of Biological Sciences, University of Minnesota, Minneapolis, MN, United States

Sivaraj Sivaramakrishnan
Department of Genetics, Cell Biology, and Development, College of Biological Sciences, University of Minnesota, Minneapolis, MN, United States

Viktor Stein
Fachbereich Biologie; Centre for Synthetic Biology, Technische Universität Darmstadt, Darmstadt, Germany

Fei Sun
Department of Chemical and Biological Engineering, The Hong Kong University of Science and Technology, Hong Kong, SAR, People's Republic of China

Chiao-Yu Tseng
Institute of Physics, Academia Sinica, Taipei, Taiwan

Ri Wang
Department of Chemical and Biological Engineering, The Hong Kong University of Science and Technology, Hong Kong, SAR, People's Republic of China

Yong Wang
Department of Physics, Cell and Molecular Biology Program, Materials Science and Engineering Program, University of Arkansas, Fayetteville, AR, United States

Anastasia Weyrich
Fachbereich Biologie, Technische Universität Darmstadt, Darmstadt, Germany

Yannick Woudstra
Molecular Enzymology Group, University of Groningen, Groningen, The Netherlands

Anna I. Wurz
Department of Chemistry, East Carolina University, Greenville, NC, United States

Xin-Hui Xing
Department of Chemical Engineering, Institute of Biochemical Engineering; Key Laboratory for Industrial Biocatalysis, Ministry of Education; Center for Synthetic and Systems Biology, Tsinghua University, Beijing; Institute of Biopharmaceutical and Health Engineering, Tsinghua Shenzhen International Graduate School; Institute of Biomedical Health Technology and Engineering, Shenzhen Bay Laboratory, Shenzhen, China

Chong Zhang
Department of Chemical Engineering, Institute of Biochemical Engineering; Key Laboratory for Industrial Biocatalysis, Ministry of Education; Center for Synthetic and Systems Biology, Tsinghua University, Beijing, China

Giovanni Zocchi
Department of Physics and Astronomy, University of California Los Angeles, Los Angeles, CA, United States

Weiyan Zuo
DWI-Leibniz Institute for Interactive Materials; Institute of Technical and Macromolecular Chemistry, RWTH Aachen University, Aachen, Germany

Preface

I think it is fair to say that linkers have thus far not received the attention that they deserve. At the same time, there is a strong desire among biochemists and protein engineers to better understand the effect of linkers on protein properties and a clear need for guidance in linker design. The first time I realized this was when we published a study that showed that the energy transfer efficiency between donor and acceptor fluorescent domains connected by a flexible SerGlyGly linker can be quantitatively explained using random coil models and experimentally derived the parameter that described the stiffness of these popular linkers. We had actually started this study because we were puzzled by the behavior of the Zn^{2+}-binding FRET sensor protein CALWY, which showed a decrease in FRET upon binding Zn^{2+}, even though Zn^{2+} binding promoted intramolecular complex formation. In our naivety we had assumed that because the sensor contained a long, 60 amino-acid linker between the two metal-binding domains, the distance would decrease upon binding Zn^{2+}. What we had not realized is that the random coil structures adopted by flexible linkers actually represent a distribution of conformations in which the average distance between the ends of linkers was much shorter than the total linker length, and also shorter than the distance of the linker in the Zn^{2+}-bound state.

Taking a step back and trying to understand the conformational properties of SerGlyGly linkers turned out to be very fruitful and provided insights that have helped our designs of protein switches and sensors ever since. Apparently, we had not been the only ones wishing to get a better grasp of understanding and tuning of linker properties. The original Biochemistry paper is one of our best cited papers and a more recent follow-up study that determined the influence of Gly/Ser content on linker flexibility has been one of the most downloaded Biochemistry papers in recent years. The reason to share this personal perspective is not to highlight these papers, but because I think it illustrates that there is a widely felt need to understand and tune linker properties better in protein engineering. I am therefore pleased to present to you this volume of *Methods in Enzymology* dedicated to the role of linkers in biomacromolecules. My aim was to bring together in this volume expert reviews focusing on different aspects of linkers, varying from thermodynamic models and fundamental studies on linker conformational behavior to experimental methods that allow rapid linker optimization, and reviews that focus on specific classes of linkers and discuss the role of linkers in specific application areas.

Chapters 1 and 6 describe the use of thermodynamic models to quantitatively describe the effect of linker lengths and linker stiffness on intramolecular interactions and multivalent binding. In addition, the authors of Chapter 6 also developed a very useful tool for linker design that can be used to calculate effective concentrations as a function of linker length, linker stiffness, and the end-to-end distance. Whereas certain properties of multidomain proteins can be effectively understood using these models, not all linker behavior can be easily predicted, in particular for smaller linkers or for linkers involved in subtle allosteric conformational changes. In these cases, methods that allow efficient screening of many different linkers are very important. ProtaTETHER and iFLinkC, two methods that were developed to allow the construction of libraries with a diversity of linker lengths and flexibilities, are described in Chapters 4 and 9, respectively. In Chapter 2, Xing and coworkers report an alternative approach to construct linkers with different ratios of flexible (GGGGS) and more rigid motifs (EAAAK). Three chapters focus on specific linkers motifs. Chapter 1 describes the properties of linkers consisting of serine and glycine residues. In Chapter 7, Sivaramakrishnan and coworkers describe the structure, stability, and function of ER/K α-helical linkers and their application as linkers in native proteins. Another peptide-based linker with the α-helix-forming properties, the EAAAK motif is described by Arai in Chapter 8.

Several chapters discuss the role of linkers for specific application areas. Boersma and coworkers (Chapter 3) detail how the conformational behavior of flexible linkers can be used to develop fluorescent sensors to monitor intracellular crowding, both in vitro and in situ. In Chapter 5, Fraaije and coworkers describe their work on a fusion protein of alcohol dehydrogenase and cyclohexanone monooxygenase as a case study for linker engineering in fusions of enzymes with complementary cofactor dependency. The tunability of DNA-based linkers and the large difference in persistence length between single-strand and double-strand DNA has been harnessed to allow mechanical control of enzyme activity in DNA–enzyme hybrids. The synthesis and characterization of these systems are explained by Zocchi and coworkers and can be found in Chapter 10. Finally, Sun and coworkers introduce the use of the SpyTag/SpyCatcher system to synthesize branched multidomain structures that go beyond the linear architecture found in genetically encoded proteins (Chapter 11).

I would like to express my appreciation for the support by the series editors to dedicate a volume of *Methods of Enzymology* to the role of linkers in biomolecular assemblies. Most importantly, I would like to thank all the authors for their hard work. I hope you will enjoy reading their contributions as much as I have enjoyed putting together this volume of *Methods of Enzymology*.

Maarten Merkx
Laboratory of Chemical Biology
and Institute for Complex Molecular Systems
Eindhoven University of Technology
Eindhoven, The Netherlands

CHAPTER ONE

Understanding and applications of Ser/Gly linkers in protein engineering

Eric Ceballos-Alcantarilla and Maarten Merkx*
Laboratory of Chemical Biology and Institute for Complex Molecular Systems, Eindhoven University of Technology, Eindhoven, The Netherlands
*Corresponding author: e-mail address: m.merkx@tue.nl

Contents

Abstract

Peptide linkers consisting of repeats of glycine and serine residues are commonly chosen by protein engineers to introduce flexible and hydrophilic spacers between protein domains. Given the popularity of these linkers, gaining a quantitative insight in their conformational behavior is important to understand the effect on functional properties of fusion proteins, including energy transfer efficiency in luminescent sensor proteins, intramolecular domain interactions and (multivalent) binding. In this chapter, we discuss how the conformational behavior of Ser/Gly linkers can be described using random coil models, and how measuring FRET as a function of linker length can be used to obtain empirical values for the stiffness of linkers containing different Ser-to-Gly ratios. Subsequently, we show how these models and the experimentally determined linker stiffness can be used to explain and predict the functional properties of multidomain proteins, providing useful rules-of-thumb and design tools for optimal linker engineering.

Methods in Enzymology, Volume 647
ISSN 0076-6879
https://doi.org/10.1016/bs.mie.2020.12.001

1. Introduction

The generation of fusion proteins consisting of multiple protein domains is an attractive and modular approach for engineering proteins with new functions. Often little consideration is given to the optimal choice of linker length and sequence, however. While in some applications the linker between the domains merely separates two protein domains and allows their independent folding, in many cases linker properties directly affect the functional properties of the fusion protein. Well-known examples include genetically encoded fluorescent or bioluminescent sensor proteins based on energy transfer (FRET and BRET, respectively). Since the performance of these sensors is determined by the change in distance between the donor and acceptor domains in the free and ligand-bound states, it is important to understand the effect of linker length and linker flexibility on the conformational behavior of these fusion proteins (van Dongen et al., 2007). In addition, fusing two domains that can interact with each other through a linker affects the probability of their interaction, so that each domain "perceives" a local effective concentration C_{eff} of the other (Fig. 1). This means that the kinetic and thermodynamic properties of the interaction change according to the properties of the linker, which are reflected in the C_{eff} value.

Repeats of glycine and serine residues represent the most commonly used type of linker to construct multidomain fusion proteins. These linkers

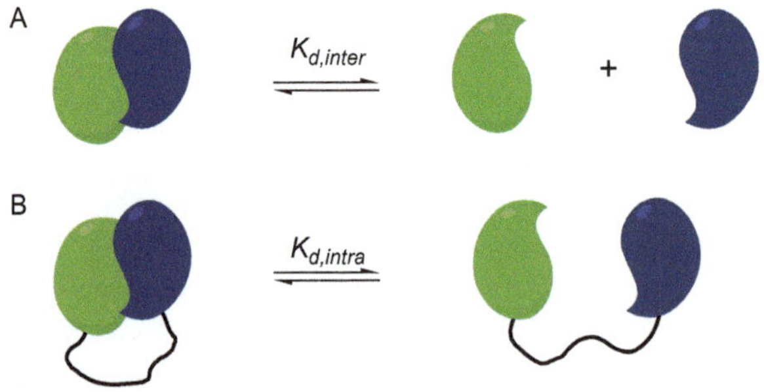

Fig. 1 (A) Two protein domains reversibility interact with each other to form a complex with an intermolecular dissociation constant $K_{d,inter}$. (B) The same interaction but with the domains fused through a flexible peptide linker, so an analogous but different intramolecular dissociation constant $K_{d,intra}$ applies. The relationship between these constants and the local effective concentration C_{eff} is given by $K_{d,inter} = K_{d,intra}C_{eff}$. C_{eff} is determined by linker length, linker stiffness and the distance that the linker needs to bridge in the complex form.

(hereinafter Ser/Gly linkers) are believed to be flexible and hydrophilic, and are assumed not to interfere with the main protein domains, such as hampering their folding or blocking their normal function. Ser/Gly linkers have been used for instance to investigate the enhancement of the expression of recombinant antibodies (Trinh, Gurbanaxi, Morrison, & Seyfzadeh, 2004), to develop enzyme variants with increased activity (Kim, Jung, Lee, Song, & Bae, 2015), to gain insights into mechanisms of enzymatic reactions (Salaun, Greaves, Tomkinson, & Chamberlain, 2020) and to afford various types of sensor proteins (Komatsu et al., 2011; Vinkenborg et al., 2009).

Our interest in quantitatively describing and understanding the conformational behavior of Ser/Gly linkers was triggered by a desire to better understand the effect of linker lengths on the sensor properties of a series of FRET-based Zn^{2+} sensor proteins (Evers, Appelhof, de Graaf-Heuvelmans, Meijer, & Merkx, 2007; Evers, van Dongen, Faesen, Meijer, & Merkx, 2006; van Dongen et al., 2007). To do so, we systematically studied energy transfer efficiency between donor and acceptor fluorescent domains as a function of linker length and linker constitution, and used these experimental data to quantitatively describe the conformational properties of various Ser/Gly linkers. The insights thus obtained not only allowed us to better understand the properties of our FRET sensor proteins, but also proved useful in the subsequent design of other proteins sensors and protein switches. In this chapter, we will first introduce the models to describe the conformational properties of flexible linkers and explain how these models can be used to quantitatively describe the amount of FRET in ECFP–linker–EYFP proteins as a function of linker length and Ser/Gly content. Fitting these models to the experimentally determined FRET values yields experimental values for the stiffness of the linkers, which can subsequently be used to calculate the effect of linker length and stiffness on the effective concentration for intramolecular domain interactions. Finally, we will discuss some examples from our own work to show how the properties of fusion proteins can be understood and tuned by modifying the linker length and/or stiffness.

2. Theoretical description of Ser/Gly linkers using random coil models

Since Ser/Gly linkers are linear copolymers of two types of monomers (serine and glycine) having moderate-to-high rotational freedom, they can adopt a large number of different conformations. Therefore, and like many other polymers, they can be seen as random coils and there is a certain probability for each of the possible conformations to be adopted. An accurate description of these kinds of systems requires the consideration of their key properties as stochastic variables, so that probability density and distribution functions can be deducted and then used as mathematical tools to better understand the linker behavior and/or predict other relevant properties. This reasoning was previously applied in polymer science, and has subsequently been adapted to describe polypeptide chains. The most important property that can be modeled as a stochastic variable is the distance between two given points of the random coil r, and more specifically the end-to-end distance r_e. For convenience, these systems are better described using spherical polar coordinates, so that one of the given points is placed in the pole and the considered distance is equal to the polar axis (Fig. 2).

Eq. (1) shows the relationship between the density and radial distribution functions for r_e using this reference system. In this equation, $p(r_e)$ is the probability density function for r_e, which gives the probability to find the ends of the coil at a distance r_e for a specified orientation (fixed values of θ and φ). On the other hand, $P(r_e)$ is the radial distribution function, which gives the probability to find the termini at the same distance r_e but regardless of the orientation (the density function is partially integrated over all possible values of θ and φ, which are equally likely because the function depends only on r_e). The same relationship would be applicable to any two other points of the chain.

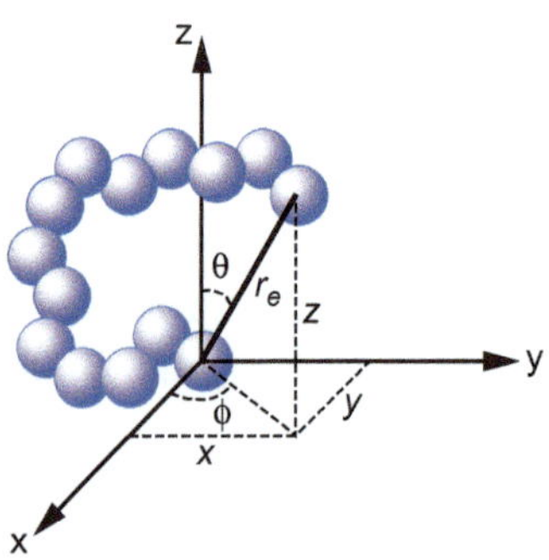

$x = r_e \sin\theta \cos\phi$

$y = r_e \sin\theta \sin\phi$

$z = r_e \cos\theta$

$dx\, dy\, dz = r_e^2 \sin\theta\, dr_e\, d\theta\, d\phi$

Fig. 2 Reference system based on spherical polar coordinates to describe the relative position of the ends of a random coil (shown as a joint of *blue circles* that represent monomers) when one of them is placed in the pole. The end-to-end distance is equal to r_e, whereas the spatial direction from the pole is given by the angles θ and φ. The relation between these polar coordinates and the respective Cartesian ones (x, y, z) are shown on the right.

$$P(r_e) = \int_0^{\pi} \int_0^{2\pi} p(r_e)\, r_e^2 \sin\theta d\phi d\theta = 4\pi r_e^2 p(r_e) \tag{1}$$

Considering fusion proteins in which two domains (or tethered ligands) connected by a Ser/Gly linker can intramolecularly interact with each other, random coil models can also be used to calculate the effective concentration C_{eff}. The density function value for a given value of r_e can be interpreted as the average number of linker molecules per volume unit (Å^3 if all the distances are expressed in Å) whose ends are separated by a distance r_e. Since the intramolecular interaction is produced at a fixed interaction distance r_{int}, which depends on the nature of each specific system, C_{eff} is proportional to the value of the density function when the two ends of the chain are separated by a distance r_{int}, $p(r_{int})$, as shown in Eq. (2). In this expression, N_A is the Avogadro number, whereas f is a dimensionless correction factor to account for excluded volume effects. For protein-ligand intramolecular binding, a value of 2 has been proposed for f by Lees and coworkers to consider the twofold increase in C_{eff} after loss of half hemisphere of access for the ligand because of the presence of the protein (Gargano, Ngo, Kim, Acheson, & Lees, 2001). However, here a value of 1 for f will be used to simplify analyses.

$$C_{eff} = f\frac{p(r_{int})}{N_A} = f\frac{P(r_{int})}{4\pi r_{int}^2 N_A} \tag{2}$$

Although several theoretical models inherited from polymer science have been proposed in the past years to quantitatively describe peptide linkers as random coils, two of them are most commonly used: the Gaussian chain model, and the wormlike chain model.

2.1 Gaussian chain model

The Gaussian chain (GC) model is the simplest way to describe a peptide linker as a random coil. It was first applied in this way by Flory and considers the linker as a chain of freely jointed alpha carbon atoms, in such a way that the distance between two given alpha carbon atoms is a stochastic variable which follows a normal Gaussian distribution (Flory, 1969). The radial distribution function $P_G(r_e)$ for the end-to-end distance r_e according to this model is expressed in Eq. (3). The function is normalized so that the total accumulated probability is 1.

$$P_G(r_e) = 4\pi r_e^2 \left(\frac{3}{2\pi\langle r_e^2\rangle_G}\right)^{3/2} e^{-\frac{3r_e^2}{2\langle r_e^2\rangle_G}} \tag{3}$$

$$\langle r_e^2\rangle_G = C_\infty b_0^2 n \tag{4}$$

In Eq. (3), $\langle r_e^2\rangle_G$ is the mean square end-to-end distance, which is calculated from the average distance between residues b_0 (3.8 Å) and the number of amino acids of the linker n (Eq. 4). A dimensionless correction factor called characteristic ratio takes in account nonideal effects (Brant & Flory, 1965). When linkers are not too short, this factor tends to a constant value C_∞, which depends on the nature of the amino acids in the chain (Brant, Miller, & Flory, 1967; Schimmel & Flory, 1967; Wang, Rivera, Benavides-Garcia, & Nall, 2005). For instance, C_∞ values for poly(Gly), poly(Ala) and poly(Pro) have been calculated as 2.16, 9.27 and 116, respectively. The main differences among these polypeptides are due to steric and/or rotational effects associated to the amino acids. Thereby, C_∞ can be regarded as a quantitative value for linker stiffness.

2.2 Wormlike chain model

Whereas the GC model provides a good quantitative description for long peptide linkers as random coils, results are not so accurate for short and stiff linkers. In these cases, the wormlike chain (WLC) model developed by Kratky and Porod allows a better description of linker behavior (Kratky & Porod, 1949). The model takes in account two important properties that are not considered in the GC model. The first one is that the peptide chain length cannot change, so a fixed contour length l_c given by Eq. (5) must be incorporated into the model. The second one is related to the stiffness of the linker. Rotational freedom is not the same in every point due to the variable steric and electronic properties of the amino acids. As a result, the model considers that the chain direction has at any point a persistent memory within a persistence length l_p, so the linker can be seen as a continuous semiflexible tube.

$$l_c = b_0 n \tag{5}$$

With these considerations, the radial distribution function $P_{WL}(r_e)$ for the end-to-end distance r_e that has been found to describe this behavior is shown in Eq. (6) (Zhou, 2004), in which w is a dimensionless function of r_e that takes the nonclosed form of Eq. (7). In addition, using Eq. (6) in combination with the theoretical definition of effective concentration (C_{eff}) that was indicated previously (Eq. 2) for random coils, a specific expression for C_{eff} according to this model can be deducted (Eq. 8).

$$P_{WL}(r_e) = 4\pi r_e^2(1 - w(r_e))\left(\frac{3}{4\pi l_p l_c}\right)^{3/2} e^{-\frac{3r_e^2}{4l_p l_c}} \tag{6}$$

$$w(r_e) = \frac{5l_p}{4l_c} - \frac{2r_e^2}{l_c^2} + \frac{33r_e^4}{80l_p l_c^3} + \frac{79l_p^2}{160l_c^2} + \frac{329r_e^2 l_p}{120l_c^3} - \frac{6799r_e^4}{1600l_c^4} + \frac{3441r_e^6}{2800l_p l_c^5} - \frac{1089r_e^8}{12800l_p^2 l_c^6} \tag{7}$$

$$C_{eff} = f\frac{P_{WL}(r_{int})}{4\pi r_{int}^2 N_A} = \frac{f}{N_A}(1 - w(r_{int}))\left(\frac{3}{4\pi l_p l_c}\right)^{3/2} e^{-\frac{3r_{int}^2}{4l_p l_c}} \tag{8}$$

In the limit of long and flexible chains ($l_c >> l_p$), $w(r_e)$ tends to 0 and thus $P_{WL}(r_e)$ equals the radial distribution function for the GC model, $P_G(r_e)$, which means that predictions made by the two random coil models become the same. Under these conditions, the relationship between the stiffness parameters of both models is given by Eq. (9).

$$l_p = \frac{C_\infty}{2} b_0 \tag{9}$$

3. Quantitative understanding of Ser/Gly linker behavior

Whereas the work by Flory yielded theoretical values of C_∞ for polypeptide homopolymers and various protein unfolding studies reported l_p values for different natural polypeptide chains, no experimental values were available for the stiffness (C_∞ or l_p) of Ser/Gly repeat linkers. We therefore designed and expressed a series of fusion proteins consisting of donor and acceptor fluorescent domains connected by a linker containing different numbers of Ser/Gly repeats. Both the linker length and the Ser-to-Gly ratio were systematically varied and experimental FRET efficiencies for each protein were determined. These data were then compared to values calculated from GC and WLC models, which allowed us to obtain quantitative information about linker stiffness and effective concentration. In this section, our main results and conclusions about the quantitative behavior of Ser/Gly linkers will be discussed.

3.1 Effect of linker length

FRET is a special type of electronic relaxation pathway that occurs when an excited fluorophore (donor) and a second fluorophore (acceptor) are spatially close enough, so that the donor is able to transfer its energy to the acceptor, which subsequently emits fluorescence. The energy-transfer efficiency E depends on the distance r_c between the two chromophores according to Eq. (10), which is known as the Förster equation (Lakowicz, 1999). In this expression, R_0 is the Förster distance, i.e., the value of r_c at which FRET efficiency is 50%. R_0 is a parameter that depends on the spectral overlap between the emission spectrum of the donor and the excitation spectrum of the acceptor, the relative spatial orientation of the fluorophores, the refraction index of the medium and the quantum yield of the donor (Lakowicz, 1999).

$$E = \frac{R_0^6}{R_0^6 + r_c^6} \tag{10}$$

We first generated a set of recombinant fusion proteins named CLYx in which enhanced cyan fluorescent protein (ECFP) and enhanced yellow fluorescent protein (EYFP) domains were fused through Ser/Gly linkers of different length. For each protein, the GGSGGS sequence was systematically repeated in the linker between 1 and 9 times (value of x in CLYx), and the observed FRET efficiencies were then calculated using the emission intensities of donor (ECFP) and acceptor (EYFP) and molar extinction coefficients. In this way, fusion proteins containing linkers between 23 and 71 residues were studied (Evers et al., 2006).

Fig. 3 shows the observed FRET efficiencies (E_{obs}) for the CLYx series as a function of linker length. As expected, the values of E decreased for longer linkers, but the decrease in E was relatively small considering the large increase in linker length and the strong dependence of E on the inter-chromophore distance (r_c^{-6}). This suggests that the linkers adopt a relatively compact random coil conformation in which the average end-to-end distance of the linker changes much less than the length of the linker. To determine whether the experimentally observed FRET efficiencies could indeed be quantitatively described assuming a random coil conformation of the linker, we used both the GC and the WLC models to calculate the average energy transfer efficiency as a function of linker length and linker stiffness.

For each CLY protein, Eq. (11) was used to calculate an average FRET efficiency value ($\langle E \rangle_{\text{ensemble}}$) by numerical integration of the product between

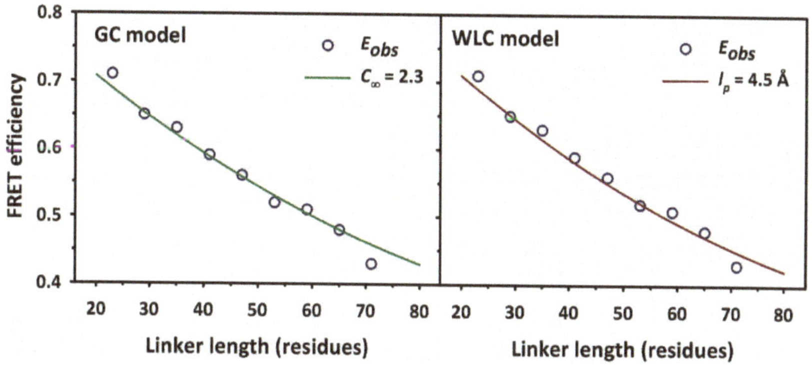

Fig. 3 Comparison between the observed FRET efficiencies (E_{obs}, *blue hollow circles*) and predicted values (*curves*) using GC model (*left*) and WLC model (*right*) for the CLY proteins. *Curves* are the best fits of the data predicted using the indicated random coil model and stiffness parameter to experimental values. *Adapted from Evers, T. H., van Dongen, E. M. W. M., Faesen, A. C., Meijer, E. W., & Merkx, M. (2006). Quantitative understanding of the energy transfer between fluorescent proteins connected via flexible peptide linkers.* Biochemistry, 45, *13183–13192. https://doi.org/10.1021/bi061288t. Copyright 2006 American Chemical Society*

the average FRET efficiency for a given value of r_e ($\langle E\rangle_{r_e}$), and the radial distribution function $P(r_e)$ deducted for the chosen model, which included the adjustable stiffness parameter (C_∞ for GC model and l_p for WLC model). The values of $\langle E\rangle_{r_e}$ were numerically estimated as an average of all the possible E values calculated with the Föster equation, using as r_c values calculated for a discrete and homogeneous set of conformations with equal r_e, after assuming both a full rotational freedom of the fluorescent domains around the ends of the linker and fixed end-to-fluorophore distances (20 Å for ECFP and 24 Å for EYFP). A calculated value of 48 Å for the ECFP-EYFP pair was used for R_0 in the Förster equation.

$$\langle E\rangle_{\text{ensemble}} = \int_0^\infty \langle E\rangle_{r_e} P(r_e)dr_{\text{e}} \tag{11}$$

As seen in Fig. 3, the excellent correspondence observed between the experimental and theoretical values from both GC and WLC models strongly supports the consideration of Ser/Gly linkers as random coils. The values of C_∞ (2.3) and l_p (4.5 Å) that gave the best fit between the experimental data and the models were only slightly higher than those predicted for polyglycine ($C_\infty = 2.16$ and $l_p = 4.1$ Å) (Brant et al., 1967), indicating that the flexibility of GGSGGS linkers is dominated by the 66.7% of glycine residues.

3.2 Influence of Ser-to-Gly ratio in linker stiffness

Although side chains of Ser residues in polypeptides are not excessively bulky, they provide sufficient steric hindrance to introduce more restrictions in rotational freedom than Gly residues would do. As a result, Flory and coworkers predicted an increase in the stiffness of Ser/Gly linkers with a decrease in Gly content (Brant & Flory, 1965). Polyserine linkers would thus be expected to be more effective than flexible polyglycine linkers in spanning long distances (>60 Å) due to a longer average end-to-end distance (see also Fig. 5). Similarly, the use of polyserine linkers could be useful to reduce energy transfer efficiency in the so-called low-FRET state of many FRET sensors.

Following a similar approach to that described above, our group more recently also studied the influence of the Ser-to-Gly ratio on the stiffness of Ser/Gly linkers (van Rosmalen, Krom, & Merkx, 2017). In addition to the already prepared set of CLY proteins with GGSGGS repeats (renamed to CLY-$(G_4S_2)_x$), three additional series of CLY proteins were produced in which the Ser-to-Gly ratio was systematically increased: CLY-$(G_2S_4)_x$ (GSSGSS repeats), CLY-$(GS_5)_x$ (GSSSSS repeats) and CLY-$(S_6)_x$ (SSSSSS repeats). All fusion proteins were readily expressed in *Escherichia coli* and purified using Ni-affinity chromatography, except for the proteins with the longest polyserine linkers, which showed some tendency to form oligomers and required size-exclusion chromatography to obtain protein monomers. To assess the stiffness of the four different Ser/Gly linkers present in these proteins, experimentally determined FRET efficiencies E_{obs} were again compared to values predicted using GC and WLC models. Molar extinction coefficients and emission intensities for donor an acceptor taken from each protein's fluorescent spectrum were used to calculate the values for E_{obs}, which are shown in Fig. 4. Overall similar E_{obs} values were obtained for G_2S_4 and GS_5 proteins, which were lower than E_{obs} values for G_4S_2. Energy transfer efficiency was lowest for the S_6 proteins.

After experimental E_{obs} values were calculated, WLC and GC models were applied to each series of CLY proteins to get quantitative information about linker stiffness. Theoretical ensemble FRET efficiency values ($\langle E \rangle_{\text{ensemble}}$) were again calculated for each protein using Eq. (11). However, the model was slightly adjusted by excluding conformations with an r_c value lower than 22 Å (interchromophore distance in the GFP dimer) from the averaging to calculate $\langle E \rangle_{r_c}$. These conformations were left out, because they correspond to conformations in which ECFP and EYFP occupy the same

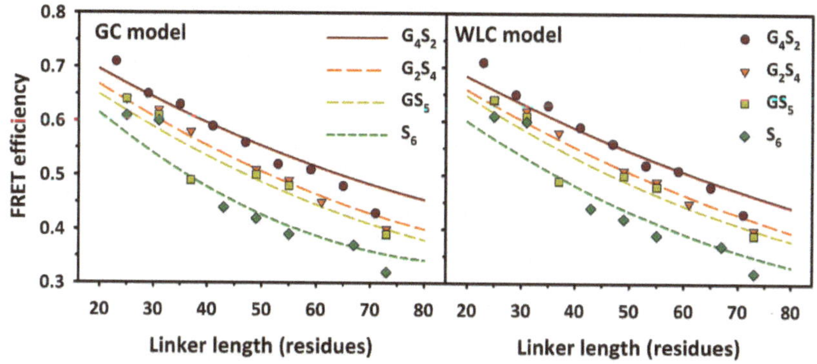

Fig. 4 Comparison between the observed FRET efficiencies (●, ▼, ■ and ◆ for G_4S_2, G_2S_4, GS_5 and S_6 CLY proteins, respectively) and predicted values (*curves*) using GC model (*left*) and WLC model (*right*) for the four series of CLY proteins with different Ser/Gly linkers. *Curves* are the best fits of the data predicted using the indicated random coil model and the stiffness parameters shown in Table 1. *Adapted from van Rosmalen, M., Krom, M., & Merkx, M. (2017). Tuning the flexibility of glycine-serine linkers to allow rational design of multidomain proteins.* Biochemistry, 56, *6565–6574. https://doi.org/10.1021/acs.biochem.7b00902. Copyright 2017 American Chemical Society.*

volume (Hink et al., 2000; Yang, Moss, & Phillips, 1996). Using this adjusted model to fit the data for the GGSGGS-containing CLY proteins yielded somewhat lower values of C_∞ and l_p of 1.9 and 3.7 Å, which were even closer to the values predicted for polyglycine linkers.

Fig. 4 shows a graphical comparison between the observed and predicted values of FRET efficiency for all of the CLY proteins. The best fits indicated a moderate increase on the linker stiffness parameters with higher Ser-to-Gly ratios. According to the GC model, the value of C_∞ varied from 1.9% for a 66% Gly content to 3.4 for poly(Ser) linkers, whereas the use of the WLC model afforded values of l_p between 3.7 and 6.2 Å. In other words, a 1.8-fold increase in linker stiffness was observed when all the Gly residues of the G_4S_2 linkers were replaced by Ser. Both models explain the observed FRET efficiencies well, with small discrepancies for proteins with the shortest poly (Ser) linkers. These differences can be attributed to a significant amount of Gly residues (18%) within N-and C-terminal ends of the fluorescent domains. These ends can be considered to be part of the linkers, resulting in a higher FRET efficiency than predicted by the model.

Table 1 contains both the values of the stiffness parameters obtained in our experimental studies and those theoretically predicted by the GC model of Flory and coworkers (Brant & Flory, 1965). Whereas the values

Table 1 Comparison of the stiffness parameters calculated for Ser/Gly linkers and values predicted by Flory and co-workers for peptide linkers with the same amount of Gly.

			C_∞[a]		l_p (Å)[b]	
Linker	**Repeating sequence**	**Gly content (%)**	**Exp.**[c]	**Pred.**[d]	**Exp.**[c]	**Pred.**[e]
G_4S_2	GGSGGS	66.7	1.9	2.2	3.7	4.2
G_2S_4	GSSGSS	33.3	2.4	3.8	4.5	7.2
GS_5	GSSSSS	16.7	2.6	5.6	4.8	10.6
S_6	SSSSSS	0.0	3.4	9.3	6.2	17.7

[a]Characteristic ratio for the GC model.
[b]Persistence length for the WLC model.
[c]Calculated by fitting the predicted FRET efficiencies with the random coil models to the experimentally observed values for CLY proteins.
[d]Value predicted by Brant and Flory (1965).
[e]Calculated using Eq. (9) and the value of C_∞ that was predicted by Flory and coworkers.

for G_4S_2 linkers are similar, the discrepancy between the experimental and the predicted value is substantial for the S_6 linkers. Apparently, the decrease in glycine content does not affect linker flexibility as much as the GC model predicts. Our experimental results are consistent with other studies that tried to quantify the effect of glycine content in linkers. A good example is work by Bowler and coworkers, who studied the effect of varying amounts of Ala and Gly in linkers on the stability of the interaction between a His residue and a linked heme-coordinated Fe^{3+}. When the amount of Gly in the linker was changed from 60% to 4%, a 1.6-fold increase in the value of C_∞ was observed (Finnegan & Bowler, 2012; Tzul, Kurchan, & Bowler, 2007). This increase is similar to the 1.8-fold increase that we observed when the linker was changed from G_4S_2 to S_6.

Based on the experimentally determined values of C_∞ and l_p, the Flory model underestimates the effective flexibility of C_α-substituted amino acids, suggesting that some assumptions behind the original Flory model are incorrect. The model assumes that only specific dihedral angles are allowed and others not, but in reality some conformations are more likely than others (Beck, Alonso, Inoyama, & Daggett, 2008; Childers, Towse, & Daggett, 2016). In addition, the model does not take in account the influence of neighboring amino acids on these dihedral angles (Hollingsworth, Lewis, & Karplus, 2016; Pappu, Srinivasan, & Rose, 2000) and the fact that intralinker noncovalent interactions conditioned by the nature of the solvent can lead to a partial or total collapse of the linker (Haran, 2012).

4. Understanding and tuning binding properties of multidomain proteins containing Ser/Gly linkers

The quantitative description of Ser/Gly peptide linkers described above for CFP-YFP fusion proteins can also be applied to other multidomain proteins. These models and the experimentally determined linker stiffness can be used to explain and predict important properties such the average end-to-end distance $\langle r_e \rangle$ of the linker and C_{eff}. $\langle r_e \rangle$ can be calculated as the integrated product of r_e and the radial distribution function of the chosen random coil model $P(r_e)$ over the range of r_e values (Eq. 12).

$$\langle r_e \rangle = \int_0^\infty r_e \, P(r_e) \, dr_e \qquad (12)$$

Fig. 5 shows how the stiffness of Ser/Gly linkers affect both the probability of finding a specific end-to-end distance and the average value of this distance. For given values of Ser-to-Gly ratio and contour length, there is a value of r_e for which $P(r_e)$ reaches a maximum, which corresponds to the most likely end-to-end distance (note that this is not the same as the average end-to-end distance). As expected, the WLC model predicts that a higher Ser content leads to an increase in the most likely end-to-end distance, as well as in the average end-to-end distance $\langle r_e \rangle$, for a given contour length.

As shown in Eq. (8), C_{eff} depends on both the contour length (l_c) and the interaction end-to-end distance (r_{int}) for a given linker stiffness (l_p). Fig. 6 graphically depicts this dependence for G_4S_2 linkers (upper left) and S_6 linkers (upper right), and also compares the C_{eff} values obtained for the two linkers as a

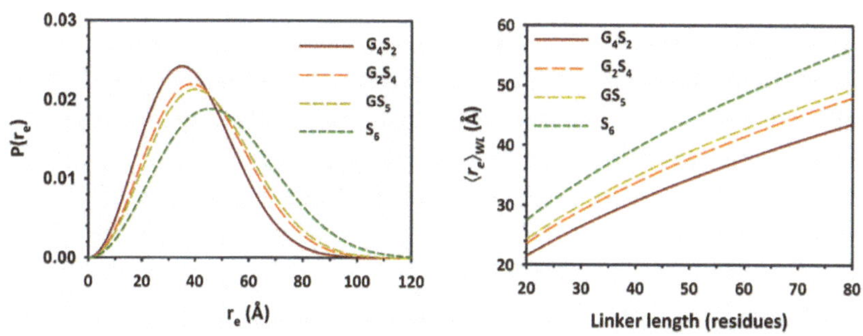

Fig. 5 Radial distribution functions $P(r_e)$ for a linker length of 64 residues (*left*) and average end-to-end distances as a function of linker length (*right*) for various types of Ser/Gly linkers. Values were calculated with the WLC model using Eqs. (6) and (12) with the experimental values of l_p for each type of linker listed in Table 1.

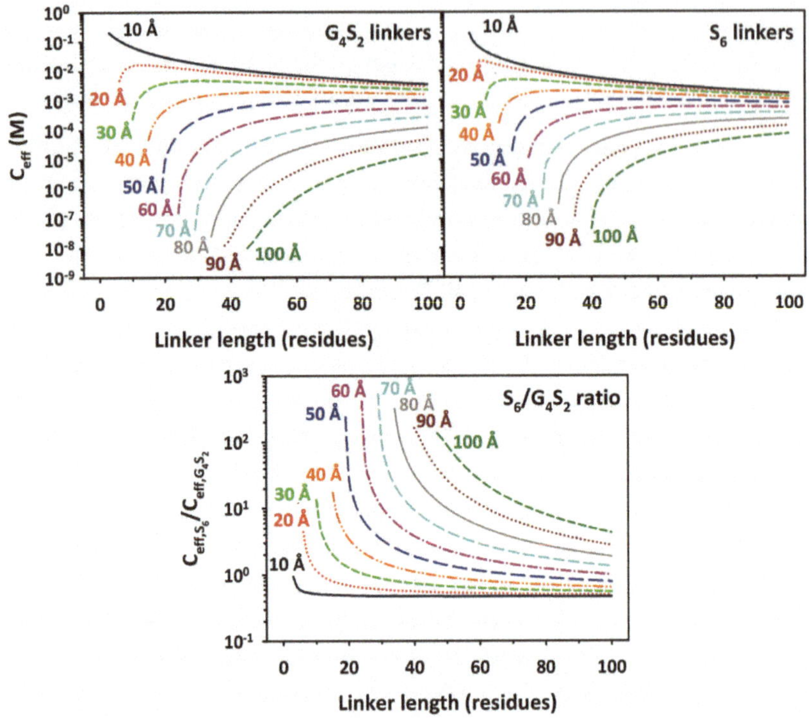

Fig. 6 Dependence of C_{eff} on linker length predicted with the WLC model for Ser/Gly linkers at interaction distances (r_{int}) ranging from 10 to 100 Å. *Upper left*: G_4S_2 linkers ($l_p = 3.7$ Å). *Upper right*: S_6 linkers ($l_p = 6.2$ Å). *Down*: Values for S_6 linkers ($C_{eff,S6}$) divided by those of G_4S_2 linkers ($C_{eff,G4S2}$) with the same r_{int} and linker length. C_{eff} values were calculated using Eq. (8).

function of linker length (below). Examination of this figure reveals some useful guidelines for linker design. A first important observation is that C_{eff} decreases rapidly for linker lengths that are "too short" (shorter than the optimal linker length), whereas "too long" linkers have only a minor effect on C_{eff}. In other words, in case of doubt it is better to be safe than sorry and choose long linkers. For interaction end-to-end distances (r_{int}) up to 30 Å, C_{eff} ranges between 1 and 10 mM, with G_4S_2 linkers providing slighter higher C_{eff} values then S_6 linkers. In this regime the effective concentration is relatively insensitive to the linker length for both linkers. However, when the linker needs to bridge a distance of 40 Å or more, polyserine linkers become more effective. Indeed, the relative effectiveness of polyserine linkers over G_4S_2 linkers increases as the value or r_{int} becomes higher and also for short linkers, which reflects the lower conformational freedom of polyserine chains. However, both linkers become relatively inefficient for large distances, with $\sim 10\,\mu M$ being the highest value of C_{eff} that can be achieved for a distance of 100 Å.

To illustrate how quantitative models of the conformational behavior of Ser/Gly linkers can help to understand and tune the properties of multi-domain sensor proteins, we first discuss the CALWY and eCALWY family of FRET-based Zn^{2+} sensor proteins developed by our group (Fig. 7). The initial version of this sensor (CALWY) consisted of two small metal binding domains (ATOX and WD4), with each domain providing two cysteine

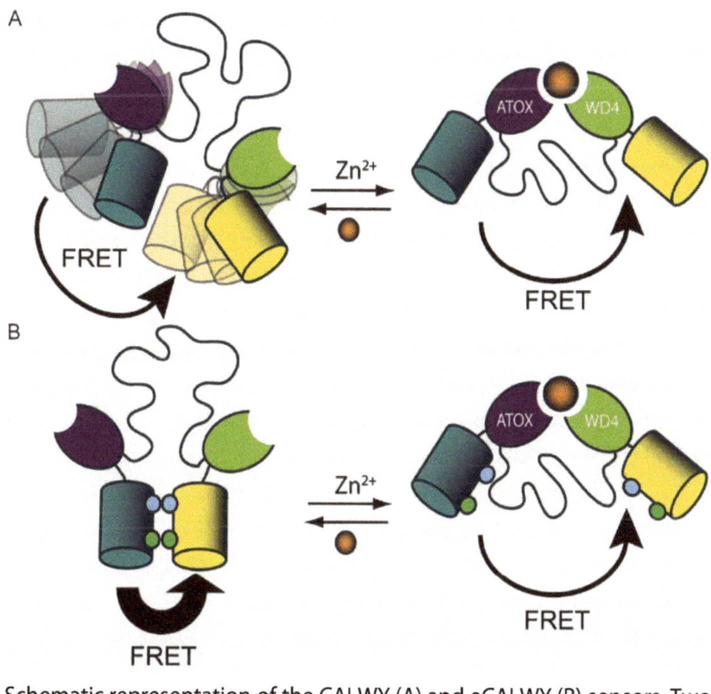

Fig. 7 Schematic representation of the CALWY (A) and eCALWY (B) sensors. Two metal-binding domains (*purple* and *green lunes*) are fused through a Ser/Gly peptide linker, and each one is tightly joined to a different fluorescent domain derived from GFP (*blue* and *yellow cylinders*). Addition of Zn^{2+} (*orange circle*) to the sensor induces the intramolecular dimerization of the metal-binding domains, which leads to a conformational rearrangement that pulls the fluorescent domains apart. As a result, FRET decreases. In eCALWY, two mutations are introduced in each fluorescent domain (*blue* and *green circles*) to promote an interaction between them in the Zn^{2+}-unbound state. *Adapted with permission from Vinkenborg, J. L., Nicolson, T. J., Bellomo, E. A., Koay, M. S., Rutter, G. A., & Merkx, M. (2009). Genetically encoded FRET sensors to monitor intracellular Zn^{2+} homeostasis.* Nature Methods, 6(10), *737–740. https://doi.org/10.1038/nmeth.1368.*

residues to form a tetrahedral Zn^{2+} complex (van Dongen et al., 2007). These domains were fused through a GGSGGS linker of variable length, with ECFP and EYFP domains flanking the Atox1 and WD4 domains, to afford a functional FRET-sensor. CALWY variants containing G_4S_2 linkers with lengths ranging from 22 (2 GGSGGS repeats) to 64 (9 GGSGGS repeats) residues were obtained, showing high affinity Zn^{2+} binding with K_d values ranging between 1.4 pM for the variant with the shortest linker and 170 fM for the protein containing the longest linker. In all cases, Zn^{2+} binding resulted in a decrease in FRET, with the change in emission ratio ranging from 15% for the variant with the longest linker to 37% for the protein containing a linker with only 2 GGSGGS repeats.

To improve the change in emission ratio, a second generation of sensors was developed (eCALWY, enhanced CALWY) in which the ECFP and EYFP domains were replaced by variants of cerulean and citrine with mutations that promote intramolecular association of the domains in the absence of Zn^{2+} (resulting a significant increase in FRET), while still allowing the protein to switch to a low-FRET, Zn^{2+}-bound state (Vinkenborg, Evers, Reulen, Meijer, & Merkx, 2007). A series of six eCALWY variants was obtained that differed from each other in the length of their G_4S_2 linkers and/or mutation of one of the Zn^{2+}-binding cys residues to a serine (C416S in WD4), which yielded a broad range of K_d values between 1.8 pM and 2.9 nM (Vinkenborg et al., 2009). Again a decrease in affinity (increase in K_d) was observed when changing the linker length from 9 GGSGGS repeats (eCALWY1/4) to 5 (eCALWY2/5) and 3 repeats (eCALWY3/6).

The WLC model can be used to explain both the changes in the emission ratio observed for the CALWY sensors and the dependence of K_d on linker length. For the CALWY-sensors the change in emission ratio is related to the difference between the average FRET efficiencies in the Zn^{2+}-bound and unbound states, which is highly dependent on the interchromophore distance, as shown in Section 3. Fig. 5A shows that the average end-to-end distance $\langle r_e \rangle$ predicted by the WLC model ranges between 22.6 Å for the linker with 2 GGSGGS repeats to 38.9 Å for the linker with 9 GGSGGS repeats. These distances are all smaller than the end-to-end distance of 50 Å that is estimated for the Zn^{2+}-bound state, which is based on an X-ray structure for the analogous Cd^{2+} complex (Wernimont, Huffman, Lamb, O'Halloran, & Rosenzweig, 2000). Thereby, both the decrease in emission ratio and the magnitude of the changes can be readily explained by the WLC model.

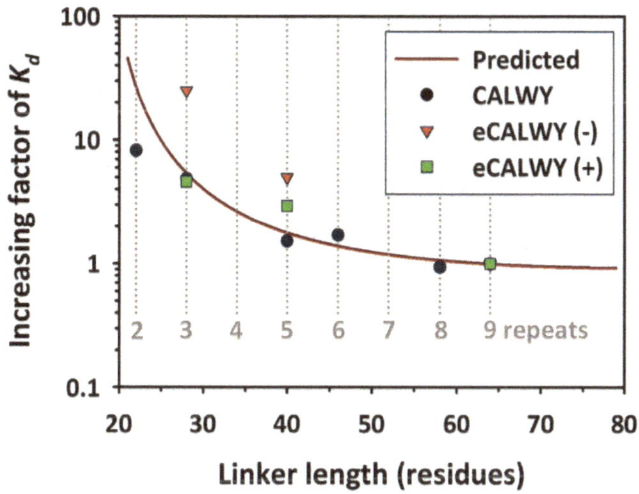

Fig. 8 Expected increases in K_d upon linker shortening calculated with the WLC model for the CALWY and eCALWY sensor proteins ($l_p = 3.7$ Å, $r_{int} = 50$ Å). Values in each series are normalized to the protein with the longest linker (CA-L9-WY for CALWY, eCALWY-1 for eCALWYs without the C416S mutation (−) and eCALWY-4 for mutated eCALWYs (+)) and are calculated as the quotient $C_{eff,0}/C_{eff}$, in which $C_{eff,0}$ is the value calculated for the protein of reference. C_{eff} values were calculated using Eq. (8). Experimentally determined values are also shown for comparison.

A second interesting aspect of the CALWY sensor is that it provides a well-understood experimental system to study the relation between linker length and effective concentration. Fig. 6 shows that an increase in linker length can give rise to either an increase or a decrease in effective concentration, depending on the distance that the linker needs to bridge in the complex. In this case, where the linker needs to bridge a distance of 50 Å, C_{eff} is expected to increase with increasing linker length, which is consistent with the higher Zn^{2+} affinity that is observed experimentally. Moreover, the C_{eff} value predicted by the model can be compared by the value of C_{eff} that can be derived by dividing the association constant for Zn^{2+} binding to the CALWY sensor by the association constant for formation of the ternary complex between CFP-ATOX1, WD4-YFP, and Zn^{2+} (no linker) that was determined independently ($\beta = 4.5 \times 10^{15}$ M^{-2}). For CALWY9, this experimentally derived value ($C_{eff} = 1.3$ mM) agrees well with the value predicted by the WLC model ($C_{eff} = 1.0$ mM) assuming an end-to-end distance of 50 Å in the complex form (van Dongen et al., 2007). Fig. 8 compares the relative increase in K_d values for the CALWY and eCALWY sensors as function of linker length with that predicted by the WLC model.

Excellent agreement is observed for the CALWY-series of sensors, whereas for the high-affinity eCALWY sensors the increase in K_d is higher than predicted. The latter may be explained by the fact that the linker also affects the stability of the interaction between the fluorescent domains in the Zn^{2+}-unbound state, which is likely to be stabilized by the shorter linkers.

As graphically depicted in Fig. 6, the effect of linker length on C_{eff} is critically dependent on the distance that the linker has to bridge for the intramolecular interaction to take place (r_{int}). A good example of what this means for multidomain proteins was observed in our ZinCh sensors (Fig. 9), which are variants of the CLY-$(G_4S_2)_x$ proteins (Section 3) that contain two mutations (Y39H and S208C) on the surface of both fluorescent domains to make them capable of chelating two Zn^{2+} ions with a different K_d value (Evers et al., 2007). It was found that the values of K_d for the high-affinity coordination site ranged from 108 nM for the protein with the shortest linker (1 GGSGGS repeat, 23 residues) to 213 nM for the variant with the longest spacer (9 GGSGGS repeats, 71 residues), i.e., a twofold decrease in the affinity when the linker is enlarged. This opposite and much more modest effect of linker length on Zn^{2+} affinity can be explained by the smaller distance the linker needs to bridge in the Zn^{2+}-bound state of the ZinCh sensors. Assuming an end-to-end linker distance of 30 Å in the Zn^{2+}-bound complex, the WLC model predicts C_{eff} values ranging from 3.1 mM for the protein with the longest spacer to 4.3 mM for the ZinCh variant with the shortest linker. This modest 1.4-fold increase in C_{eff} is thus consistent with twofold decrease in K_d observed experimentally.

Whereas Ser/Gly linkers can effectively span the distances of 10–50 Å observed in many protein–protein interaction, the effective concentration rapidly decreases for distance above 60 Å (Fig. 6). We encountered this phenomenon when developing FRET sensor proteins (AbSense) for antibody

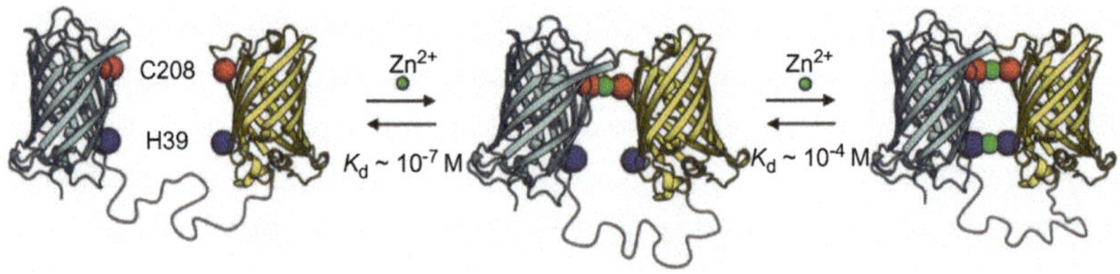

Fig. 9 Schematic overview of ZinCh sensors and their working principle. *Reproduced with permission from The Royal Society of Chemistry from Hessels, A. M., & Merkx, M. (2014). Genetically encoded fluorescent probes for intracellular Zn^{2+} imaging. In Fukada, T., & Kambe, T. (Eds.), Zinc signals in cellular functions and disorders. Japan: Springer. pp. 135–159. https://doi.org/10.1007/978-4-431-55114-0_7.*

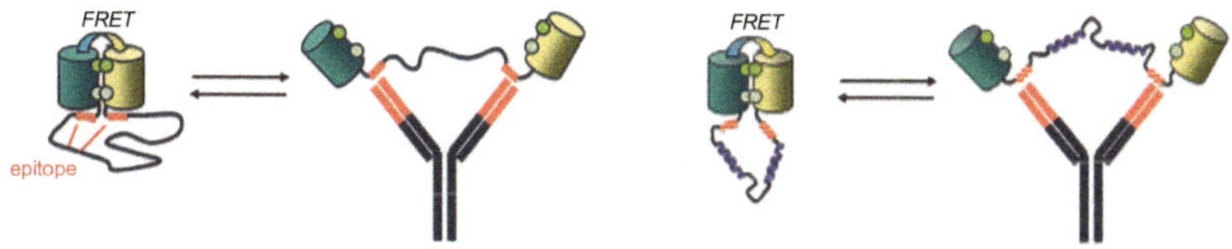

Fig. 10 Schematic overview of the AbSense protein sensor for antibody detection containing a flexible G_4S_2 linker (*left*) or a semiflexible linker (*right*). *Adapted with permission from Golynskiy, M. V., Rurup, W. F., & Merkx, M. (2010). Antibody detection by using a FRET-based protein conformational switch.* Chembiochem, 11, *2264–2267. https://doi.org/10.1002/cbic.201000143.*

detection (Fig. 10) (Golynskiy, Rurup, & Merkx, 2010). The design of the AbSense sensor consists of self-associating variants of Cerulean and Citrine (the same as were used in the eCALWY sensor), separated by a long and semiflexible linker with two flanking antibody-binding epitope sequences, one adjacent to each of the two fluorescent domains. Bivalent binding of the sensor to each of the epitope sequences disrupts the interaction between the fluorescent domains, resulting in a fivefold decrease in emission ratio. Interestingly, an initial design that used a linker consisting of 9 GGSGGS repeats did not show efficient bivalent binding, which can be explained by the large distance of ~120 Å that the linker needs to bridge in the antibody-bound state (Saphire et al., 2002). Even though this linker can bridge this distance when fully extended (3.8 Å × 54 = 205 Å), the entropic penalty for doing so is very large, which is reflected in $C_{eff} < 1\ \mu M$. Indeed, effective bivalent binding was observed only after introduction of two 45 Å α-helical blocks, and these linkers have also been used in subsequent bioluminescent antibody sensors (LUMABS) (Arts et al., 2016; Ni, Arts, & Merkx, 2019).

5. Summary and conclusions

Flexible linkers consisting of repeats of serine and glycine residues are widely used in the construction of multidomain proteins. Systematic study of the energy transfer efficiency between donor and acceptor fluorescent domains as a function of linker length have revealed that energy transfer efficiency can be quantitatively understood using models describing the linkers as random coil structures, yielding experimental values for linker stiffness as a function of Ser/Gly content. While an increase in persistence length is observed with a decrease in glycine content, polyserine linkers were found to be more flexible than predicted by the classical model of Flory and coworkers. In addition to allowing a quantitative understanding and rational tuning of the energy transfer efficiency in FRET sensor proteins, the experimentally determined persistence lengths also allow one to calculate the effect of linker length and linker constitution on the strength of intramolecular and multivalent interactions via its effect on effective concentration. The models predict that this effect critically depends on the distance that the linker needs to bridge in the complex form, which has been confirmed in studies on FRET sensor proteins with different sensor architectures. Whereas the availability of these models now allows rational design of Ser/Gly linkers, similar models should be developed for linkers containing different amino-acid repeats (e.g., the stiffer polyproline linkers) and alternating blocks of flexible and more structured domains (e.g., EAAAK repeats). The latter would be particularly useful for designing linkers that can effectively bridge distances of >8 nm.

Acknowledgments

The work by ECA has received funding from the European Union's Horizon 2020 research and innovation program under the Marie Skłovodowska-Curie grant agreement no. 754462 (EuroTechPostdoc).

References

Arts, R., den Hartog, I., Zijlema, S. E., Thijssen, V., van der Beelen, S. H., & Merkx, M. (2016). Detection of antibodies in blood plasma using bioluminescent sensor proteins and a smartphone. *Analytical Chemistry*, *88*, 4525–4532. https://doi.org/10.1021/acs.analchem.6b00534.

Beck, D. A. C., Alonso, D. O. V., Inoyama, D., & Daggett, V. (2008). The intrinsic conformational propensities of the 20 naturally occurring amino acids and reflection of these propensities in proteins. *Proceedings of the National Academy of Sciences of the United States of America*, *105*, 12259–12264. https://doi.org/10.1073/pnas.0706527105.

Brant, D. A., & Flory, P. J. (1965). The configuration of random polypeptide chains. II. Theory. *Journal of the American Chemical Society*, *87*(13), 2791–2800. https://doi.org/10.1021/ja01091a003.

Brant, D. A., Miller, W. G., & Flory, P. J. (1967). Conformational energy estimates for statistically coiling polypeptide chains. *Journal of Molecular Biology*, *23*(1), 47–65. https://doi.org/10.1016/S0022-2836(67)80066-4.

Childers, M. C., Towse, C.-L., & Daggett, V. (2016). The effect of chirality and steric hindrance on intrinsic backbone conformational propensities: Tools for protein design. *Protein Engineering, Design and Selection*, *29*(7), 271–280. https://doi.org/10.1093/protein/gzw023.

Evers, T. H., Appelhof, M. A. M., de Graaf-Heuvelmans, P. T. H. M., Meijer, E. W., & Merkx, M. (2007). Ratiometric detection of Zn(II) using chelating fluorescent protein chimeras. *Journal of Molecular Biology*, *374*, 411–425. https://doi.org/10.1016/j.jmb.2007.09.021.

Evers, T. H., van Dongen, E. M. W. M., Faesen, A. C., Meijer, E. W., & Merkx, M. (2006). Quantitative understanding of the energy transfer between fluorescent proteins connected via flexible peptide linkers. *Biochemistry*, *45*, 13183–13192. https://doi.org/10.1021/bi061288t.

Finnegan, M. L., & Bowler, B. E. (2012). Scaling properties of glycine-rich sequences in guanidine hydrochloride solutions. *Biophysical Journal, 102*(10), 1969–1978. https://doi.org/10.1016/j.bpj.2012.03.049.

Flory, P. J. (1969). *Statistical mechanics of chain molecules*. New York: Interscience Publishers.

Gargano, J. M., Ngo, T., Kim, J. Y., Acheson, D. W. K., & Lees, W. J. (2001). Multivalent inhibition of AB_5 toxins. *Journal of the American Chemical Society, 123*(51), 12909–12910. https://doi.org/10.1021/ja016305a.

Golynskiy, M. V., Rurup, W. F., & Merkx, M. (2010). Antibody detection by using a FRET-based protein conformational switch. *Chembiochem, 11*, 2264–2267. https://doi.org/10.1002/cbic.201000143.

Haran, G. (2012). How, when and why proteins collapse: The relation to folding. *Current Opinion in Structural Biology, 22*(1), 14–20. https://doi.org/10.1016/j.sbi.2011.10.005.

Hink, M. A., Griep, R. A., Borst, J. W., van Hoek, A., Eppink, M. H. M., Schots, A., et al. (2000). Structural dynamics of green fluorescent protein alone and fused with a single chain Fv protein. *Journal of Biological Chemistry, 275*(23), 17556–17560. https://doi.org/10.1074/jbc.M001348200.

Hollingsworth, S. A., Lewis, M. C., & Karplus, P. A. (2016). Beyond basins: φ,ψ preferences on a residue depend heavily on the φ,ψ values of its neighbors. *Protein Science, 25*(9), 1757–1762. https://doi.org/10.1002/pro.2973.

Kim, H. M., Jung, S., Lee, K. H., Song, Y., & Bae, H.-J. (2015). Improving lignocellulose degradation using xylanase–cellulase fusion protein with a glycine–serine linker. *International Journal of Biological Macromolecules, 73*, 215–221. https://doi.org/10.1016/j.ijbiomac.2014.11.025.

Komatsu, N., Aoki, K., Yamada, M., Yukinaga, H., Fujita, Y., Kamioka, Y., et al. (2011). Development of an optimized backbone of FRET biosensors for kinases and GTPases. *Molecular Biology of the Cell, 22*, 4647–4656. https://doi.org/10.1091/mbc.e11-01-0072.

Kratky, O., & Porod, G. (1949). Röntgenuntersuchung gelöster Fadenmoleküle. *Recueil des Travaux Chimiques des Pays-Bas, 68*(12), 1106–1122. https://doi.org/10.1002/recl.19490681203.

Lakowicz, J. R. (1999). *Principles of fluorescence spectroscopy*. New York: Kluwer Academic/Plenum Publishers.

Ni, Y., Arts, R., & Merkx, M. (2019). Ratiometric bioluminescent sensor proteins based on intramolecular split luciferase complementation. ACS Sensors *4*(1), 20–25. https://doi.org/10.1021/acssensors.8b01381.

Pappu, R. V., Srinivasan, R., & Rose, G. D. (2000). The Flory isolated-pair hypothesis is not valid for polypeptide chains: Implications for protein folding. *Proceedings of the National Academy of Sciences of the United States of America, 97*(23), 12565–12570. https://doi.org/10.1073/pnas.97.23.12565.

Salaun, C., Greaves, J., Tomkinson, N. C. O., & Chamberlain, L. H. (2020). The linker domain of the SNARE protein SNAP25 acts as a flexible molecular spaces that ensures efficient *S*-acylation. *Journal of Biological Chemistry, 295*(21), 7501–7515. https://doi.org/10.1074/jbc.RA120.012726.

Saphire, E. O., Stanfield, R. L., Crispin, M. D. M., Parren, P. W. H. I., Rudd, P. M., Dwek, R. A., et al. (2002). Contrasting IgG structures reveal extreme asymmetry and flexibility. *Journal of Molecular Biology, 319*(1), 9–18. https://doi.org/10.1016/S0022-2836(02)00244-9.

Schimmel, P. R., & Flory, P. J. (1967). Conformational energy and configuration statistics of poly-L-proline. *Proceedings of the National Academy of Sciences of the United States of America, 58*(1), 52–59. https://doi.org/10.1073/pnas.58.1.52.

Trinh, R., Gurbanaxi, B., Morrison, S. L., & Seyfzadeh, M. (2004). Optimization of codon pair use within the $(GGGGS)_3$ linker sequence results in enhanced protein expression. *Molecular Immunology, 40*(10), 717–722. https://doi.org/10.1016/j.molimm.2003.08.006.

Tzul, F. O., Kurchan, E., & Bowler, B. E. (2007). Sequence composition effects on denatured state loop formation in iso-1-cytochrome *c* variants: Polyalanine *versus* polyglycine inserts. *Journal of Molecular Biology, 371*(3), 577–584. https://doi.org/10.1016/j.jmb.2007.04.060.

van Dongen, E. M. W. M., Evers, T. H., Dekkers, L. M., Meijer, E. W., Klomp, L. W. J., & Merkx, M. (2007). Variation of linker length in ratiometric fluorescent sensor proteins allows rational tuning of Zn(II) affinity in the picomolar to femtomolar range. *Journal of the American Chemical Society, 129*, 3494–3495. https://doi.org/10.1021/ja069105d.

van Rosmalen, M., Krom, M., & Merkx, M. (2017). Tuning the flexibility of glycine-serine linkers to allow rational design of multidomain proteins. *Biochemistry, 56*, 6565–6574. https://doi.org/10.1021/acs.biochem.7b00902.

Vinkenborg, J. L., Evers, T. H., Reulen, S. W. A., Meijer, E. W., & Merkx, M. (2007). Enhanced sensitivity of FRET-based protease sensors by redesigning of the GFP dimerization interface. *Chembiochem, 8*, 1119–1121. https://doi.org/10.1002/cbic.200700109.

Vinkenborg, J. L., Nicolson, T. J., Bellomo, E. A., Koay, M. S., Rutter, G. A., & Merkx, M. (2009). Genetically encoded FRET sensors to monitor intracellular Zn^{2+} homeostasis. *Nature Methods, 6*(10), 737–740. https://doi.org/10.1038/nmeth.1368.

Wang, L., Rivera, E. V., Benavides-Garcia, M. G., & Nall, B. T. (2005). Loop entropy and cytochrome *c* stability. *Journal of Molecular Biology, 353*(3), 719–729. https://doi.org/10.1016/j.jmb.2005.08.038.

Wernimont, A. K., Huffman, D. L., Lamb, A. L., O'Halloran, T. V., & Rosenzweig, A. C. (2000). Structural basis for copper transfer by the metallochaperone for the Menkes/Wilson disease proteins. *Nature Structural Biology*, 7, 766–771. https://doi.org/10.1038/78999.

Yang, F., Moss, L., & Phillips, G. N. (1996). The molecular structure of green fluorescent protein. *Nature Biotechnology, 14*, 1246–1251. https://doi.org/10.1038/nbt1096-1246.

Zhou, H.-X. (2004). Polymer models of protein stability, folding and interactions. *Biochemistry, 43*(8), 2141–2154. https://doi.org/10.1021/bi036269n.

CHAPTER TWO

Design and construction of chimeric linker library with controllable flexibilities for precision protein engineering

Ziliang Huang[a,b,c], Chong Zhang[b,c,d], and Xin-Hui Xing[b,c,d,e,f,*]
[a]Department of Bioengineering, University of California San Diego, La Jolla, CA, United States
[b]Department of Chemical Engineering, Institute of Biochemical Engineering, Beijing, China
[c]Key Laboratory for Industrial Biocatalysis, Ministry of Education, Beijing, China
[d]Center for Synthetic and Systems Biology, Tsinghua University, Beijing, China
[e]Institute of Biopharmaceutical and Health Engineering, Tsinghua Shenzhen International Graduate School, Shenzhen, China
[f]Institute of Biomedical Health Technology and Engineering, Shenzhen Bay Laboratory, Shenzhen, China
*Corresponding author: e-mail address: xhxing@tsinghua.edu.cn

Contents

Abstract

Linkers play essential roles in the engineering of fusion proteins, and have been extensively demonstrated to affect protein properties such as expression level, solubility, and biological functions. For linker design and optimization, one of the key factors is the flexibility or rigidity of linkers, which describes the tendency of a linker to maintain a stable conformation when expressed, and can directly contribute to the physical distance between domains of a fusion protein. In this chapter, we discuss the design and engineering of linkers in fusion proteins, and describe a library-based method for optimization of linker flexibility. This approach is based on chimeric linkers, which

Methods in Enzymology, Volume 647
ISSN 0076-6879
https://doi.org/10.1016/bs.mie.2020.12.004

are composed of both flexible and rigid (helix-forming) linker motifs. We demonstrate that the chimeric linker library capable of controlling the flexibility in a wide range can fill the gap between flexible and rigid linkers by molecular dynamics simulation and fluorescence resonance energy transfer experiments, as well as its applications in fusion protein optimization.

1. Introduction

Fusion proteins have been widely used in both research and applications in the fields of biology, biotechnology, and biomedicine (Chen, Zaro, & Shen, 2013; Hussain, Heppenstall, Kampmeier, Meinhold-Heerlein, & Barth, 2019; Lei et al., 2020; Liu, Wang, Shao, Zaro, & Shen, 2020). By construction of fusion proteins, multiple desired functions originally from different proteins can be integrated into the same molecules, which provide a simple approach to engineering the functions of existing proteins and enable possible synergistic effects between different proteins (Albertsen et al., 2011; Moon, Lee, Lee, & Lee, 2020; Tokuhiro et al., 2009; Zhong, Zhang, Wu, Liu, & Chen, 2019). Based on the fusion approach, there are mainly three types of fusion proteins (Fig. 1): end-to-end fusion, insertional fusion, and branched fusion. In the end-to-end fusion, two or more domains or proteins (considered as domains for the rest of this chapter) are linked with the N-terminal end of the downstream domain linked to the C-terminal end of the upstream domain, forming a linear fusion protein with continuous domains (Doi & Yanagawa, 1999; Huang, Li, Zhang, & Xing, 2016). The insertional fusion includes the insertion of the full length of one domain (guest domain) into the other domain (host domain) at a tolerant location without affecting the structure and functions, which can be determined either rationally or experimentally (Pavlenko et al., 2019; Shah, Charlton, & Kim, 2018). Compared to the end-to-end fusion, the domains in an insertional fusion protein are more sensitive to changes of the other domain, and thus insertional

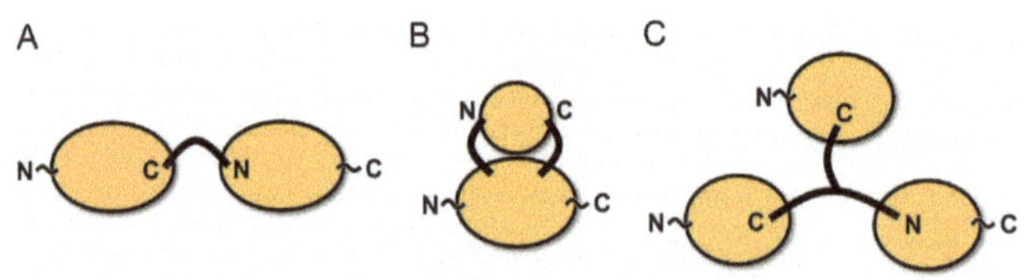

Fig. 1 Three types of fusion proteins. (A) end-to-end fusion, (B) insertional fusion, and (C) branched fusion. N and C denote the N- and C-terminus of each protein domain, respectively. The linkers between domains are shown in *bold black* lines.

fusion has advantages in developing regulatory machinery and biosensors, and requires more engineering efforts. First, a tolerant insertion site within the host domain needs to be identified, which should allow the correct folding of the host domain after translation with the guest domain inserted in the middle, while the insertion location should also allow the correct folding of the guest domain. Second, the structure of the host protein, with the guest domain inserted, should allow its original biological functions. Accordingly, as a general guideline, a successful insertion fusion usually requires the guest domain to have its N- and C-termini in close proximity in the folded structure (Ostermeier, 2005). Branched fusion is a nonlinear fusion approach in which more than two domains are linked together surrounding a center (Hirakawa, Kamiya, Kawarabayasi, & Nagamune, 2010; Hirakawa, Kamiya, Tanaka, & Nagamune, 2007). Because branched fusion proteins cannot be genetically encoded at the DNA level, the fusion process is carried out at the protein level, which can be catalyzed by enzymes such as transglutaminase, sortase, and laccase (Mattinen et al., 2006; Minamihata, Goto, & Kamiya, 2011; Nagamune, 2017). Compared to other fusion approaches, the branched fusion allows more balanced distribution of domains, thus reducing the interference and steric hindrance. However, due to the requirement of additional in vitro modification steps, the applications are still limited.

As a convenient and effective approach, the end-to-end fusion is most widely used for fusion protein construction. Unlike natural multidomain proteins which have evolved for millions of years, the artificially engineered multidomain proteins may encounter problems when expressed in cells, such as low expression, incorrect folding, and impaired functions, most of which are due to interference between the adjacent domains (Chen et al., 2013). To overcome these problems, one of the most often adopted strategy is to introduce a linker between different domains, which can act as a buffer to provide certain distancing between domains while keeping them connected. In the construction of end-to-end fusion proteins, since the choice of domains to be fused together is largely determined by the desired functions of the fusion protein, the engineering efforts mostly focus on the linker peptide, which as a bridge can control the spatial and functional relationships between the domains. For this purpose, properties of linkers have been extensively studied, such as length (number of residues), composition, and hydrophobicity, with the goal to identify a linker which can provide stable linkage between domains while the functions of different domains are not impaired (Jiang et al., 2020; Lager, Looger, Hilpert, Lalonde, & Frommer, 2006; Shen, Yan, Zhang, Mernaugh, & Zeng, 2008; van Rosmalen, Krom, & Merkx, 2017).

2. Engineering of protein linkers

Protein linkers are short peptides with the typical length of 3–30 amino acid residues. So far, there is no single linker which can satisfy most of the fusion proteins, and studies from different aspects have demonstrated the importance of linkers on the expression and function of fusion proteins (Argos, 1990; Doi & Yanagawa, 1999; Evers, van Dongen, Faesen, Meijer, & Merkx, 2006; Nagamune, 2017). Therefore, the design and optimization of fusion protein linkers is an essential step toward the success of fusion protein engineering.

Generally, linkers are introduced into fusion proteins through molecular cloning methods. Typical linker peptides are less than 30 residues, while the size of protein domains can easily reach over 200 residues (e.g., GFP) or even larger. This discrepancy in size may cause issues during the molecular cloning steps of a fusion protein construct. For instance, the small size of the linker sequence may not allow normal PCR amplification of the linker DNA fragment from an existing DNA template.

2.1 Introducing the linker sequence

One direct approach to introduce a linker sequence in a fusion protein is to include the full-length sequence of the linker in the primer, either the reverse primer of the N-terminal domain or the forward primer of the C-terminal domain of the fusion proteins. As such, the amplified DNA fragment would contain either the N-terminal or C-terminal domain with the full-length linker. To connect this fragment with the other domain of the fusion protein, there are typically three options: ligation, overlap PCR, or Gibson assembly. Here we discuss the situation in which the linker is amplified together with the N-terminal domain.

For ligation, although the N-terminal domain-linker fragment can directly ligate with the DNA fragment encoding the C-terminal domain via blunt-end ligation, the specificity and overall efficiency is typically low. Instead, a restriction enzyme cutting site can be engineered after the linker sequence, as well as at the 5′-end of the C-terminal domain, both of which can be easily included in the PCR primers. This allows the ligation with compatible sticky ends (after digestion by restriction enzyme) between the two DNA fragments with higher specificity and efficiency.

Alternatively, the N-terminal domain-linker fragment can fuse with the C-terminal domain fragment via overlap PCR. For this method, an overlapping sequence between the two fragments is necessary. This can be realized by adding extra nucleotides from the 5′-end of the C-terminal domain (~15–25 nucleotides, or with T_m over 50°C) to the end of the reverse primer for the N-terminal domain. Sometimes, the additional nucleotides may make the reverse primer too long to perform PCR properly, because of the long sequence of the linker and the overlap which could interfere with the annealing of the primer to the DNA template. In this case, the overlap sequence can be added in the forward primer for the C-terminal domain instead. The overlap sequence can be from the linker, or linker and the N-terminal domain, depending on the size of the primer. Then the overlap PCR can be performed using the forward primer of the N-terminal domain, reverse primer of the C-terminal domain, and the two amplified DNA fragments as template (0.1–1 ng in a 50 μL reaction), resulting in the full-length fragment of the (N-terminal domain)-linker-(C-terminal domain). Gibson assembly is a convenient method to assemble multiple DNA fragments in one shot. The Gibson assembly reaction includes three different enzymes, 5′-exonuclease, DNA polymerase, and DNA ligase, which can work together under the same condition. Specifically, the exonuclease can remove nucleotides from 5′-end of DNA fragments and create single-stranded 3′-overhangs which allow the annealing of fragments that share complementary sequences (i.e., overlap before 5′-exonuclease digestion). The DNA polymerase then can fill in gaps in the annealed fragments, and the DNA ligase can seal nicks within the assembled DNA. Similar to the PCR method, it requires overlap sequence between the two fragments to be connected, which can be realized in the same way as described for the PCR method. Compared to the methods mentioned above, the most obvious advantage is that the two fragments as well as the plasmid vector can be assembled simultaneously, generating the target construct without intermediate steps such as restriction enzyme digestion and gel purification.

When trying to incorporate a long linker (such as over 30 nucleotides) in the fusion protein, including the whole linker in one of the primers of the N- or C-terminal domain may cause issues for PCR. In this case, the whole linker sequence can be split into two parts and added to the N-terminal domain reverse primer and the C-terminal domain forward primer, respectively. With the split linker on the N- and C-terminal fragments, the full-length fusion protein sequence can be obtained through the methods described above.

Besides being included in the primers of the protein domains, the linker sequence can also be introduced into the fusion protein as a separate fragment. This is particularly useful when the linker is too long to incorporate into the N- or C-terminal's PCR primers (e.g., over 100 base pairs), or when the linker contains repeating sequence. If the linker length is within the range of oligo synthesis capacity (typically less than 200 base pairs), a forward and a reverse DNA oligos encoding the linker sequence can be synthesized. These two oligos can be annealed to generate the double-stranded DNA fragment for the full-length linker by following procedure.

Besides the coding sequence of the linker, additional nucleotides which complement the overhangs of the digested restriction recognition site (e.g., 5′-AATT-3′ for *Eco*RI site) can also be included in the oligos (typically 5′-end). As such, after annealing, sticky ends would be generated, which can be directly used for restriction recognition site-based ligation.

Lastly, for linkers with the size of over 200 base pairs, the oligo method can still be useful to generate the desired linker. Since one annealed DNA fragment would not be enough to cover the whole linker, two or more such annealed DNA fragments can be generated following the same design principles. However, in this case, sticky ends would be designed for ligation or assembly between linker fragments, so that a longer linker can be formed. Of course, the best scenario would be that the DNA template for the long linker is available and it can be amplified by PCR. Alternatively, as the size of the linker is already within the range for double-stranded DNA block synthesis, direct synthesis of the whole linker would be possible, although it may not be cost-effective.

2.2 Considerations in linker design

Assembly of the fusion protein construct is not enough, of course; one must obtain the soluble expression of the fusion protein with the expected biological activities. Since the choice of protein domains is largely based on the desired functions, the variable when designing a fusion protein is usually the linker. For a given fusion protein, the choice of linker can lead to very different outcomes. There are two considerations to keep in mind when selecting or designing the linker for a fusion protein: linker length and linker flexibility. The former refers to the number of residues in the linker sequence, which indicates the distance between the functional domains in the primary structure of the fusion protein, and to some extent it can also affect the physical distance of the domains in the three-dimensional structure (Evers et al., 2006).

The linker flexibility is related to the conformation or secondary structure of the linker. Linkers with high flexibility (e.g., glycine-rich linkers) tend to form random coils, without stable conformation, while linkers with low flexibility have more defined and stable conformations, such as helixes (Arai, Ueda, Kitayama, Kamiya, & Nagamune, 2001; Tan, Zhang, Karcher, & Bock, 2019; van Rosmalen et al., 2017).

The linker length can affect the properties of the fusion protein. For instance, if the linker is short, it may not sufficiently separate the domains, and interference between adjacent domains can happen during folding of the newly translated peptide, resulting in incorrect folding or low expression level (Carlsson, Ljung, & Bulow, 1996; Chen et al., 2013), or the biological functions of the fusion protein may be impaired (Grunberg et al., 2013; Nagamune, 2017). On the other hand, if the linker is too long, especially with repetitive sequences such as the EV linkers developed for FRET biosensors (Komatsu et al., 2011), first, it would be more difficult to engineer; second, the chance for the linker itself to form some secondary structure that can affect the biological functions may increase. It should be noted that linker length and stiffness can also affect the strength of intramolecular (between two domains in the same fusion protein) or intermolecular interactions (e.g., bivalent interactions). Regarding the range of linker length, Argos analyzed the linkers from 50 natural multidomain proteins, and showed that the linkers of natural proteins are generally between 3 and 15 residues, with the majority being ~5 residues (Argos, 1990). For artificially engineered fusion proteins, significantly longer linkers have also been applied successfully in many cases. For instance, Carlsson et al. compared fusion proteins of beta-galactosidase and galactose dehydrogenase with linkers of 3, 9, and 13 residues, respectively. Compared to 3-residue linker, the fusion proteins with long linkers (9 and 13 residues) showed enhanced specific activity and substrate affinity of galactose dehydrogenase, suggesting less intramolecular strain and interference in the fusion proteins with a longer linker (Carlsson et al., 1996). For proteins/domains that require oligomer formation for their biological functions, the length of linker can be crucial and needs to be carefully optimized. Currently, it is still challenging to accurately predict the linker length for a given fusion protein, except for linkers based on Ser/Gly repeats (van Rosmalen et al., 2017), but according to the reported studies, the range of 5–25 residues should work for the majority of fusion proteins (Cameron et al., 2019; Fontana et al., 2020; Grunberg et al., 2013; Morita et al., 2016; Yi, Kang, Lee, Braubach, & Baker, 2018).

For linker flexibility, typically, there are two options: the flexible linkers and the rigid linkers. The most often used flexible linkers are the glycine-rich linkers, which are hydrophilic due to the glycine residues and thus can increase the solubility of fusion proteins. As the side chain of glycine is small, this type of linkers can have flexible conformations, and their interference on the structure and activities of the functional domains is minimal. Besides, these linkers can provide certain distancing for the adjacent domains in the primary structure of fusion proteins, avoiding direct connection of residues from these domains which may impair the folding of the nascent peptides, as well as spatial separation at the three-dimensional structure. The $(GGGGS)_n$ (typically, $n \leq 6$) linkers have been broadly applied in fusion protein construction, and have been shown to be able to reduce the interference between domains. For instance, Lu et al. reported that direct end-to-end fusion (without linker) of beta-glucanase and xylanase lead to 31% decrease of xylanase activity compared to the nonfusion xylanase, whereas with $(GGGGS)_2$ linker the fusion protein showed a 43% increase of xylanase activity, as well as similar increase for the glucanase activity. Besides $(GGGGS)_n$, other flexible linkers have also been developed such as $(GGSGGT)_n$, KESGSVSSEQLAQFRSLD, EGKSSGSGSESKST, and GSAGSAAGSGEF (Bird et al., 1988; Suzuki et al., 2016; Waldo, Standish, Berendzen, & Terwilliger, 1999), which have been applied successfully in fusion protein construction especially scFv engineering. However, as the flexible linkers are usually hydrophilic and exposed to the solvent environment, they or the adjacent residues can be easily accessible for potential proteases in the cells, which can cause protein degradation. For example, Kavoosi et al. characterized a series of family 9 carbohydrate binding module (CBM9)-tagged GFP fusion proteins with different linkers (CBM9-linker-GFP), and found out that the widely used glycine-rich linkers, $(Gly)_3$ and $(Gly)_5$, are susceptible to hydrolysis by peptidases when expressed in *Escherichia coli*, with multiple cleavage sites within the linker region predicted by MEROPS (Rawlings, Morton, & Barrett, 2006) and confirmed experimentally by tandem mass spectrometry (Kavoosi, Creagh, Kilburn, & Haynes, 2007). It was reported that proteolysis occurs far more frequently within flexible or less compact regions (Fontana et al., 1997) which are more susceptible to endopeptidase (Huber & Bennett Jr., 1983; Huyton et al., 2003). The high flexibility also means that these linkers cannot maintain a stable conformation, therefore the physical distance between the domains may have large variation, and may not be effectively manipulated by simply changing the linker length although successful examples have also been reported (Evers et al., 2006; van Rosmalen et al., 2017).

Rigid linkers, on the contrary, are those with defined secondary structure when expressed in the fusion protein, mostly forming α-helices. Maeda et al. reported that the insertion of a three α-helix bundle domain between protein G and luciferase can successfully restore the binding capacity of protein G in the fusion protein (Maeda et al., 1997), suggesting that rigid linkers containing α-helices can be superior to flexible linkers in some cases. Later, Arai et al. demonstrated the helix-forming peptide $A(EAAAK)_nA$ ($n=2$–5) as linker for fusion protein construction (Arai et al., 2001), which is stabilized by the Glu-Lys salt bridges, with ~80% helicity based on circular dichroism studies (Marqusee & Baldwin, 1987). Although the $(EAAAK)_n$ peptide may still maintain some flexibility under physiological conditions, because of the stable conformation of the helical structure, the $(EAAAK)_n$ linker is more predictable for controlling the physical distance between domains in the folded structure of the protein compared to flexible linkers.

One of the main goals of introducing a linker is to achieve certain physical distancing between domains in the fusion protein. Although further separation can be expected with increased linker length, the effects can be different for flexible linkers and helical or rigid linkers. Fluorescence resonance energy transfer (FRET) between fluorescent proteins provides an elegant way to study this relationship, because the FRET efficiency is extremely sensitive to the distance between two fluorophores. Evers et al. constructed a series of fusion proteins of enhanced cyan fluorescent protein and enhanced yellow fluorescent protein with 1–9 repeats of GGSGGS as the linker (ECFP-linker-EYFP). By measuring the intramolecular FRET efficiency between ECFP and EYFP in different fusion proteins, the accurate distance between ECFP and EYFP can be calculated when different linkers were used. As expected, their results showed a gradual decrease in FRET efficiency with more repeats or increased linker lengths, with the shortest linker of 23 residues showing a distance of 41.3 Å between the chromophores of ECFP and YFP, and the longest linker of 71 residues showing 50.3 Å (Evers et al., 2006). It is important to note that the distance values reported here are apparent distance obtained from applying the Forster equation on FRET efficiency, but in reality, FRET measurements are the results of average of many different conformations, and therefore not a single distance, but a distribution of distances. On the other hand, a series of fusion proteins of EBFP and EGFP with different repeats of $(EAAAK)_n$ as linker showed that with increased repeats of EAAAK ($n=2$–5), the distance between the fluorophores can increase from 51.2 to 61.8 Å, suggesting linker length can have more dramatic effects on domain separation for helical linkers (Arai et al., 2001). With that said, it is still necessary to stress that, compared to the flexible linkers, the helical linkers have higher chance of interfering the folding of domains due to their intrinsic secondary structure (Chen et al., 2013; Fitzkee & Rose, 2004; Huston et al., 1988), therefore engineering of the linker flexibility or rigidity is still necessary and important for design of fusion proteins.

3. Design and construction of chimeric linker library

Although there are many successful examples for fusion protein design, the selection of linker is still largely based on the trial-and-error strategy (Fontana et al., 2020; Hires, Zhu, & Tsien, 2008; Ibraheem, Yap, Ding, & Campbell, 2011; Jiang et al., 2020; Piljic, de Diego, Wilmanns, & Schultz, 2011). The GGGGS motif is a universal flexible unit for linker design, while for situations where domain separation is desired the helix-forming EAAAK unit can be used. In addition, in some cases linkers containing flexible (Ser/Gly) and rigid (EAAAK) units have been successfully used and found to be superior to fully flexible linkers (Banala, Aper, Schalk, & Merkx, 2013; Boersma, Zuhorn, & Poolman, 2015; Golynskiy, Rurup, & Merkx, 2010). However, in these linkers it is difficult to characterize and quantify the effects of linker flexibility systematically. To solve this problem and expand the linker flexibility palette, we explored the recombination of flexible and rigid units and developed the chimeric linkers (Huang et al., 2016; Li et al., 2016). A library of linkers with the full flexibility spectrum from flexible $(GGGGS)_n$ to rigid $(EAAAK)_n$ was designed, constructed, and characterized. We demonstrated that the fine-tuning of flexibility between $(GGGGS)_n$ and $(EAAAK)_n$ can be achieved with the chimeric linker library, which greatly enhanced fusion protein functions compared to conventional linker types. In the following section, we describe the design and construction of chimeric linker library, which can be further used for the library-based strategy of linker optimization.

3.1 Design of chimeric linker library

The parameter of linker flexibility describes empirically the ability for a linker to maintain a specific conformation in a protein, which has been adequately shown to be a critical factor for the function of fusion proteins (Arai et al., 2001; Argos, 1990; Nagamune, 2017). As an empirical description, linker flexibility is difficult to be quantified or directly measured, although much progress has been made especially for Ser/Gly linkers (Evers et al., 2006;

van Rosmalen et al., 2017). For the optimization of linker flexibility during the construction of a specific fusion protein, the trial-and-error strategy with limited choice is still being used for the most time, and the results and conclusions are difficult to compare because usually different linker sets were tested. We aimed to develop a universal linker library for optimization which consists of a full spectrum of linker flexibility from flexible to rigid. Inspired by the idea of chimeric proteins, we developed the chimeric linkers based on the well-recognized glycine-rich GGGGS motif and the helix-forming EAAAK motif. Since a linker consists of only GGGGS repeats is flexible, while that consists of only EAAAK repeats is rigid, we reasoned that chimeric inkers which consist of both GGGGS and EAAAK motifs would have flexibility between flexible and rigid. The modular design of the chimeric linkers allows us to calculate the percentage of flexible or rigid units within the whole chimeric linker, which is difficult, although possible in some cases, to achieve for traditional linkers. As such, the flexibility of the chimeric linkers can be easily quantified (i.e., by the percentage of GGGGS repeats in the entire linker), which provides an index for the optimization of linker flexibility. In addition, since this quantifiable universal linker library can be readily applied to different fusion protein systems, these results can be directly compared with the same flexibility index, which can provide more insight of our understanding of linker flexibility, such as the relationship between domain interaction and linker flexibility.

To demonstrate the proof of concept for chimeric linker library, we designed chimeric linkers with the length of 25 residues consisting 5 repeat units of either GGGGS (designated as F unit) or EAAAK (designated as R unit). Any linker in this library is composed of these two basic units (Fig. 2A). For instance, the most flexible linker in the library is FFFFF, which only consists of GGGGS repeats. Similarly, the most rigid linker is RRRRR. Chimeric linkers which consist of both R and F are considered to have flexibility between FFFFF and RRRRR, such as FFRRF, RRFRR, and RFRFR. As such, the total library size is $2^5 = 32$.

To investigate the conformation of these linkers and thus verify the flexibility spectrum and the concept of chimeric linkers, we performed molecular dynamics simulation for the designed chimeric linkers. Briefly, molecular dynamics simulation was performed with Discovery Studio 2.5 (Accelrys, USA) based on the force field CHARMM (Brooks et al., 2009). The implicit solvent model Generalized Born was employed, and the target temperature was set as 300 K. The initial status of each linker was set as fully extended conformation. For the simulation process, the program was run until the system

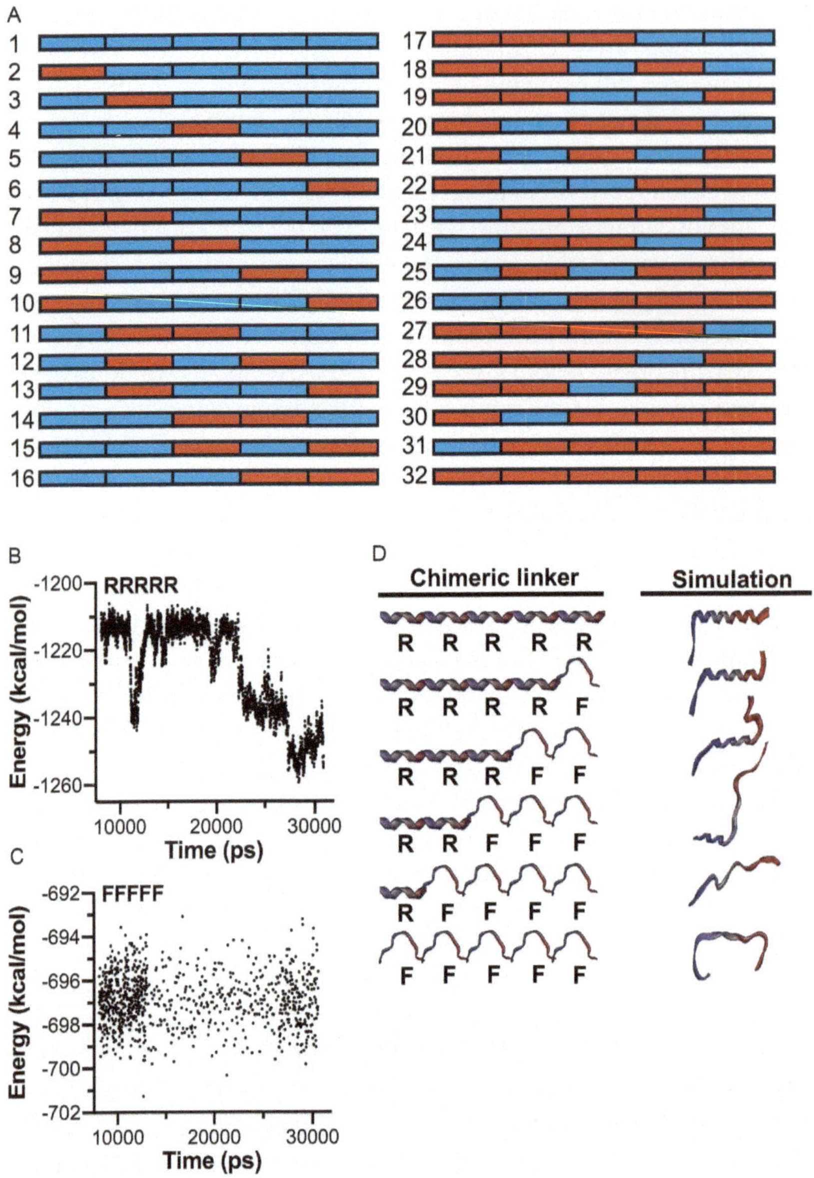

Fig. 2 Design of chimeric linkers and library. (A) A total of 32 variants (numbered) in a chimeric linker library can be generated for linkers with the length of 5 motifs. The *blue* and *red blocks* represent the flexible motif GGGGS and the helix-forming rigid motif EAAAK, respectively. Each number corresponds to a specific chimeric linker. (B–C) Change of total energy during the molecular dynamics simulation of rigid (B) and flexible (C) linkers. RRRRR, the helix-forming rigid linker $(EAAAK)_5$. FFFFF, the Gly-rich flexible linker $(GGGGS)_5$. (D) The expected conformations (left) and simulation results (right) for a group of five-motif chimeric linkers with different number of rigid (R) units. *Blue* and *red* in the simulation results represent N- and C-termini of the linkers.

total energy showed a significant decrease followed by a period where no significant energy change was observed (referred to as steady state). For the situation of flexible linkers, since no significant energy decrease was expected when the system reached a steady state, we assumed that the steady state was obtained after a simulation time of ~20 ns with energy fluctuation level similar to that of other linkers at their steady states and stable end-to-end distance. All simulations were sufficiently equilibrated before production and the simulation results are consistent with the experimental FRET measurements.

The RRRRR linker, which is known to form an α-helix (Arai et al., 2001; Marqusee & Baldwin, 1987), was used to verify the reliability of the simulation in predicting the steady state conformation. As expected, the RRRRR showed a conformation of α-helix after 28 ns of simulation, with a significant drop of total energy, indicating that the system indeed reached a much more stable status compared to the initial extended conformation (Fig. 2B). On the contrary, the FFFFF linker did not show any significant energy change during the whole simulation time, indicating that there is no preferred specific conformation for this linker (Fig. 2C). The results for both RRRRR and FFFFF are consistent with the reported conformations for $(GGGGS)_n$ and $(EAAAK)_n$ linkers. To test whether flexibility of the chimeric linkers can gradually change as we designed, we performed simulations for a group of chimeric linkers with gradually changed compositions (FFFFF, RFFFF, RRFFF, RRRFF, RRRRF, RRRRR, see Fig. 2D). Our simulation results showed that as the number of R units increased, the chimeric linkers had more helical conformation and hydrogen bonds, which facilitate the linker to maintain a specific conformation and less flexible, suggesting that the chimeric linker design can indeed generate linkers with heterogeneous flexibility ranging from fully flexible and fully rigid.

3.2 Construction of chimeric linker library

In this section we describe the construction method for the chimeric linker library of 5 units or 25 residues, with 32 variants as described in Section 3.1. Obviously, the most direct way to introduce these different linkers into the target fusion protein is to include the different linker sequences in the PCR primers. However, this would be difficult for a linker with the length of 75 base pairs, and constructing all of the 2^n variants would require much time and efforts. For this purpose, we developed a systematic approach for generating the chimeric linker library.

Briefly, the full length of a specific linker is generated by the ligation of two "halflinker" fragments (the N-terminal and C-terminal halves). The half linker at the N terminus contains two complete units of either GGGGS (F) or EAAAK (R), followed by an incomplete unit encoding either F or R, with a specifically designed restriction endonuclease recognition site in the coding sequence at the 3′-end (denoted as Fx or Rx, respectively, see Fig. 3A–B). Similarly, the half linker at the C-terminus contains an incomplete unit of F or R, with a restriction endonuclease recognition site within the coding

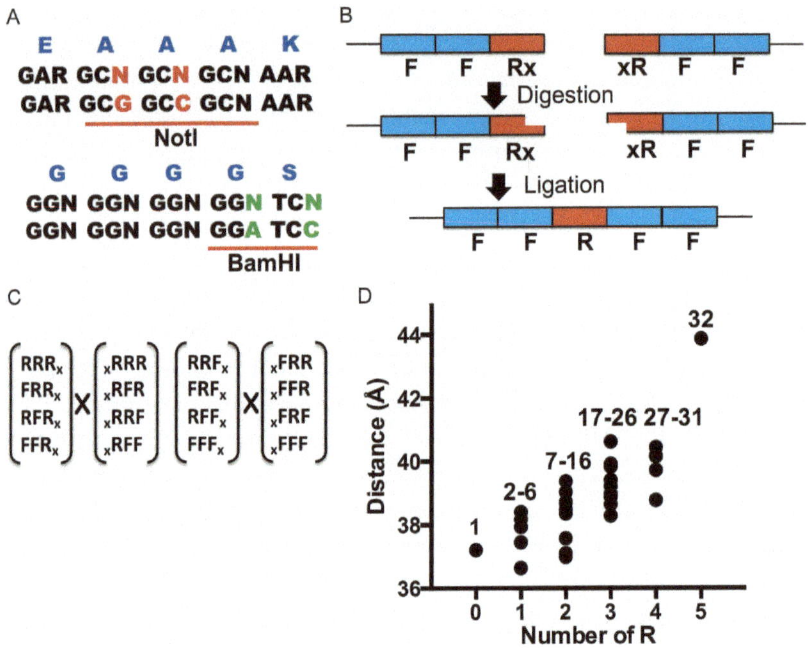

Fig. 3 Construction of chimeric linkers and library. (A) Design of restriction endonuclease recognition sites within the EAAAK and GGGGS motifs located at the third motif of the linker. The degenerate codons are shown below each residue of the motif (row 1), the designed sites (*Not*I for EAAAK, *Bam*HI for GGGGS) are underlined and indicated (row 2). (B) Strategy for chimeric linker construction. The FFRFF linker is demonstrated here as an example. Rx and xR refer to the EAAAK motif with *Not*I site at the 3′-end and 5′-end, respectively. (C) Strategy for generation of chimeric linker library. Fx and xF refer to the GGGGS motif with *Bam*HI site at the 3′-end and 5′-end, respectively. The half linkers to the left of the cross (X) can ligate with any of the half linkers to the right of the cross. A total of 32 chimeric linkers can be generated based on the random combination of half linkers. (D) Physical distance between ECFP and EYFP in fusion proteins with different chimeric linkers, calculated from the intramolecular FRET between ECFP and EYFP. Each data point represents one specific chimeric linker, which is labeled with the numbering system shown in Fig. 2A (labeled as a group when R = 1,2,3,4). Error bars, standard deviations ($n = 3$).

sequence at the 5′-end (denoted as xF or xR, respectively), followed by two complete units (either F or R). To generate the full-length linker without introducing additional amino acid residues in the middle, the Rx and xR were designed to have the *Not*I site within the coding sequence, while Fx and xF were designed to have the *Bam*HI site (Fig. 3A). As such, only when the N-terminal and C-terminal half linkers have the compatible incomplete units (either Rx with xR or Fx with xF), can they be ligated, thus generating the full-length chimeric linker. For example, the half linker FFRx (containing two F units followed by an Rx unit) can ligate with the half linker xRFF to generate the FFRFF linker, while it cannot ligate with the half linker xFFR (Fig. 3B). Because both the N-terminal and the C-terminal half linkers have 8 variants (different combinations of F and R, Fx, Rx, xF, and xR), the combination of these 8 N-terminal half linkers and 8 C-terminal half linkers can generate all of the 32 variants of the chimeric linker library (Fig. 3C).

Below we describe the general procedures for generating the chimeric linker library aiming at the design of fusion proteins. A more detailed example is provided in the next section. We designate the target fusion protein as A-linker-B, with A and B being the protein domains flanking the linker sequence.

(1) *Generate DNA fragments encoding A and the N-terminal half linker.* Different N-terminal half linkers can be introduced at the 3′-end of A during the cloning of A. The small size of the half linker would allow the reverse primer for amplification of A to include this additional sequence at the 5′ of the annealing sequence. As such, a total of 8 primers are needed to generate all the variants of the N-terminal half linkers.

(2) *Generate DNA fragments encoding B and the C-terminal half linker.* Similarly, different C-terminal half linkers can be introduced to B during the cloning step by including the C-terminal half linker sequence in the forward primer for PCR amplification of B at the 5′ of the annealing sequence. A total of eight primers are needed to generate all the variants of the C-terminal half linkers.

(3) *Generate A-linker-B through ligation.* Digest the generated DNA fragments in (1) and (2) with *Bam*HI for fragments containing Fx or xF, and *Not*I for fragments containing Rx or xR. After digestion and purification, the 8 different DNA fragments in (1) can ligate with any of the DNA fragments with the compatible ligation site in (2), thus generating 32 different combinations with each representing a specific A-linker-B, which can be incorporated into the desired vector for protein expression.

3.3 Construction of PhoC-RRFRR-GFP as an example of fusion proteins

For characterization of the chimeric linkers, the linkers need to be applied in a fusion protein system. In this section, as an example, we describe the construction of a fusion protein of acid phosphatase PhoC from *Enterobacter aerogenes* and green fluorescent protein (GFP) with the chimeric linker $(EAAAK)_2$-(GGGGS)-$(EAAAK)_2$ (PhoC-RRFRR-GFP) based on the approach described above.

Equipment

- Thermal cycler (Bio-Rad C1000 or similar)
- Water bath (37, 42, 55°C)
- Agarose gel apparatus (Bio-Rad Sub-Cell GT Cell or similar)

Buffers and reagents

- DNA polymerase (e.g., Q5 High-Fidelity DNA Polymerase from New England Biolabs)
- Gel purification kit (Zymoclean Gel DNA Recovery Kit D4001 or similar)
- DNA clean-up kit (Zymo DNA Clean & Concentrator Kit D4033 or similar)
- Restriction endonucleases *Eco*RI, *Bam*HI, *Hin*dIII (New England Biolabs)
- T4 ligase (New England Biolabs)
- One Shot Stbl3 Chemically Competent Cells (Thermo Fisher) or similar cloning *E. coli* strain
- Miniprep kit (Qiagen or Sigma)

Protocol

(1) Design the forward and reverse primers for PCR amplification of *phoC* gene. An *Eco*RI site (underlined) is included the 5′-end of the forward primer for the ligation with the vector plasmid. The sequence encoding the first R unit (in bold) is also included at the 5′-end of the reverse primer. The designed primers are.

Forward primer: 5′-TTA<u>GAATTC</u>ATGTTTGCGCTGGTTC CCGCCGGCAAT-3′

Reverse primer: 5′-**CTTGGCCGCGGCCTC**ATTTCGCTG TTTCGCAAACTC GTCTT-3′

(2) Perform PCR using the primers designed in (1) with the template of encoding PhoC. Purify the target DNA band which contains *phoC* and one R unit (phoC-R, ~720 bp) on 1% agarose gel following the directions provided by the gel purification kit.

(3) Design another reverse primer which can anneal to the 3′-end of the phoC-R fragment and contains the sequence encoding the second R unit and the Fx unit (in bold) with *Bam*HI site (underlined). The designed primer is

5′-TTA**GGATCCACCGCCACCTTTCGCCGCGGCTTC**TTTCGCCGCGGCTTCATTTCGCTGTT-3′

(4) Perform PCR using the purified DNA in (2) as template, the forward primer in (1) and the reverse primer in (3). Purify the target DNA band which encodes PhoC-RRFx (with *Bam*HI site at the 3′-end, ~750 bp) on 1% agarose gel.

(5) Similarly, design the forward and reverse primers for PCR amplification of GFP. A *Hin*dIII site (underlined) is included the 3′-end of the reverse primer for the ligation with the vector plasmid. The sequence encoding the last R unit of the linker (in bold) is also included at the 5′-end of the forward primer. The designed primers are

Forward primer: 5′-**GGTGGCGGTGGCTCA**ATGGTGAGCAAGGGCGAGGAGCTG-3′

Reverse primer: 5′-TCCAAGCTTTTACTTGTACAGCTCGTCCATGCCGAG-3′

(6) Perform PCR using the primers designed in (5) with the template of encoding GFP. Purify the target DNA band which contains GFP and one R unit (R-GFP, ~730 bp) on 1% agarose gel following the directions provided by the gel purification kit.

(7) Design another forward primer which can anneal to the 5′-end of the R-GFP fragment and contains the sequence encoding xFR with *Bam*HI site (underlined). The designed primer is 5′-TTA**GGATCCGAAGCCGCGGCGAAA**GAAGCCGCGGCGAAAATGGTGAGCAAGGGCG-3′

(8) Perform PCR amplification using the purified DNA in (6) as template, the forward primer in (7) and the reverse primer in (5). Purify the target DNA band which encodes xFRR-GFP (with *Bam*HI site at the 5′-end, ~750 bp) on 1% agarose gel.

(9) Digest the purified PhoC-RRFx with *Eco*RI and *Bam*HI, digest the purified xFRR-GFP with *Bam*HI and *Hin*dIII, and digest the vector plasmid with *Eco*RI and *Hin*dIII. Purify all the digested products using the DNA clean-up kit.

(10) For ligation, add 20 ng of digested PhoC-RRFx, 20 ng of digested xFRR-GFP, 60 ng of digested vector, 1 μL 10x T4 DNA ligase buffer, and 0.5 μL of T4 DNA ligase in a total volume of 10 μL. Incubate at room temperature for 10 min.

(11) Transform 2 μL of ligation reaction into 50 μL aliquot of competent cells following the manufacturer's instructions. Plate cells on LB media supplemented with the appropriate antibiotic. Negative controls (ligation with digested vector only or without DNA) can be run in parallel to determine the amount of background colonies. Prepare plasmids from three colonies using standard miniprep kits (e.g., Qiagen or Sigma). Ensure the correct insertion sequence of PhoC-RRFRR-GFP by Sanger sequencing using compatible sequencing primers.

4. Measuring the physical length of chimeric linkers with FRET

We described the construction of chimeric linker library, and our molecular dynamics simulation study showed that the conformation of the chimeric linkers has more α-helix with more EAAAK (R) units or less GGGGS (F) units in the chimeric linker, suggesting these chimeric linkers, with the same length but different flexibility, can have different effects in separating the protein domains when applied in fusion proteins. Fluorescence resonance energy transfer (FRET) between fluorescent proteins provide an extremely sensitive method for studying the distance between protein domains. The donor chromophore, when in the excited status, can transfer energy to the acceptor chromophore if both chromophores are in close proximity (typically 1–10 nm), through non-radiative dipole–dipole coupling. Because the efficiency of this energy transfer is inversely proportional to the sixth power of the distance between donor and acceptor, FRET is extremely sensitive to small changes in distance. Another consideration for FRET to happen is the spectral overlap between the donor emission spectrum and the acceptor excitation spectrum, such as the ECFP-EYFP pair, which has been used in many FRET studies (Arai et al., 2001; van Rosmalen et al., 2017). To study the separation distance of the chimeric linkers using FRET, we constructed a total of 32 variants of ECFP-linker-EYFP fusion proteins containing the chimeric linker library described in Section 3.1. The FRET measurement results of the CFP-linker-YFP fusion proteins clearly showed that the chimeric linker library can accurately control linker flexibility and the separation between protein domains. With the same number of total residues, the chimeric linkers can cause larger distance between CFP and YFP when more R units were included in the chimeric linker (Fig. 3D). The fully rigid linker (RRRRR) showed a separation distance of 44 Å, while the fully flexible linker (FFFFF) showed a separation distance of 37 Å, which are consistent with other reported studies using these linkers. The chimeric linkers showed a wide spectrum of separation distance between that of FFFFF and RRRRR linkers. Interestingly, with this approach, we observed that the chimeric linkers with the same number of R and F units (for instance, three R units and two F units) also demonstrated variations in CFP-YFP distance within a smaller range compared with that caused by different numbers of R and F units, suggesting that the arrangement of flexible and rigid units can further fine-tune the overall flexibility and the separation effect of the chimeric linkers. In the following section, we describe the methods used for measuring the FRET efficiencies of the chimeric linker library.

Equipment

- Spectrophotometer (Hitachi F2500 or similar)
- Shaking incubator (37°C)
- Water bath (42°C)
- Centrifuge
- Cell sonicator

Buffer and reagents

- *E. coli* BL21 (DE3) competent cells (New England Biolabs)
- LB media with 50 mg/L kanamycin
- Phosphate Buffered Saline (50 mM, pH 7.4)
- Quick Start Bradford Protein Assay (Bio-Rad)

Protocol

(1) Plasmids encoding the different ECFP-linker-EYFP variants can be constructed into the pET-28a vector following the protocol described in Section 3.3. For protein expression, transform each of the verified constructs into *E. coli* BL21 (DE3) competent cells following manufacturer's instructions.

(2) Create starter cultures by inoculating 2 mL LB media containing 50 mg/L kanamycin with a single colony. Grow cultures in a 37°C shaking incubator at 180 rpm overnight.

(3) Subculture 300 μL of the starter cultures into 30 mL LB media containing 50 mg/L kanamycin. Continue incubation in a 37°C shaking incubator at 180 rpm.

(4) Monitor the optical density of the media at 600 nm (OD_{600}). When OD_{600} reaches 0.6–0.8, add isopropyl β-D-1-thiogalactopyranoside (IPTG) to the LB media to a final concentration of 0.04 mM to induce expression of the fusion protein. Continue to grow the cultures at 16°C for 20 h.

(5) After the culture, centrifuge the cells at 6000 × *g*, 4°C, for 10 min. Discard the supernatant, and wash the cells with PBS. After centrifuge, resuspend the cells in 10 mL PBS.

(6) Disrupt the cells by sonication (200 W, 10 min, on ice). Centrifuge the disrupted cells at 13000 × *g*, 4°C, for 30 min. Transfer each supernatant to a clean tube and keep on ice for all the following procedures.

(7) Purify ECFP-linker-EYFP proteins from supernatants using His-tag affinity chromatography. Verify the size of the purified proteins by protein gel electrophoresis.

(8) Determine the protein concentration of each sample using the Bradford protein assay kit. Adjust the concentrations of samples to the same with PBS.

(9) To measure FRET efficiency between ECFP and EYFP, fluorescence intensity of ECFP in the presence of EYFP (F_{DA}) and fluorescence intensity of ECFP in the absence of EYFP (F_D) are measured (the concentrations of the purified fusion proteins and ECFP are adjusted to be the same before measurement), and efficiency of FRET (*E*) can be calculated as $E = 1 - F_{DA}/F_D$ (Arai et al., 2001). Make sure that the reading is below the saturation level by diluting the sample with PBS. In our experiments, the reading is kept under 10,000 RFU. The distance (r) between ECFP and EYFP can be calculated following $r = R_0\ (1/E - 1)^{1/6}$, where R_0 is the Forster distance between ECFP and EYFP.

5. Application of chimeric linker library in fusion proteins

With the chimeric linker library, it is convenient to characterize the effects of linker flexibility systematically for a fusion protein of interest. We previously applied the library to the fusion protein of acid phosphatase PhoC from *Enterobacter aerogenes* and GFP (PhoC-GFP). The phosphotransferase activity of PhoC can phosphorylate inosine to produce inosine-5′-monophosphate (5′-IMP), while its phosphatase activity can dephosphorylate 5′-IMP back to inosine (Asano, Mihara, & Yamada, 1999; Mihara, Utagawa, Yamada, &

Asano, 2001). To maximize the production of 5′-IMP, it is critical to inhibit the phosphatase activity while improving the phosphotransferase activity. To generate a fusion protein of PhoC-GFP with the desired properties, two groups of PhoC-GFP fusion proteins were constructed with chimeric linkers of different flexibility as demonstrated in Section 3.3. The linkers used in group I include RRRRR, RRRRF, RRRFF, RRFFF, RFFFF, and FFFFF, while the linkers used in group II include RRRRR, FRRRR, FFRRR, FFFRR, FFFFR, and FFFFF, both of which change from rigid to flexible. The phosphotransferase and phosphatase activities of PhoC and fluorescence of GFP of the purified fusion proteins (through His-tag mediated affinity chromatography) were measured and compared (Fig. 4A). The folding and structure, and hence fluorescence, of GFP may be affected due to fusion with PhoC, and the extent of interference can vary when different linkers are used between GFP and PhoC. Interestingly, the highest specific activity for either phosphotransferase activity or phosphatase activity was not obtained with the conventional flexible or rigid linker (FFFFF or RRRRR), but with the chimeric linker FFFRR for both activities. By fine-tuning the linker flexibility, the specific activities of phosphotransferase and phosphatase showed changes of 5.26-fold and 10.7-fold, respectively, suggesting that linker flexibility indeed can have significant impacts on fusion protein functions. As an important index for 5′-IMP production, we also compared the ratio between phosphotransferase and phosphatase activities among different PhoC-GFP fusion proteins, which showed a change of ~threefold (highest ratio obtained with FRRRR, Fig. 4B), suggesting that different functions in a protein domain may be affected by linker flexibility to different extents. Our results provide the proof-of-concept to demonstrate that applying the chimeric linker library with controllable flexibility to the fusion protein system will be an efficient way to optimize the functions of fusion proteins. To further facilitate the application of linker library, unique endonuclease recognition sites can be introduced flanking the linker region during the library construction, thus the linker library from an established fusion protein system can be easily applied to another fusion protein system. Here we describe the methods used for the transferring of the chimeric linker library.

Equipment

- Thermal cycler (Bio-Rad C1000 or similar)
- Water bath (37, 42, 55°C)
- Agarose gel apparatus (Bio-Rad Sub-Cell GT Cell or similar)

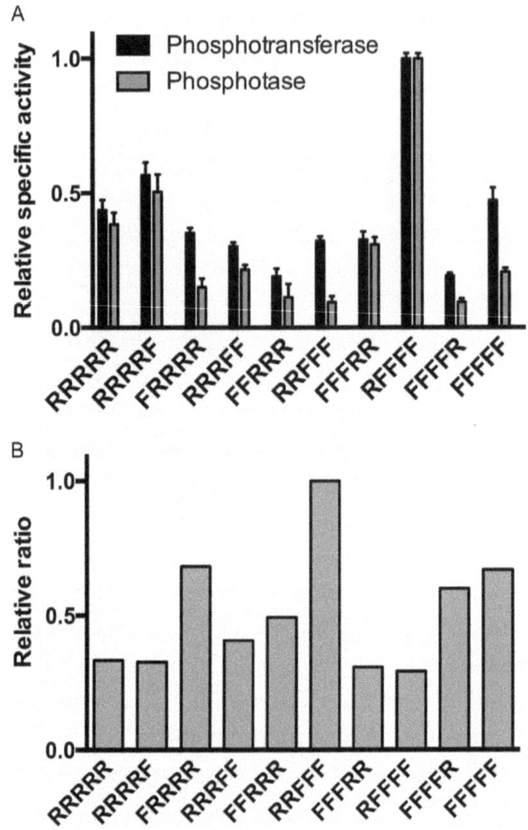

Fig. 4 Application of chimeric linker library in PhoC-GFP fusion protein. (A) Characterization of phosphotransferase and phosphatase activities and GFP fluorescence of PhoC-GFP fusion proteins with the chimeric linkers as shown. The specific activity/fluorescence values are normalized to the highest values (RFFFF for both activities and fluorescence). Error bars, standard deviation ($n = 3$). (B) Comparison of the ratio between phosphotransferase activity and phosphatase activity for PhoC-GFP fusion proteins with different chimeric linkers. The ratio values are normalized to the highest ratio.

Buffers and reagents

- DNA polymerase (e.g., Q5 High-Fidelity DNA Polymerase from New England Biolabs)
- Gel purification kit (Zymoclean Gel DNA Recovery Kit D4001 or similar)
- DNA clean-up kit (Zymo DNA Clean & Concentrator Kit D4033 or similar)
- Restriction endonucleases (New England Biolabs)
- T4 ligase (New England Biolabs)

- One Shot Stbl3 Chemically Competent Cells (Thermo Fisher) or similar cloning *E. coli* strain
- Miniprep kit (Qiagen or Sigma)

Protocol

(1) Introduce unique endonuclease sites flanking the linker region during library construction. For the PhoC-RRFRR-GFP example described in Section 3.3, a *Kpn*I site (italicized) can be introduced in the reverse primer for *phoC* gene PCR (used in step 1), thus the modified reverse primer is.

Reverse primer:
5′-**CTTGGCCGCGGCCTC***GGTACC*ATTTCGCTGTTTCG CAAACTC-3′

Similarly, a *Pst*I site (italicized) can be introduced in the forward primer for *gfp* gene PCR (used in step 5):

Forward primer:
5′-**GGTGGCGGTGGCTCA***CTGCA*GATGGTGAGCAAGG GCGAGGAGCTG-3′

(2) Perform PCR with the modified primers in (1) following the protocol described in Section 3.3 for the construction of chimeric linker library in the PhoC-linker-GFP system.

(3) Digest the PhoC-linker-GFP constructs with *Kpn*I and *Pst*I. Purify the target DNA band which encodes the chimeric linker (~75 bp) on 1% agarose gel.

(4) Introduce the same recognition sites flanking the linker to the functional domains of the new fusion protein system. Since *Kpn*I and *Pst*I sites are used here, the *Kpn*I site should be included at the 3′-end of the N-terminal functional domain, while the *Pst*I site included at the 5′-end of the C-terminal functional domain.

(5) Digest the DNA fragments amplified in (3) with *Kpn*I and *Pst*I. The DNA fragments encoding the functional domains may also contain other recognition sites for ligation with the vector plasmid, in this case digestion of these sites can be performed together with *Kpn*I and *Pst*I. Purify the target bands (depending on the size of the domain) on a 1% agarose gel.

(6) Ligate the N-terminal and C-terminal domains and linker into the vector. Based on the design described, the N-terminal domain can be ligated with the 5′-end of any of the chimeric linkers, while the C-terminal domain can be ligated with the 3′-end of any of the chimeric linkers. For ligation, add 20 ng of digested N-terminal domain, 20 ng of digested C-terminal domain, 5 ng of the digested linker, 60 ng of digested vector,

1 μL 10 × T4 DNA ligase buffer, and 0.5 μL of T4 DNA ligase in a total volume of 10 μL. Incubate at room temperature for 10 min.

(7) Transform the ligation products into competent cells following the Section 3.3. Ensure the correct insertion sequence by Sanger sequencing using the sequencing primers for the target vector.

(8) As such, the chimeric linker library can be transferred from PhoC-linker-GFP system to the new fusion protein system. If for screening purpose, where separation of individual fusion proteins is not necessary, different constructs from the original fusion protein system can be pooled and digested together as one component, which can be ligated into the new fusion protein system.

6. Summary

In this chapter we described the design and the methods for the engineering of chimeric linkers with controllable flexibility or rigidity, which are composed of the well-established flexible and helix-forming rigid linker motifs. Due to their modular characteristics, the chimeric linkers can be generated in a systematic manner and thus constitute a library with the size of n^2 with n being the number of motifs in the linkers. The flexibility of the chimeric linker library has been demonstrated to range from fully flexible to fully rigid by molecular dynamics simulation as well as characterization of intramolecular FRET between ECFP and EYFP in fusion proteins with different variants from the chimeric linker library. We also demonstrated the application of the chimeric linker library in a fusion protein system of acid phosphatase PhoC and GFP, which showed improved biological activities compared to the conventional flexible or rigid linkers. The chimeric linker library can be used as a systematic and effective method for the linker optimization of fusion proteins.

Acknowledgments

This work was supported by the National Key R&D Program of China (Grant No. 2018YFA0901500), the National Nature Science Foundation of China (No. 20836004), and Tsinghua University Initiative Scientific Research Program (2013Z02-1).

References

Albertsen, L., Chen, Y., Bach, L. S., Rattleff, S., Maury, J., Brix, S., et al. (2011). Diversion of flux toward sesquiterpene production in Saccharomyces cerevisiae by fusion of host and heterologous enzymes. *Applied and Environmental Microbiology*, 77(3), 1033–1040.

Arai, R., Ueda, H., Kitayama, A., Kamiya, N., & Nagamune, T. (2001). Design of the linkers which effectively separate domains of a bifunctional fusion protein. *Protein Engineering*, *14*(8), 529–532.

Argos, P. (1990). An investigation of oligopeptides linking domains in protein tertiary structures and possible candidates for general gene fusion. *Journal of Molecular Biology, 211*(4), 943–958.

Asano, Y., Mihara, Y., & Yamada, H. (1999). A novel selective nucleoside phosphorylating enzyme from Morganella morganii. *Journal of Bioscience and Bioengineering, 87*(6), 732–738.

Banala, S., Aper, S. J., Schalk, W., & Merkx, M. (2013). Switchable reporter enzymes based on mutually exclusive domain interactions allow antibody detection directly in solution. *ACS Chemical Biology, 8*(10), 2127–2132.

Bird, R. E., Hardman, K. D., Jacobson, J. W., Johnson, S., Kaufman, B. M., Lee, S. M., et al. (1988). Single-chain antigen-binding proteins. *Science, 242*(4877), 423–426.

Boersma, A. J., Zuhorn, I. S., & Poolman, B. (2015). A sensor for quantification of macromolecular crowding in living cells. *Nature Methods, 12*(3), 227–229. 221 p following 229.

Brooks, B. R., Brooks 3rd, C. L., Mackerell, A. D., Jr., Nilsson, L., Petrella, R. J., Roux, B., et al. (2009). CHARMM: the biomolecular simulation program. *The Journal of Computational Chemistry, 30*(10), 1545–1614.

Cameron, P., Coons, M. M., Klompe, S. E., Lied, A. M., Smith, S. C., Vidal, B., et al. (2019). Harnessing type I CRISPR-Cas systems for genome engineering in human cells. *Nature Biotechnology, 37*(12), 1471–1477.

Carlsson, H., Ljung, S., & Bulow, L. (1996). Physical and kinetic effects on induction of various linker regions in beta-galactosidase/galactose dehydrogenase fusion enzymes. *Biochimica et Biophysica Acta, 1293*(1), 154–160.

Chen, X., Zaro, J. L., & Shen, W. C. (2013). Fusion protein linkers: Property, design and functionality. *Advanced Drug Delivery Reviews, 65*(10), 1357–1369.

Doi, N., & Yanagawa, H. (1999). Design of generic biosensors based on green fluorescent proteins with allosteric sites by directed evolution. *FEBS Letters, 453*(3), 305–307.

Evers, T. H., van Dongen, E. M., Faesen, A. C., Meijer, E. W., & Merkx, M. (2006). Quantitative understanding of the energy transfer between fluorescent proteins connected via flexible peptide linkers. *Biochemistry, 45*(44), 13183–13192.

Fitzkee, N. C., & Rose, G. D. (2004). Steric restrictions in protein folding: An alpha-helix cannot be followed by a contiguous beta-strand. *Protein Science, 13*(3), 633–639.

Fontana, J., Dong, C., Kiattisewee, C., Chavali, V. P., Tickman, B. I., Carothers, J. M., et al. (2020). Effective CRISPRa-mediated control of gene expression in bacteria must overcome strict target site requirements. *Nature Communications, 11*(1), 1618.

Fontana, A., Zambonin, M., Polverino de Laureto, P., De Filippis, V., Clementi, A., & Scaramella, E. (1997). Probing the conformational state of apomyoglobin by limited proteolysis. *Journal of Molecular Biology, 266*(2), 223–230.

Golynskiy, M. V., Rurup, W. F., & Merkx, M. (2010). Antibody detection by using a FRET-based protein conformational switch. *Chembiochem, 11*(16), 2264–2267.

Grunberg, R., Burnier, J. V., Ferrar, T., Beltran-Sastre, V., Stricher, F., van der Sloot, A. M., et al. (2013). Engineering of weak helper interactions for high-efficiency FRET probes. *Nature Methods, 10*(10), 1021–1027.

Hirakawa, H., Kamiya, N., Kawarabayasi, Y., & Nagamune, T. (2010). Artificial self-sufficient P450 in reversed micelles. *Molecules, 15*(5), 2935–2948.

Hirakawa, H., Kamiya, N., Tanaka, T., & Nagamune, T. (2007). Intramolecular electron transfer in a cytochrome P450cam system with a site-specific branched structure. *Protein Engineering, Design & Selection, 20*(9), 453–459.

Hires, S. A., Zhu, Y., & Tsien, R. Y. (2008). Optical measurement of synaptic glutamate spillover and reuptake by linker optimized glutamate-sensitive fluorescent reporters. *Proceedings of the National Academy of Sciences of the United States of America, 105*(11), 4411–4416.

Huang, Z., Li, G., Zhang, C., & Xing, X. H. (2016). A study on the effects of linker flexibility on acid phosphatase PhoC-GFP fusion protein using a novel linker library. *Enzyme and Microbial Technology, 83*, 1–6.

Huber, R., & Bennett, W. S., Jr. (1983). Functional significance of flexibility in proteins. *Biopolymers*, *22*(1), 261–279.

Hussain, A. F., Heppenstall, P. A., Kampmeier, F., Meinhold-Heerlein, I., & Barth, S. (2019). One-step site-specific antibody fragment auto-conjugation using SNAP-tag technology. *Nature Protocols*, *14*(11), 3101–3125.

Huston, J. S., Levinson, D., Mudgett-Hunter, M., Tai, M. S., Novotny, J., Margolies, M. N., et al. (1988). Protein engineering of antibody binding sites: Recovery of specific activity in an anti-digoxin single-chain Fv analogue produced in *Escherichia coli*. *Proceedings of the National Academy of Sciences of the United States of America*, *85*(16), 5879–5883.

Huyton, T., Pye, V. E., Briggs, L. C., Flynn, T. C., Beuron, F., Kondo, H., et al. (2003). The crystal structure of murine p97/VCP at 3.6A. *Journal of Structural Biology*, *144*(3), 337–348.

Ibraheem, A., Yap, H., Ding, Y., & Campbell, R. E. (2011). A bacteria colony-based screen for optimal linker combinations in genetically encoded biosensors. *BMC Biotechnology*, *11*, 105.

Jiang, H. K., Lee, M. N., Tsou, J. C., Chang, K. W., Tseng, H. W., Chen, K. P., et al. (2020). Linker and N-terminal domain engineering of Pyrrolysyl-tRNA Synthetase for substrate range shifting and activity enhancement. *Frontiers in Bioengineering and Biotechnology*, *8*, 235.

Kavoosi, M., Creagh, A. L., Kilburn, D. G., & Haynes, C. A. (2007). Strategy for selecting and characterizing linker peptides for CBM9-tagged fusion proteins expressed in Escherichia coli. *Biotechnology and Bioengineering*, *98*(3), 599–610.

Komatsu, N., Aoki, K., Yamada, M., Yukinaga, H., Fujita, Y., Kamioka, Y., et al. (2011). Development of an optimized backbone of FRET biosensors for kinases and GTPases. *Molecular Biology of the Cell*, *22*(23), 4647–4656.

Lager, I., Looger, L. L., Hilpert, M., Lalonde, S., & Frommer, W. B. (2006). Conversion of a putative agrobacterium sugar-binding protein into a FRET sensor with high selectivity for sucrose. *The Journal of Biological Chemistry*, *281*(41), 30875–30883.

Lei, C., Qian, K., Li, T., Zhang, S., Fu, W., Ding, M., et al. (2020). Neutralization of SARS-CoV-2 spike pseudotyped virus by recombinant ACE2-Ig. *Nature Communications*, *11*(1), 2070.

Li, G., Huang, Z., Zhang, C., Dong, B. J., Guo, R. H., Yue, H. W., et al. (2016). Construction of a linker library with widely controllable flexibility for fusion protein design. *Applied Microbiology and Biotechnology*, *100*(1), 215–225.

Liu, Y., Wang, H. Y., Shao, J., Zaro, J. L., & Shen, W. C. (2020). Enhanced insulin receptor interaction by a bifunctional insulin-transferrin fusion protein: An approach to overcome insulin resistance. *Scientific Reports*, *10*(1), 7724.

Maeda, Y., Ueda, H., Kazami, J., Kawano, G., Suzuki, E., & Nagamune, T. (1997). Engineering of functional chimeric protein G-Vargula luciferase. *Analytical Biochemistry*, *249*(2), 147–152.

Marqusee, S., & Baldwin, R. L. (1987). Helix stabilization by Glu-...Lys + salt bridges in short peptides of de novo design. *Proceedings of the National Academy of Sciences of the United States of America*, *84*(24), 8898–8902.

Mattinen, M. L., Hellman, M., Permi, P., Autio, K., Kalkkinen, N., & Buchert, J. (2006). Effect of protein structure on laccase-catalyzed protein oligomerization. *Journal of Agricultural and Food Chemistry*, *54*(23), 8883–8890.

Mihara, Y., Utagawa, T., Yamada, H., & Asano, Y. (2001). Acid phosphatase/phosphotransferases from enteric bacteria. *Journal of Bioscience and Bioengineering*, *92*(1), 50–54.

Minamihata, K., Goto, M., & Kamiya, N. (2011). Protein heteroconjugation by the peroxidase-catalyzed tyrosine coupling reaction. *Bioconjugate Chemistry*, *22*(11), 2332–2338.

Moon, J. H., Lee, K., Lee, J. H., & Lee, P. C. (2020). Redesign and reconstruction of a steviol-biosynthetic pathway for enhanced production of steviol in Escherichia coli. *Microbial Cell Factories*, *19*(1), 20.

Morita, S., Noguchi, H., Horii, T., Nakabayashi, K., Kimura, M., Okamura, K., et al. (2016). Targeted DNA demethylation in vivo using dCas9-peptide repeat and scFv-TET1 catalytic domain fusions. *Nature Biotechnology*, *34*(10), 1060–1065.

Nagamune, T. (2017). Biomolecular engineering for nanobio/bionanotechnology. *Nano Convergence*, *4*(1), 9.

Ostermeier, M. (2005). Engineering allosteric protein switches by domain insertion. *Protein Engineering, Design & Selection*, *18*(8), 359–364.

Pavlenko, O. S., Gra, O. A., Mustafaev, O. N., Kabarbaeva, K. V., Sadovskaya, N. S., Tyurin, A. A., et al. (2019). Thermostable Lichenase from Clostridium thermocellum as a host protein in the domain insertion approach. *Biochemistry (Mosc)*, *84*(8), 931–940.

Piljic, A., de Diego, I., Wilmanns, M., & Schultz, C. (2011). Rapid development of genetically encoded FRET reporters. *ACS Chemical Biology*, *6*(7), 685–691.

Rawlings, N. D., Morton, F. R., & Barrett, A. J. (2006). MEROPS: The peptidase database. *Nucleic Acids Research*, *34*(Database issue), D270–D272.

Shah, V., Charlton, T., & Kim, J. R. (2018). Laboratory evolution of Bacillus circulans xylanase inserted into Pyrococcus furiosus maltodextrin-binding protein for increased xylanase activity and thermal stability toward alkaline pH. *Applied Biochemistry and Biotechnology*, *184*(4), 1232–1246.

Shen, Z., Yan, H., Zhang, Y., Mernaugh, R. L., & Zeng, X. (2008). Engineering peptide linkers for scFv immunosensors. *Analytical Chemistry*, *80*(6), 1910–1917.

Suzuki, K., Kimura, T., Shinoda, H., Bai, G., Daniels, M. J., Arai, Y., et al. (2016). Five colour variants of bright luminescent protein for real-time multicolour bioimaging. *Nature Communications*, 7, 13718.

Tan, J., Zhang, F., Karcher, D., & Bock, R. (2019). Engineering of high-precision base editors for site-specific single nucleotide replacement. *Nature Communications*, *10*(1), 439.

Tokuhiro, K., Muramatsu, M., Ohto, C., Kawaguchi, T., Obata, S., Muramoto, N., et al. (2009). Overproduction of geranylgeraniol by metabolically engineered Saccharomyces cerevisiae. *Applied and Environmental Microbiology*, *75*(17), 5536–5543.

van Rosmalen, M., Krom, M., & Merkx, M. (2017). Tuning the flexibility of Glycine-serine linkers to allow rational Design of Multidomain Proteins. *Biochemistry*, *56*(50), 6565–6574.

Waldo, G. S., Standish, B. M., Berendzen, J., & Terwilliger, T. C. (1999). Rapid protein-folding assay using green fluorescent protein. *Nature Biotechnology*, *17*(7), 691–695.

Yi, B., Kang, B. E., Lee, S., Braubach, S., & Baker, B. J. (2018). A dimeric fluorescent protein yields a bright, red-shifted GEVI capable of population signals in brain slice. *Scientific Reports*, *8*(1), 15199.

Zhong, W., Zhang, Y., Wu, W., Liu, D., & Chen, Z. (2019). Metabolic engineering of a Homoserine-derived non-natural pathway for the De novo production of 1,3-Propanediol from glucose. *ACS Synthetic Biology*, *8*(3), 587–595.

CHAPTER THREE

Engineering crowding sensitivity into protein linkers

Theodoros Pittas[a,b,†], Weiyan Zuo[a,b,†], and Arnold J. Boersma[a,b,*]
[a]DWI-Leibniz Institute for Interactive Materials, Aachen, Germany
[b]Institute of Technical and Macromolecular Chemistry, RWTH Aachen University, Aachen, Germany
[*]Corresponding author: e-mail address: boersma@dwi.rwth-aachen.de

Contents

[†] These authors contributed equally.

Methods in Enzymology, Volume 647
ISSN 0076-6879
https://doi.org/10.1016/bs.mie.2020.09.007

Abstract

The intracellular environment contains a high concentration of biomacromolecules that present steric barriers and ample surface area for weak chemical interactions. Consequently, these forces influence protein conformations and protein self-assembly, with an outcome that depends on the sum of the effects resulting from crowding. Linkers are disordered domains that lack tertiary structure, and this flexible nature would render them susceptible to compression or extension under crowded conditions, compared to the equilibrium conformation in a dilute buffer. The change in distance between the linked proteins can become essential where it attenuates protein activity. In this chapter, we first discuss the experimental findings in vitro and in the cell on how linkers and other relevant macromolecules are affected by crowding. We focus on the dependence on the linker's size, flexibility, and the intra- and intermolecular interactions. Although the experimental data on the systematic variation of proteins in a buffer and cells is limited, extrapolating the available insights allows us to propose a protocol on how to engineer the directionality of crowding effects in the linker. Finally, we describe a straightforward experimental protocol on the determination of crowding sensitivity in a buffer and cell.

1. Introduction

The intracellular environment is highly crowded with biomacromolecules that include proteins, RNA, and DNA, with total concentrations estimated up to 300 mg/mL for *Escherichia coli* (Cayley, Lewis, Guttman, & Record, 1991; Zimmerman & Trach, 1991). The consequence is that the average protein has a water hydration shell of only 10 water molecules that separate it from the adjacent protein hydration shell (Milo & Phillips, 2015). The intracellular macromolecules take up a significant biopolymer volume fraction of 0.25, hampering each other's movement. Among a variety of effects (Cheung & Gasic, 2018; Fonin, Darling, Kuznetsova, Turoverov, & Uversky, 2018; Ostrowska, Feig, & Trylska, 2019; Rivas & Minton, 2018), the steric repulsion between the biomacromolecules increases the free energy of the system: The decrease in translational degrees of freedom is a decrease in entropy. The decrease in entropy raises the osmotic pressure in the cell. The osmotic pressure can be reduced by reducing biopolymer volume fractions through aggregation, protein folding, and reducing effective radii of disordered/compressible proteins. Protein linkers would fall in the latter category. In the case of protein linkers, the Asakura Oosawa (AO) depletion theory suggests that the depletion of crowders from the interstitial voids of the linker induces an osmotic pressure difference that compresses the polymer-like linker (Asakura & Oosawa, 1954; Kang, Pincus, Hyeon, & Thirumalai, 2015). Simultaneously, the polymer's decreased size provides more translational degrees of freedom for the crowders, in bulk reducing the osmotic pressure. Thus, intracellular crowding may influence the equilibrium conformation of protein linkers.

Macromolecular crowding manifests itself through various physical and chemical pathways. The effect depends, for a large part, on the size of the protein and the crowder and is most significant when the crowders are smaller than the protein under investigation (Minton, 2005; Rivas & Minton, 2016). However, when the crowders are close to the size of water, they are not excluded from the interstitial voids anymore and will not induce an osmotic pressure difference. When, on the other hand, crowders are much larger than the protein, more confinement-type effects are expected, where the smaller protein is moving in a cage between the larger crowders. Confinement should only affect the protein if the cavities are sufficiently small, e.g., similar to the radius of the protein (Zhou, Rivas, & Minton, 2008). In a living cell, proteins and the other macromolecules are not equally spread throughout the cytoplasm. Spatial heterogeneity will emerge because proteins self-sort due to size-dependent depletion forces but also because they interact with each other through nonsteric interactions (Feig, Yu, Wang, Nawrocki, & Sugita, 2017; Spitzer & Poolman, 2013). Nonsteric interactions include, for example, electrostatic, van der Waals, hydrophobic, and H-bonding interactions. Also, further complexity arises because active processes push the intracellular environment out of equilibrium. Hence, the high concentration of biomacromolecules in a cell induces a multitude of effects relevant to protein linkers.

Crowding effects are challenging to predict. Theory and simulation provide understanding from simplified systems, but the extensive coarse-graining can oversimplify the intracellular environment. Buffer solutions that contain artificial macromolecular crowders mimick aspects of the intracellular environment allowing the isolation of the individual effects, but also here oversimplifications are stark, and crowder-specific artifacts arise. While the real intracellular environment provides direct information, clear trends are rare, and it is not easy to identify its source. Therefore, we need to combine theory, simulation, experiments in a buffer, and cell experiments. Arguably the most useful method to determine crowding effects on linkers in these various in vitro and in vivo media is Förster Resonance Energy Transfer (Fig. 1). Because this method is fluorescence-based, it only required a low nM amount of protein, and the advent of genetically encoded fluorescent proteins allows expression and measurement inside cells with

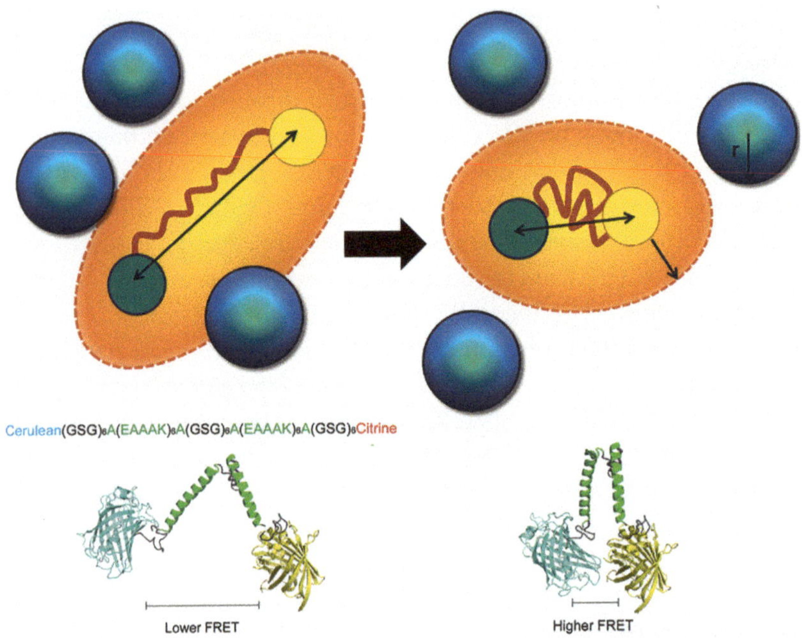

Fig. 1 Schematic overview of the effect of macromolecular crowders on inert linkers. The crowders (blue) cannot enter the excluded volume of the protein, which is defined as the volume of the protein plus the radius of the crowder (yellow/orange ovals). Exclusion of the center of the crowders from the excluded volume leads to depletion zones between the linker or protein segments where crowders cannot enter. This osmotic pressure difference with the bulk is relieved by compaction (right). Compaction results in a decrease in the distance between the proteins that are linked. Below: Linker compression can be followed by FRET using, for example, two genetically-encoded fluorescent proteins that form a FRET pair. Here, the linker contains a $(GSG)_n$ domain that is a random coil and an $(EAAAK)_n$ domain that forms an α-helix. *Bottom figures are adapted from Boersma, A. J., Zuhorn, I. S., & Poolman, B. (2015). A sensor for quantification of macromolecular crowding in living cells.* Nature Methods, 12*(3), 227–229.*

minimal perturbations. The FRET efficiency is related to the distance with the highest sensitivity in the 2–8 nm range for GFP-type fluorescent proteins: This falls ideally in the range of linkers. FRET measurements can have some limitations. The FRET efficiency is highly sensitive to the orientation between the fluorescent proteins, which requires them to move freely. Besides, FRET measurements induce small inaccuracies in distance determinations because short distances contribute more to FRET than long distances do (Liu et al., 2017), which is experimentally unavoidable with the rapid interchange of linker conformations. Because GFP and its derivatives have a typical Förster radius of about 5 nm, and the chromophore is buried in the fluorescent proteins, linkers that generate a radius of 2 nm between the fluorophores are outside the dynamic range and would require, e.g., EPR or NMR. Moreover, such small linkers would not allow free movement of the 2 nm-sized fluorophores due to steric repulsion, which is needed to approximate the average orientation when calculating the distance from FRET. Together, FRET is an excellent method to determine the effective size of a linker under relatively unperturbed conditions allowing to extrude the effects of crowding in artificially crowded media and comparison to the intracellular environment.

In this chapter, we focus on providing a protocol for designing crowding sensitivity into linkers that evolves from FRET-based measurements. Only crowding effects on linkers are discussed and, where needed, findings from other systems will be used to illustrate a relevant principle that applies to linkers. We aim to compare the different parameters that affect the influence of macromolecular crowding, both from artificially crowded and intracellular crowded media. A comparison between different studies is performed on a qualitative level because different fluorescent proteins, linkers, and cell lines are used (Grailhe et al., 2006; Hoppe, Christensen, & Swanson, 2002; Rizzo, Springer, Granada, & Piston, 2004). This overview leads us to extrapolate towards a roadmap on how to design a linker to be crowding sensitive, or not crowding sensitive, in cells or with artificial crowders.

2. The dependence of crowding effects on linker size

The effect of purely sterically interacting crowders on the linker can be estimated by considering the linker as a sphere (Fig. 2). Although a rather crude simplification, this viewpoint has been successfully applied to intrinsically disordered proteins (IDPs), serving as a model for linkers when the linkers are long enough and their behavior not determined by the proteins

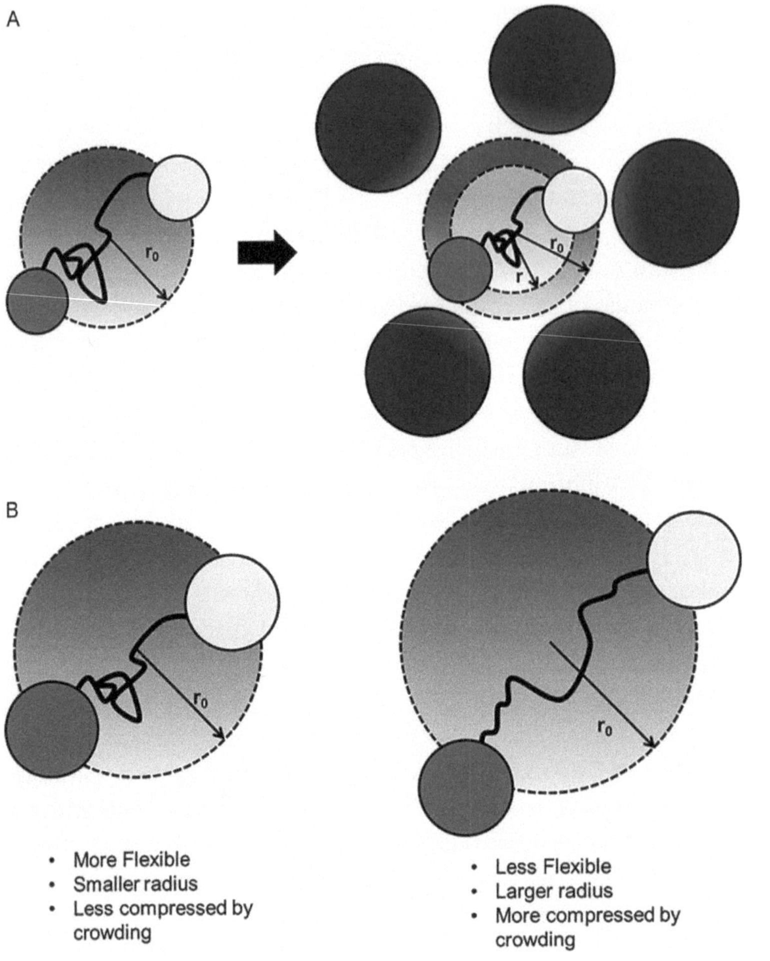

Fig. 2 Steric effects of crowders on linkers. (A) The linker adopts a smaller radius of gyration (or end-to-end distance when measured by FRET) due to the presence of crowders. A larger radius without crowders, r0, leads to a relatively higher compression with crowders, r/r0. (B) The radius without crowders can be increased by using less flexible linkers, and can, therefore, be compressed more by crowders.

to which they are linked. The AO depletion theory suggests that the depletion of crowders from the polymer induces an osmotic pressure difference that compresses the polymer. Simultaneously, the polymer's decreased size provides more translational degrees of freedom for the crowders in the bulk solution, reducing the osmotic pressure. Because longer polymers take up more space, longer linkers experience a more substantial osmotic pressure difference due to crowding and are compressed more. Although the effect strongly depends on the radii of the crowder and the linker, the prediction can be made qualitatively.

To understand the effect of linker size, we designed linkers that were systematically varied in size while retaining the linkers' chemical properties (Liu, Åberg, et al., 2017). Previously, using Ficoll 70, a synthetic sugar-based branched macromolecule often used to induce crowding in buffer with minimal chemical interactions, it was already shown that unstructured proteins are compressed by crowding (Le Coeur, Teixeira, Busch, & Longeville, 2010; Ohashi, Galiacy, Briscoe, & Erickson, 2007). Other mechanisms can also influence the distance between the donor and acceptor (see below); therefore, the chemical properties need to be similar when comparing different linkers. Varying the sizes of the helix $(EAAAK)_n$ and random coil $(GSG)_n$ domains in a linker that connects a FRET donor (mCerulean3, CFP) and a FRET acceptor (mCitrine, YFP) allowed deconvolution of the crowding effects to isolate the effect of linker size. All linkers were compressed by a set of crowders, which included polyethylene glycol, Dextran, Ficoll, bovine serum albumin, and ovalbumin, with an extent depending on the crowder size and concentration. Following arguments based on AO theory, the compression scaled with the effective size of the linker without crowder, which is the distance between the donor and acceptor protein, and not with the number of amino acids. Calculating the apparent distance between the fluorophores with the Förster equation from the experimentally measured FRET efficiencies, it was shown that larger linkers provided a higher relative compression: 20% w/w Ficoll provided a r/r_0 ($r =$ distance with crowder, $r_0 =$ distance without crowder) of 0.9 and 40% w/w 0.85 for a 114 aa linker (7.6 nm diameter), while the smallest linker of 36 aa (5.7 nm diameter) provided a r/r_0 of 0.96 at 20% w/w Ficoll and 0.94 at 40% w/w Ficoll. To calculate these distances, the measured FRET efficiencies were taken to correspond to a single conformation. However, the accuracy can further be improved by correcting for the distribution of conformations using polymer models (Evers, van Dongen, Faesen, Meijer, & Merkx, 2006; van Rosmalen, Krom, & Merkx, 2017). The determined FRET efficiencies were confirmed by time-resolved fluorescence lifetime measurements, time-resolved anisotropy measurements (Currie et al., 2017; Leopold et al., 2018), and single-molecule FRET experiments (Höfig et al., 2018).

In-cell compression depends on the linker length as it does in a buffer. Given that crowding compresses linkers by ~2–15% (Liu, Åberg, et al., 2017), the significant variation between different studies would preclude a precise quantitative comparison (Chen, Mills, & Periasamy, 2003; Periasamy & Day, 1998; Tsien, 1998; Xia & Liu, 2001). The small energies involved also suggests that FRET sensors designed to bind a particular analyte, for example, to quantify zinc(II) in vivo, do not provide a noticeable difference between a buffer and HEK293 cells (Vinkenborg et al., 2009). Therefore, direct measurements of the relation between intracellular crowding and the linker length are sparse. When retaining the relative GSG and EAAAK composition, we showed that the linkers followed a similar compression in *E. coli* as they do in buffer solutions (Liu, Åberg, et al., 2017). In this case, the crowding in the cells was increased by placing the cells in a hyperosmotic medium. A hyperosmotic medium reduces the water content in the cell. The compression of the linker in cells growing in a regular medium is equivalent to 18% w/w Ficoll while with a 500 mM NaCl upshift this reaches an equivalent of ~30% w/w. Especially those sequences that contained the α-helical $(EAAAK)_n$ domains followed the same dependence on the linker size and volume fraction of crowder as with Ficoll. The linker with the highest diameter arrived at an r/r_0 of 0.91 for regular growing cells to a 0.87 ratio for 500 mM NaCl upshift. In the absence of $(EAAAK)_n$, the linker consisted of exclusively random coil $(GSG)_n$ and provided a marginally higher FRET efficiency in cells than buffer solutions (r/r_0 0.98) with almost no length dependence. This dependence of the intracellular crowding effects on the primary sequence is in contrast to the in vitro findings. Hence, while the compression scales with the radius of gyration of the linker and the volume fraction of the crowder both in cells and buffer, the linker's primary structure can contribute in cells, and sequences of similar structural properties should be compared to determine the effect of linker size in cells.

3. The dependence of crowding effects on linker flexibility

In principle, the linker flexibility, or rigidity, can be approximated by a random walk model where the step size is a measure for the flexibility. In polymer-terms, this would be the Kuhn length. The Kuhn length relates to the persistence length, which is the length below which the linker behaves as a rigid rod, and above it behaves like a random walk. Decreasing the linker flexibility increases the radius of gyration of a linker because larger steps in the random walk are taken (an elephant that walks 4 m in two random steps gets further than an ant that walks 4 m in many more small random steps) (Fig. 2). Thus, decreasing the flexibility should increase the compressibility of a linker by crowding if a sphere approximated it because larger spheres are compressed more. This approximation breaks down in the limit of a rigid rod because the linker will no longer be dynamic and compressible.

In artificially crowded buffer solutions, linkers of varying flexibility have been tested. To achieve this, we generated a series of sensors by adjusting the length of rigid $(EAAAK)_n$ and flexible $(GSG)_n$ domains of the linker and compared FRET efficiency in Ficoll solution (Liu, Åberg, et al., 2017). In the absence of crowders, lower FRET efficiency of helix-containing linkers indicated that replacing flexible domains by rigid structures increased the linker's radius of gyration. This finding correlates with the calculated hydrodynamic radius using time-resolved fluorescence anisotropy and fluorescence correlation spectroscopy (FCS) (Lee et al., 2018). In solutions with artificial crowders, all constructs were compressed, while those with larger radii were more susceptible to the crowding effect by showing a more considerable compression. Hence the effect of crowding on linkers with different flexibilities can, in principle, be accounted for by only assessing the radius without crowder. As expected, when taking very rigid linkers instead, the compressibility due to crowding disappears: An oligonucleotide labeled with a FRET donor and acceptor dye at either end was only sensitive to crowding when a flexible single-strand nucleotide hinge region was placed between rigid double-strand nucleotide arms. When instead, this single-stranded region was shortened, crowding sensitivity decreased accordingly, and no change occurred with fully double-stranded constructs (Murade & Shubeita, 2019). Hence, there is an optimum in linker flexibility: On the one hand, the crowding effect is increased by reducing the flexibility due to an increase in the radius of gyration and, on the other hand, reducing the flexibility too much will reduce the intrinsic compressibility, decreasing the crowding effect.

Inside *E. coli* cells, the linkers based on $(EAAAK)_n$ and $(GSG)_n$ domains have a dependence on the α-helix content, as described above (Liu, Åberg, et al., 2017). Here, the compression depended more on the ratio of (EAAAK)/(GSG) repeat units than on the distance between the fluorophores, in contrast to experiments in a buffer with crowders. Adding helices reduces the linker's overall flexibility, but because this dependence of (EAAAK) content was not observed in crowded buffer solutions, we postulate that this dependence may have other causes (see below). Other effects can arise due to the chemical nature of the amino acids that induce rigidity. In the absence of other studies investigating linker flexibility in cells, which could for example be probed by varying the G/S ratio (van Rosmalen et al., 2017), we assume that the dependence on flexibility in cells is similar to crowding in buffer solutions, hence mainly influencing compression by increasing the radius of gyration in the absence of crowder.

4. How do interactions within the linker influence the effect of crowding?

When a linker possesses weak attractive chemical interactions between two of its residues or segments, it equilibrates between two states: The random coil and the state where the segments interact (Fig. 3). The state where the segments interact will have a lower volume. Hence, the random coil conformation will be affected more by macromolecular crowding and depopulates in favor of the state where the segments bind each other. Because these linkers are energetically easier to compress, being aided by the associative interactions, the crowders' effect should be more substantial compared to the absence of intramolecular interactions where steric repulsion between the linker segments operates. Sources of intramolecular interactions are based on hydrophobic, electrostatic, H-bonding, and Van Der Waals interactions derived from the amino acid composition. In contrast to associative interactions, a linker with high internal repulsion is expected to be compressed less due to the unfavorable energy contribution.

Intramolecular interactions are well known to occur in IDPs, for example, between hydrophobic residues (Miller, Kim, & Mittal, 2016) or charged residues (Uversky, 2009). These can be either repulsive or attractive. From a series of IDPs in the presence of PEG crowding, it was observed that the compression was determined by the radius of the protein but modulated by the internal chemical composition, where hydrophilic proteins with internal charge repulsion that render them disordered were compressed most (Kang et al., 2015; Soranno et al., 2014). In contrast, when a protein contained more attractive interactions, it was close to folding and, due to its decreased size, compressed less. Simulations carried out that varied the repulsive/attractive interactions between the segments showed that, on the one hand, the intrachain attraction between hydrophobic residues might help to stabilize more compact conformations that are entropically disfavored. On the other hand, compression of the purely repulsive excluded-volume chain without intrachain interaction is lower due to the entropic penalty (Miller et al., 2016). Hydrophobic chains that are already

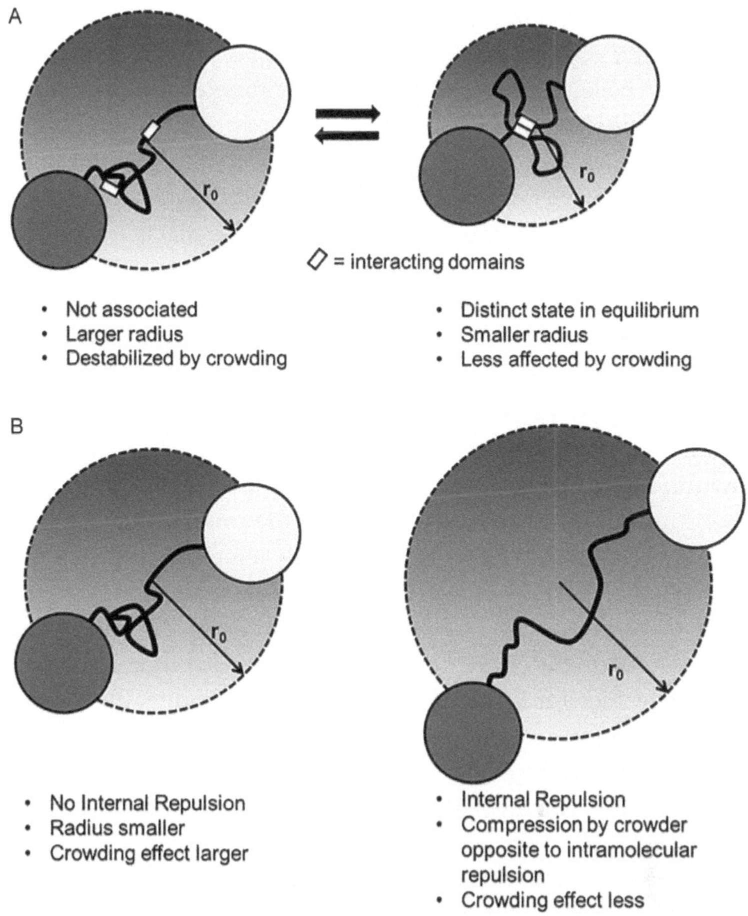

Fig. 3 The role of intramolecular interactions on crowding effects. (A) Attractive interaction by, for example, hydrophobic interactions or opposite charges, induces an equilibrium between the conformation where the domains interact and where they do not. In the conformation where the domains do not interact, the radius of gyration is higher and, therefore, more destabilized by crowding. (B) Internal repulsion by, for example, like-charges increases the radius of gyration. Although a larger radius of gyration would provide more significant crowding effects, this is balanced by the intramolecular repulsive force.

compact are also not compressed. A recent screen of four natural disordered proteins coupled to fluorescent proteins showed a wide array of behaviors depending on the solution properties, including macromolecular crowding induced by PEG and Ficoll (Moses et al., 2020). This protein-specific behavior can be explained by a complex balance between intramolecular interactions and the interactions with the solute-containing solvent. Because these disordered proteins possess a heterogeneous primary and secondary structure, it is challenging to predict how the intramolecular interactions will develop under various solution conditions. Together these simulations and in vitro data suggest that crowding effects are most significant for linkers of intermediate attractive intramolecular interactions, of which the net contribution is not straightforward to determine, while the effects are less at either end of the scale.

Even less clear data is available on the relation between intramolecular interactions in linkers and crowding effects in cells. The effect of crowding on IDPs or linkers that contain charged residues, either attractive or repulsive, is dominated by the ionic strength to screen the electrostatic interactions. In the case of an ionic strength sensor based on a helix-coil-helix-motif, where one helix was positively charged and the other negatively charged (Liu, Poolman, & Boersma, 2017), the ionic strength dominated over the effect of crowding in HEK293 cells. Here, hyperosmotic stress, where both the macromolecular crowding and ionic strength increase, increased the linker radius, while from crowding, a decrease would be expected. Although the linker that contains opposite charges is still sensitive to crowding by Ficoll as a consequence of its flexible nature, the electrostatic effect strongly outweighs the crowding effect. In the case of the highly charged IDP ProTα, which is one of the most highly negatively charged peptides, salt decreased its radius of gyration, relieving internal repulsion (König et al., 2015). However, intracellular crowding itself did not alter its radius compared to buffer. In this case, the internal repulsion is possibly too high to overcome for the moderate in vivo crowding effects. Together, in comparison with neutral linkers in cells as described above and the findings in vitro, intramolecular interactions strongly modulate crowding sensitivity and can easily overcome them.

A consequence of the presence of intramolecular interactions is that linker dimerization or oligomerization could occur. Dimerization reduces the total volume of the dimerizing proteins compared to the sum of the volumes of the monomeric linkers. Self-association occurs in archetypical aggregating peptides as linkers such as Huntingtin exon 1, containing the polyQ motif that misfolds into a β-strand that seeds aggregation (Bates, 2003; Büning et al., 2017). Next to misfolding, many native disordered proteins are involved in the formation of dynamic liquid-like condensates by complexing proteins, oligonucleotides, or themselves. In general, the propensity to undergo these events is accelerated by crowders as condensation reduces the total volume of the condensate forming proteins (Mitrea & Kriwacki, 2016). In general, these effects depend on the concentration of the protein itself. They may require high-affinity interactions to achieve oligomerization/condensation in vivo, either obtained through multivalent interactions by stickers and spacer domains or a specific fold that functions as a seed for aggregation. Thus, linkers without sticky domains or specific misfolding/aggregation behavior, are not expected to undergo these effects at reasonable physiological concentrations (that is, not highly overexpressed).

5. How do associative interactions of the linker with sticky crowders modulate the steric crowding effect?

When a linker interacts attractively with the crowders, in contrast to the purely steric repulsion discussed above, the crowders will no longer be excluded from the linker (Fig. 4). The eventual outcome on the linker depends on the number and distribution of interaction sites, the strength of the interaction, the size of the interacting macromolecule(s), and crowder concentration. Hence, attractive interactions can both compress and extend a linker, modulating the steric compression effect.

In vitro, artificial crowders may engage in nonspecific chemical interactions with the linkers, dominating the steric effects. As an example, where a disordered peptide is elongated, urea-unfolded apoazurin adopted an even more elongated conformation due to the presence of dextran 20 (Zegarra et al., 2019). The elongated shape of dextran combined with its associative interactions with the unfolded apoazurin elongated the apoazurin. In various cases, the contribution of the crowder-specific chemical interaction component with the protein modulates the crowding effect. By applying small-angle neutron scattering, both compaction and extension of the FlgM conformation were observed due to the presence of the protein-based crowders BSA or lysozyme. In contrast, compaction was observed with the carbohydrate-based crowders Ficoll or Dextran (Banks, Qin, Weiss, Stanley, & Zhou, 2018). This crowder-dependence could be explained by the existence of attractive interactions between FlgM and the protein-based crowders. Similarly, crowder-dependent repulsive or attractive interactions with IDPs changed the magnitude of IDPs' compression (Soranno et al., 2014). By increasing the chemical inertness in linkers with, for example, the $(EAAAK)_n$ and $(GSG)_n$ motifs that are relatively hydrophilic and neutral, less crowder-specific behavior is observed. However, even in this case, an exceptional sticky crowder such as gamma-globulins, which is the antibody-containing fraction of bovine serum, can diminish the compression.

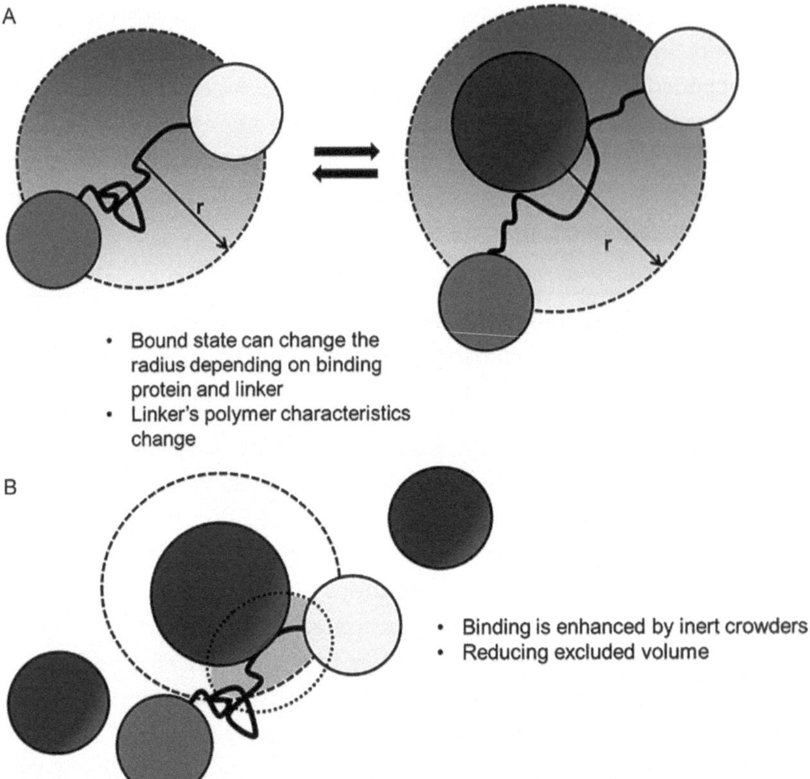

Fig. 4 Intermolecular interactions with linkers. (A) Associative interactions with crowders lead to a change in the conformation of the linker. The linker will adopt a different radius of gyration. Depending on the precise binding geometries, one can elongate or shorten the size of a linker. (B) The presence of inert crowders enhances the binding of a sticky crowder because the inert crowders are depleted from the overlap volume, which is the zone where the excluded volume of the sticky crowder and the excluded volume of the linker segment overlap, allowing for an additional attractive force between the linker and the sticky crowder.

Inert crowders increase the affinity of sticky proteins to protein linkers because the resulting complex takes up less volume. However, linkers are possibly less susceptible to this increase in affinity due to crowding compared to a folded protein (Zosel, Soranno, Buholzer, Nettels, & Schuler, 2020). Here, the volume of the linker segment that binds needs to be taken into account instead of the larger volume of the complete protein. Therefore, the volume gain upon association is less. This argument explained the binding affinity increase between an IDP and an approximately globular protein due to the presence of PEG accurately. Although inert crowders enhance the binding affinity to sticky crowders, the chemical nature of the linker and the sticky crowder determines its ultimate strength.

A dense solution of macromolecules can induce other effects than the depletion effect or direct binding of the linker to a sticky macromolecule. The large Ficoll and its monomer both stabilized chymotrypsin inhibitor 2, which is not expected based on depletion interactions. Because these molecules bound the protein only weakly (Benton, Smith, Young, & Pielak, 2012), it was postulated that preferential hydration of the protein induced by the high concentration of solutes stabilized the protein. We suggest that these effects should be less when well-hydrated amino acids are used. Together, crowder-specific chemical interactions can arise but with a magnitude that strongly depends on the surface chemistries of the linker and the crowder.

Determining the effect of intermolecular interactions on extension or compression of linkers inside a living cell is a challenge, due to the diversity of biomacromolecules and the spatiotemporal continuum of dynamic interactions. A variety of studies that use NMR to study the conformation of proteins in living cells is available (Danielsson et al., 2015; Kurnik, Hedberg, Danielsson, & Oliveberg, 2012; Sharaf, Barnes, Charlton, Young, & Pielak, 2010), where the stability of proteins is enhanced or compromised due to nonspecific interactions, also termed quinary structure. Simple mutations can reduce or enhance stability by altering the interaction potential with the proteome (Gnutt et al., 2019). Thus, in many cases, a counterintuitive result concerning crowding may be the result of nonspecific chemical interactions with the intracellular crowders. A PEG equipped with a donor and acceptor dye reported no change in FRET efficiencies between the cytoplasm of HeLa cells and a diluted aqueous buffer solution (Gnutt, Gao, Brylski, Heyden, & Ebbinghaus, 2015), suggesting compensation of excluded volume interactions with chemical interactions. The PEG was compressed by increasing crowding through hyperosmotic stress, indicating the contribution of the increasing steric component that overcomes the chemical interaction component.

Well-studied conformationally dynamic FRET probes that are based on fluorescent proteins connected via small linkers (<6 aa) to a globular substrate binding protein show, qualitatively, similar effects as the larger disordered linkers in the presence of crowders. A FRET sensor for ATP, mutated to remove ATP binding affinity (Dteam), retained the same diameter in *E. coli* as in buffer, while measurement of the sensor in cell lysates extended the sensor, in contrast to more inert crowders such as PEG and Ficoll that compressed the sensor (Groen et al., 2015). Hence also here, the chemical interactions dominate at lower intracellular crowding and the excluded volume at higher crowding. On the other hand, the globular phosphoglycerate kinase (PGK) equipped with fluorescent proteins was found to be inert in cells and compressed, and most sensitive to the excluded volume (Ebbinghaus, Dhar, McDonald, & Gruebele, 2010). The extracellular protein VlsE was extended due to increased chemical interactions in its non-native environment inside (Guzman, Gelman, Tai, & Gruebele, 2014). Of note is that both proteins that are influenced by chemical interactions, Dteam and VlsE, are both positively charged.

It is thus beneficial to increase the inertness, or repulsion, of the linker when crowding effects are desired. In the case of the GSG/EAAAK containing linkers, an increase in sensitivity of linkers containing helical EAAAK motifs vs random coil GSG motifs toward crowding was observed inside cells (Liu, Åberg, et al., 2017). The GSG motif does not contribute to the compression by crowding inside cells, which is dictated by the EAAAK motifs. A possible explanation is that the charged glutamate and lysine moieties in the helices are well hydrated, preventing interaction with the proteome. In addition, these helices are zwitterionic in nature, possibly providing some additional repulsion. The GSG motif is neutral and hydrophilic as well: lysine, glutamate, and serine are the most efficiently hydrated amino acids and therefore the most abundant amino acids on the intracellular protein surfaces (White et al., 2012). However, the amide backbone of the GSG motif is exposed for interaction, and its flexible nature may provide a larger surface area to interact with compared to a rigid helix. Together, these data show that tuning the chemical interaction potential with the other crowders in vivo is possible yet subtle.

6. Other crowding effects on linkers

Crowding can induce other effects on linker behavior, which have received somewhat less attention or have been difficult to quantify. For example, the spatiotemporal variation in crowding determines linker distribution in the cell: A linker may be excluded from a highly crowded region due to depletion effects, and report a lower crowding than the spatial average of the complete cell. Similarly, the stickiness of the linker to a specific part of the cell may direct it to the corresponding regions. Further, the magnitude of crowding effects depends not only on the linker size but also on the size of the linked proteins. When the linked proteins are more substantial, a higher crowding effect is expected. Further, the linked proteins may undergo intramolecular interactions, in which case the same conclusions would hold as discussed above in the linker-only case.

7. Protocol

7.1 Maximizing linker compressibility by crowding in vitro

1. Increase the length to achieve an as large as possible radius of gyration; the longer the linker, the more it is compressed

   In artificially crowded buffer, constructs with small linkers (for example 12 aa (Ohashi et al., 2007) or <6 aa (combined with globular proteins (Groen et al., 2015)) already show an increase in FRET by Ficoll 70. The compression increases with linker size: $(GSG)_{12}$ has a diameter of 5.7 nm and is compressed by 6% and $(GSG)_{24}$ with a diameter of 6.3 nm is compressed by 9% with 30% w/w Ficoll.

2. Obtain a linker that is as rigid as possible to increase the radius of gyration, yet retains sufficient flexibility to be compressed by crowding

   Decreased flexibility can, for example, be achieved by incorporating α-helices, which increase the radius of gyration and the crowding sensitivity. As an example, $(GSG)_{24}$ contains 72 amino acids and provides a diameter of 6.3 nm while $A(EAAAK)_6(GSG)_2(EAAAK)_6A$ contains 66 amino acids providing a diameter of 7.3 nm (Liu, Åberg, et al., 2017), assuming that the FRET efficiency relates to a single conformation. This increase in diameter is due to the helices' rigidity. Indeed, the latter is compressed by 11% and the former by 9% by 30% w/w Ficoll. In our experience, a short disordered domain of 6 amino acids between rigid α-helices is sufficient to maintain flexibility.

3. Engineer a low content of attractive intramolecular interactions. Decreasing interactions is achieved by preventing repulsive interactions of like-charges or only hydrophilic amino acids, while a too high attractive potential can be avoided by preventing too many hydrophobic amino acids

   Using a hydropathicity scale such as presented by Kyte & Doolittle (1982) to maintain an index below 0: GSG and EAAAK domains vary between 0 and −0.6 (negative is hydrophilic, positive hydrophobic). Prothymosin α, a highly negative disordered protein, is about −2, and both the C and N terminal half of this molecule are significantly compressed by crowding. Proteins with distinct domains above 0 may behave as folded proteins: ACTR, the disordered domain of steroid receptor coactivator 3, has two domains above 0 and is considered molten globule-like, and less compressed by crowding. Because the final result is difficult to predict, we recommend minimally charged species, e.g., charge neutral or a negative charge, and intermediate hydrophilicity.

4. Reduce nonspecific interactions between linkers and crowder molecules. These interactions are somewhat predictable based on the chemical nature of both linker and crowder surfaces and exposed interaction sites. Ficoll appears to be a relatively inert crowder in comparison to others, possibly because of its surface that consists of well hydrated and uncharged hydroxyl groups. As in the case of intramolecular interactions, intermolecular interactions are, for a large part, prevented by hydrophilic and neutral linkers such as $(EAAAK)_n$ and $(GSG)_n$ but specific sequences will depend strongly on the properties of the crowder. In general, the requirements are not strict when an inert crowder is used, and direct interaction with Ficoll is rare.
5. Take the size of the linked protein into account. Based on crowding theories, enlarging the entire construct would increase its sensitivity to crowding. The linked proteins itself may not be needed to drive compaction because replacing GFP derivatives with small-molecule dyes retains the crowding effect. In the absence of a direct experimental comparison, we hypothesize that compression is higher with bigger linked proteins.

8. Maximizing linker compressibility by crowding in cells

In general, the same considerations apply as in vitro. In this section, we focus on some modifications needed to address the challenging environment in the cell.

1. Most important is to increase the inertness of the linker to prevent associative intermolecular interactions

   Incorporation of the most common native intracellular surface-exposed amino acids, which are glutamate, lysine, and serine, may provide inertness because they are well hydrated. Further, a net positive charge is to be prevented because cationic proteins show high nonspecific interactions inside cells, as in the case of VlsE and Dteam:

A small negative charge as in PGK or GFP, or neutral as in the $(GSG)_n$ and $(EAAAK)_n$ based linkers, prevents interactions with the predominantly negatively charged biomacromolecules in cells. Besides, surface-exposed hydrophobic domains should be prevented, or hydrophobic domains that can become exposed in less well-folded structures.

2. Increase the linker diameter to provide higher compression in cells. In *E. coli*, linkers with a diameter of >6.6 nm were compressed as they were in a buffer. In buffer, the compression scales with the biopolymer volume fraction and the size of the linker. The relation has not yet been tested in mammalian cells.
3. Fine-tune the linker's secondary structure for crowding sensitivity in cells. Primary structure variations and the presence of secondary structures (e.g., α-helices) can alter the crowding effects in cells. For example, maximizing the EAAAK content with respect to the GSG content maximized the crowding sensitivity: Linkers with the coil-helix motif that are >6.6 nm in diameter are compressed in *E. coli* by about 10% while the more flexible coil-only constructs that span a diameter up to 6.3 nm only by 2%. In the absence of a definitive explanation, we suggest incorporating helices to prevent interactions with the GSG amide backbone.
4. Prevent intramolecular associative and repulsive interactions. Engineering interactions within the linker may induce sensitivity to, e.g., the ionic strength or small molecules. Opposite charges in the linkers render it sensitive to the ionic strength, similar to like-charges.
5. To determine the sensitivity to macromolecular crowding in cells, place the cell under an osmotic upshift. Reducing the water content leads to an increase in the concentration of cellular biomacromolecules. An osmotic upshift is done by adding a high concentration of salt or a sugar that is not taken up by the cell. For *E. coli*, up to 1 Osm NaCl or sucrose is sufficient, for *Saccharomyces cerevisiae* 2 Osm NaCl and HEK293 up to 400 mOsm sorbitol. Because cells are very adapt to recover from the stress by taking up potassium or compatible solutes, these experiments are to be executed in the absence of these ions or molecules, or rapidly executed, before recovery ensues. The precise timing depends on the species but is generally in the seconds-minute timescale.

9. Preventing crowding sensitivity

In cases where crowding effects are not desired, we present the following protocol.

1. Crowding effects on linkers are generally small in cells, up to 18% decrease in diameter in cells and much higher with some artificial crowders such as PEG. The intracellular compression and the energy involved is much smaller than those acting on larger complexes such as protein condensates, microtubule bundles, or the complexes involved in DNA replication or transcription/translation. We calculated that the effect of crowding on the Dteam sensor (not optimized for crowding sensing) equaled the energy of one hydrogen bond (Groen et al., 2015). Hence, it appears to be easier to prevent crowding sensitivity than to design it into a linker.
2. Shorten the effective diameter of the linker. The diameter can be reduced by decreasing the number of amino acids and increasing its flexibility by preventing secondary structure.
3. Maximize the rigidity of the linker. Inserting a rigid α-helix (or bundle) without any flexible domain should prevent crowding.
4. Increase the intramolecular interactions to achieve a stable conformation that is not affected by crowding. A stable conformation can be achieved with known protein interaction domains or hydrophobic patches.
5. Increase the charge of the linker so that its conformation will be determined primarily by the ionic strength, as exemplified by charged IDPs (König et al., 2015) and a specifically designed ionic strength sensor (Liu, Poolman, & Boersma, 2017).
6. Counteract crowding effects with nonspecific associative interactions. Using less well-hydrated amino acids is most likely sufficient for other interactions to dominate over the crowding effect.

10. Methods in buffer

10.1 Definition

Measurement of the effect of crowding on linkers in a buffer.

10.2 Rationale

Here we describe a quantitative and straightforward protocol on determining the effect of artificial crowder on a linker using FRET and fluorescent proteins.

10.3 Materials, equipment, and reagents

- Fluorometer
- Ficoll PM70

- Sodium Phosphate (10 mM), buffered at pH 7.4, supplemented with 100 mM NaCl and 0.2 mg/mL bovine serum albumin
- The purified linker conjugated with an N-terminal mCerulean3 and a C-terminal mCitrine
- Quartz cuvette for fluorescence, volume ~100 μL
- Balance, 15 mL capped tubes (Falcon/Greiner)
- Proteinase K (5.0 mg/mL in water)
- PMSF (100 mM in isopropanol)

10.4 Protocols

1. Weigh Ficoll PM70 and add sodium phosphate buffer up to 5 g total, resulting in 0, 10, 20, 30, and 40% w/w Ficoll solutions in 15 mL tubes. Cap the tube and mix gently by hand, leave the tube for several hours for complete dissolution, and gently mix again before the experiment.
2. Add the protein (~10 nM) and mix thoroughly by pipette to 0.5–1.0 mL of all the Ficoll concentrations.
3. Place 100 μL of the Ficoll solution of the desired concentration with the protein in the quartz cuvette.
4. Excite the sample at 420 nm and record its fluorescent emission spectrum at room temperature on a fluorometer.
5. Determine the donor emission without acceptor by adding 2.0 μL Proteinase K into the solution and mix by pipette. Proteinase K cleaves the linker but leaves the fluorescent proteins intact.
6. Incubate at 20 °C for 1 min.
7. Add 2.0 μL PMSF to quench the reaction and record the fluorescence emission spectrum.
8. Remove the content, clean the cuvette, and measure the Ficoll solution without the protein to determine the background signal.
9. Export the data and subtract the background spectrums from the protein-containing spectrums for each Ficoll concentration in Excell.
10. Calculate mCitrine/mCerulean (525 nm/475 nm fluorescence) for the different Ficoll solutions. Plot the corresponding ratios versus the Ficoll concentrations. The trend will show how the emission ratio changes with the increase of Ficoll concentration, indicating that the crowding effect on the linker.
11. The FRET efficiency is calculated using FRET efficiency $= 1 - \frac{F_{DA}}{F_D}$, where F_{DA} is the fluorescence of the donor before proteinase

K treatment, and F_D is the donor fluorescence after proteinase K treatment.

12. The distance is correspondingly determined using the Förster equation $E = \frac{1}{1-\left(r/R_0\right)^6}$, in which E is the FRET efficiency, r is the distance between the fluorophores, and R_0 is the Förster distance where the efficiency is 50%. The Förster distance is fluorescent protein-dependent and is about 5.1 nm for a randomly orienting mCerulean-mCitrine pair. Determining the r/r_0 gives the relative compression, in which r is the radius with crowder and r_0 the radius without crowder. The r is taken to correspond to a single FRET efficiency, which deviates from the true r resulting from the ensemble of conformations that each contribute to the observed FRET efficiency (Evers et al., 2006; Liu, Åberg, et al., 2017).

10.5 Precursor techniques

The gene needs to be constructed by mutagenesis to insert the desired linker $(GSG)_6A(EAAAK)_6A(GSG)_6A(EAAAK)_6A(GSG)_6$ or variations thereof between an N-terminal mCerulean3 and a C-terminal mCitrine, with a 6 His-tag fused at the N-terminus of the construct. The gene was inserted in a pRSET A vector under a T7 promoter and with a gene encoding for ampicillin resistance. Genetic constructs can be obtained from the author or obtained via standard cloning techniques.

The purified linker conjugated to the fluorescent proteins needs to be expressed in *E. coli* and purified by NiNTA affinity chromatography and size exclusion chromatography according to literature procedures (Boersma et al., 2015). Size exclusion chromatography removes any remaining aggregates. Fresh samples are to be used to prevent aggregation that may occur in old samples.

10.6 Analysis and statistics

The high viscosity of the concentrated Ficoll solution requires careful pipetting, mainly because the data needs to be subtracted by a Ficoll-only solution, which has a relatively high background signal. In general, however, the measurement leads to high precision and low error bars, especially with three individual repeats. The determination of the radius does not accurately represent the real average radius in a flexible molecule because the small distances contribute more than the long distances. The real distances are ~1–2 nm longer than those measured by FRET, as determined by our molecular dynamics simulations. In this simulation, we determined the distance from FRET in simulation as in a wet experiment: We first calculated the distances between the fluorophores for each conformation in the simulation. Next, these distances were converted to the corresponding FRET values with the Föster equation, and subsequently averaged. From this average FRET, we recalculated the distance and compared this "FRET distance" with the real average distance of all the conformations, yielding a difference of about 1–2 nm. The distance calculation is further perturbed by the orientation factor, which may not be the usual 2/3 in dense conformations with restricted movement due to steric hindrance. Also, the Förster radius is taken from literature, but a comparison between different labs shows a variation of about ~0.2 nm.

10.7 Related techniques

Radius determinations by FRET in crowded solutions can also be performed with time-resolved fluorescence measurements, time-resolved anisotropy measurements, and single-molecule FRET experiments. Linker behavior can be modeled with Gaussian Chain and Worm Like Chain models (Evers et al., 2006; Liu, Åberg, et al., 2017). Alternatively, SAXS and NMR give insight into the dynamics and radius of the linker, which have the advantage that they do not need labeling by fluorophores and provide complementary information.

10.8 Pros and cons

Pros	Cons
The measurements are precise and easy to execute	Determining the effect of crowding on linkers with FRET requires fusion with fluorescent proteins
Easy measurement and variation of linkers provides insight into crowding effects with Ficoll or any other crowder that can be obtained at high concentrations	There is some systematic deviation associated with distance determinations with FRET, and care should be taken accordingly

10.9 Alternative methods/procedures

Molecular dynamics simulations using, for example, SWISS-Model or software PYMOL is helpful for further analysis of linker structure and behavior.

10.10 Troubleshooting and optimization

Problem	Solution
A high variation between measurements	Careful pipetting circumvents imprecision due to Ficoll solution viscosity. Further, addition of BSA (~0.2 mg/mL) prevents a decrease of the concentration of FRET probe in solution (possibly due to sticking to the cuvette walls)
There is no change in FRET because the acceptor fluorescence is not visible	The linker can get cleaved upon long-term storage. Aliquot purified protein and store at -80 °C fridge. Avoid melting and freezing cycles multiple times. Use freshly thawed samples for measurements. The latter also prevents aggregate formation

10.11 Summary

This FRET method allows straightforward and rapid determination of crowding effects on linkers in buffer solutions with different concentrations of crowders.

11. Methods in *E. coli*

11.1 Definition

Quantification of the sensitivity of linkers to crowding in *E. coli*.

11.2 Rationale

Here we describe a method to determine the linker sensitivity to intracellular crowding inside a model organism, *E. coli*. The method is based on FRET with the same protein constructs as in buffer solutions and was applied to the $(EAAAK)_n$ and $(GSG)_n$ containing linkers described extensively in the text.

11.3 Materials, equipment, and reagents

- *E. coli* BL21 (DE3)
- MOPS Minimal medium. MOPS Minimal medium is an optimized medium for E coli with defined components (here with glucose, but without added amino acids and vitamins) (Neidhardt, Bloch, & Smith, 1974)
- Plasmid (pRSET A) containing the linker fused to the fluorescent protein described above (Boersma et al., 2015)
- Plasmid (pRSET A) containing a nonfluorescent protein (monomeric streptavidin) (Liu, Åberg, et al., 2017)
- Coverslip optionally modified with (3-aminopropyl) triethoxysilane
- 50 mL tubes (Falcon/Greiner)
- Incubator, 1.5 mL tubes, table-top centrifuge
- Confocal microscope equipped with a laser between 405 and 440 nm
- The materials from the buffer method section for calibration

11.4 Protocols

1. Pick a colony from a plate with freshly transformed *E. coli* strain BL21 (DE3) with the expression vector.
2. Innoculate 10 mL of filter-sterilized MOPS minimal medium (pH 7.4) supplemented with 20 mM glucose and 1 mg/mL Ampicillin. The leaky expression of the proteins is sufficient, and IPTG is not used to induce the expression.
3. Culture bacteria in MOPS Mineral medium at 30 °C, 200 rpm to an $OD_{600} = 0.1$–0.2.
4. In parallel, follow steps 1–3 for the plasmid containing a nonfluorescent protein to the same conditions (culture 2). This culture will function as a control and background.
5. Mix cultures 1 and 2 to obtain a ~70:30 ratio of fluorescent cells and control cells.
6. Spin down a total of ~100 μL cells by centrifugation (10 °C, 5000 rpm) and resuspend the pellet by pipetting in
   a. ~30 μL MOPS minimal medium, which will be the "regular" growth condition, or
   b. ~30 μL MOPS minimal medium with the desired amount of NaCl (e.g., 500 mM, retain pH), in the absence of potassium source and glucose, to prevent adaptation of the cells

7. Add 10 μL of the mixture and place it to a coverslip modified with (3-aminopropyl) triethoxysilane and place a glass slide on top.
8. Place the coverslip on a laser-scanning confocal microscope.
9. Use a wide pinhole to collect more photons. Focus on one of the fluorescent channels with low laser power and high gain to prevent bleaching.
10. Excite the donor with 405 nm and measure the emission at 450–505 nm (mCerulean) and 505–780 (mCitrine). Experiments can be compared with the two channels that had the same gain, yet the value was tuned depending on the conditions.
11. Determine the fluorescence for each fluorophore in each cell by ImageJ by selecting the cell in the brightfield channel and superimposing the selection into the corresponding fluorescent channels.
12. Plot the intensity of the donor vs the acceptor for each cell in Origin or Excell. The control cells should be included, and background subtraction is not needed. Fit the data to a linear equation. The slope is the emission ratio. Alternatively, the mCitrine/mCerulean ratio for each cell can be determined (after subtraction of the control cell intensities), and the median or average value is similar to the value obtained by plotting the mCitrine vs the mCerulean.
13. To calibrate the values obtained in cells, determine the corresponding FRET ratios in Ficoll with and without FRET on a cover slide in the microscope.
14. Follow the in buffer procedure to determine the distances from the FRET efficiency and the corresponding compression.

11.5 Precursor techniques

MOPS minimal medium should be prepared according to the protocol published by Neidhardt et al. (1974). Modification of cover slides is described in Boersma et al. (2015).

11.6 Safety considerations and standards

Standard practices regarding working safely with genetically modified organisms should be followed.

11.7 Analysis and statistics

When fitting the intensities, cells with higher intensities contribute stronger. Therefore, it is more precise to select a window of intensities. Because *E. coli* is small, it gives a reasonably high measurement error, which necessitates for at least 100 cells need to be analyzed to provide a precise average. The accuracy is highest when working fast, i.e., centrifugation and measuring within 10 min. Slower handling leads to drops of ~10% in ratio, similar to cells in the stationary phase. The inherent variation between cultures of about 3–4% in the ratio is typical, and therefore >3 biological repeats are needed.

11.8 Pros and cons

Pros	Cons
Straightforward to implement and analyze	Measurement under novel growth conditions requires optimization of promoters and plasmids to obtain sufficient FRET probe concentration that match the growth rate
A skilled experimenter can achieve highly precise measurements	Calibration with standard solutions assumes that other effects do not play a role. Calibration in a buffer is not ideal and should be prevented as much as possible. Where this is not possible, it needs to be verified extensively by testing common molecules at their native concentrations in a buffer
Fluorescent proteins allow measurement at a low concentration and can be fused to any linker, can be easily purified, and are generally inert	Fluorescent proteins are often pH-sensitive and require time to mature. Therefore, different fluorescent proteins need to be tested. Fluorescent proteins that are less sensitive to acidic pH are for example Gamillus (Shinoda et al., 2018) and tdTomato (Botman, Van Heerden, & Teusink, 2020)

11.9 Alternative methods/procedures

Various methods exist to determine FRET in vivo. Here we use cleavage of the linker in vitro and compare the obtained values in cells with standard solutions. The best alternative methods are the cleavage of the linker in cells by engineering in a protease restriction site, fluorescence lifetime imaging (FLIM), or acceptor photobleaching.

In general terms, the same methods can be applied to other cell types such as HEK293 cells, HeLa cells, or *S. cerevisiae*.

11.10 Troubleshooting and optimization

Problem	Solution
Cells fail to grow or overgrow	Due to the long lag-time in this medium, cultures inoculated at the end of day 1 reach the required OD at day 2. 1:3 dilutions of the freshly inoculated cultures ensure cultures at the correct OD at any time of day 2. Triplicate cultures and sufficient inoculation amounts ensure further success in case a culture fails to grow
Dependence on the expression method	Replace fluorescent proteins with faster-maturing proteins. A constitutive promoter also minimizes maturation artifacts. In the case of a too low intensity, employ a stronger constitutive promoter
Low precision and significant outliers are observed	The experiments must be done sufficiently fast. Slow handling speed (>10 min) changes the FRET values

11.11 Summary

This method provides the behavior of a linker with increasing macromolecular crowding induced by osmotic stress, by directly measuring the FRET donor and FRET acceptor emission after donor excitation.

Acknowledgments

This work was supported by The Netherlands Organization for Scientific Research Vidi grant (723.015.002) to A.J.B.

References

Asakura, S., & Oosawa, F. (1954). On interaction between two bodies immersed in a solution of macromolecules. *The Journal of Chemical Physics*, *22*(7), 1255–1256.

Banks, A., Qin, S., Weiss, K. L., Stanley, C. B., & Zhou, H.-X. (2018). Intrinsically disordered protein exhibits both compaction and expansion under macromolecular crowding. *Biophysical Journal*, *114*(5), 1067–1079.

Bates, G. (2003). Huntingtin aggregation and toxicity in Huntington's disease. *The Lancet*, *361*(9369), 1642–1644.

Benton, L. A., Smith, A. E., Young, G. B., & Pielak, G. J. (2012). Unexpected effects of macromolecular crowding on protein stability. *Biochemistry*, *51*(49), 9773–9775.

Boersma, A. J., Zuhorn, I. S., & Poolman, B. (2015). A sensor for quantification of macromolecular crowding in living cells. *Nature Methods*, *12*(3), 227–229.

Botman, D., Van Heerden, J. H., & Teusink, B. (2020). An improved ATP FRET sensor for yeast shows heterogeneity during nutrient transitions. *ACS Sensors*, *5*(3), 814–822.

Büning, S., Sharma, A., Vachharajani, S., Newcombe, E., Ormsby, A., Gao, M., et al. (2017). Conformational dynamics and self-association of intrinsically disordered Huntingtin exon 1 in cells. *Physical Chemistry Chemical Physics*, *19*(17), 10738–10747.

Cayley, S., Lewis, B. A., Guttman, H. J., & Record, M. T., Jr. (1991). Characterization of the cytoplasm of *Escherichia coli* K-12 as a function of external osmolarity: Implications for protein-DNA interactions in vivo. *Journal of Molecular Biology*, *222*(2), 281–300.

Chen, Y., Mills, J. D., & Periasamy, A. (2003). Protein localization in living cells and tissues using FRET and FLIM. *Differentiation*, *71*(9–10), 528–541.

Cheung, M. S., & Gasic, A. G. (2018). Towards developing principles of protein folding and dynamics in the cell. *Physical Biology*, *15*(6), 063001.

Currie, M., Leopold, H., Schwarz, J., Boersma, A. J., Sheets, E. D., & Heikal, A. A. (2017). Fluorescence dynamics of a FRET probe designed for crowding studies. *The Journal of Physical Chemistry B*, *121*(23), 5688–5698.

Danielsson, J., Mu, X., Lang, L., Wang, H., Binolfi, A., Theillet, F.-X., et al. (2015). Thermodynamics of protein destabilization in live cells. *Proceedings of the National Academy of Sciences of the United States of America*, *112*(40), 12402–12407.

Ebbinghaus, S., Dhar, A., McDonald, J. D., & Gruebele, M. (2010). Protein folding stability and dynamics imaged in a living cell. *Nature Methods*, *7*(4), 319.

Evers, T. H., van Dongen, E. M., Faesen, A. C., Meijer, E., & Merkx, M. (2006). Quantitative understanding of the energy transfer between fluorescent proteins connected via flexible peptide linkers. *Biochemistry*, *45*(44), 13183–13192.

Feig, M., Yu, I., Wang, P.-h., Nawrocki, G., & Sugita, Y. (2017). Crowding in cellular environments at an atomistic level from computer simulations. *The Journal of Physical Chemistry B*, *121*(34), 8009–8025.

Fonin, A. V., Darling, A. L., Kuznetsova, I. M., Turoverov, K. K., & Uversky, V. N. (2018). Intrinsically disordered proteins in crowded milieu: When chaos prevails within the cellular gumbo. *Cellular and Molecular Life Sciences*, *75*(21), 3907–3929.

Gnutt, D., Gao, M., Brylski, O., Heyden, M., & Ebbinghaus, S. (2015). Excluded-volume effects in living cells. *Angewandte Chemie International Edition*, *54*(8), 2548–2551.

Gnutt, D., Timr, S., Ahlers, J., König, B., Manderfeld, E., Heyden, M., et al. (2019). Stability effect of quinary interactions reversed by single point mutations. *Journal of the American Chemical Society*, *141*(11), 4660–4669.

Grailhe, R., Merola, F., Ridard, J., Couvignou, S., Le Poupon, C., Changeux, J. P., et al. (2006). Monitoring protein interactions in the living cell through the fluorescence decays of the cyan fluorescent protein. *Chemphyschem: A European Journal of Chemical Physics and Physical Chemistry*, *7*(7), 1442–1454.

Groen, J., Foschepoth, D., Te Brinke, E., Boersma, A. J., Imamura, H., Rivas, G., et al. (2015). Associative interactions in crowded solutions of biopolymers counteract depletion effects. *Journal of the American Chemical Society*, *137*(40), 13041–13048.

Guzman, I., Gelman, H., Tai, J., & Gruebele, M. (2014). The extracellular protein VlsE is destabilized inside cells. *Journal of Molecular Biology*, *426*(1), 11–20.

Höfig, H., Otten, J., Steffen, V., Pohl, M., Boersma, A. J., & Fitter, J. (2018). Genetically encoded forster resonance energy transfer-based biosensors studied on the single-molecule level. *ACS Sensors*, *3*(8), 1462–1470.

Hoppe, A., Christensen, K., & Swanson, J. A. (2002). Fluorescence resonance energy transfer-based stoichiometry in living cells. *Biophysical Journal*, *83*(6), 3652–3664.

Kang, H., Pincus, P. A., Hyeon, C., & Thirumalai, D. (2015). Effects of macromolecular crowding on the collapse of biopolymers. *Physical Review Letters*, *114*(6), 068303.

König, I., Zarrine-Afsar, A., Aznauryan, M., Soranno, A., Wunderlich, B., Dingfelder, F., et al. (2015). Single-molecule spectroscopy of protein conformational dynamics in live eukaryotic cells. *Nature Methods*, *12*(8), 773–779.

Kurnik, M., Hedberg, L., Danielsson, J., & Oliveberg, M. (2012). Folding without charges. *Proceedings of the National Academy of Sciences of the United States of America*, *109*(15), 5705–5710.

Kyte, J., & Doolittle, R. F. (1982). A simple method for displaying the hydropathic character of a protein. *Journal of Molecular Biology*, *157*(1), 105–132.

Le Coeur, C., Teixeira, J., Busch, P., & Longeville, S. (2010). Compression of random coils due to macromolecular crowding: Scaling effects. *Physical Review E*, *81*(6), 061914.

Lee, H. B., Cong, A., Leopold, H., Currie, M., Boersma, A. J., Sheets, E. D., et al. (2018). Rotational and translational diffusion of size-dependent fluorescent probes in homogeneous and heterogeneous environments. *Physical Chemistry Chemical Physics*, *20*(37), 24045–24057.

Leopold, H. J., Leighton, R., Schwarz, J., Boersma, A. J., Sheets, E. D., & Heikal, A. A. (2018). Crowding effects on energy-transfer efficiencies of hetero-FRET probes as measured using time-resolved fluorescence anisotropy. *The Journal of Physical Chemistry B*, *123*(2), 379–393.

Liu, B., Åberg, C., van Eerden, F. J., Marrink, S. J., Poolman, B., & Boersma, A. J. (2017). Design and properties of genetically encoded probes for sensing macromolecular crowding. *Biophysical Journal*, *112*(9), 1929–1939.

Liu, B., Poolman, B., & Boersma, A. J. (2017). Ionic strength sensing in living cells. *ACS Chemical Biology*, *12*(10), 2510–2514.

Miller, C. M., Kim, Y. C., & Mittal, J. (2016). Protein composition determines the effect of crowding on the properties of disordered proteins. *Biophysical Journal*, *111*(1), 28–37.

Milo, R., & Phillips, R. (2015). *Cell biology by the numbers*. Garland Science.

Minton, A. P. (2005). Models for excluded volume interaction between an unfolded protein and rigid macromolecular cosolutes: Macromolecular crowding and protein stability revisited. *Biophysical Journal*, *88*(2), 971–985.

Mitrea, D. M., & Kriwacki, R. W. (2016). Phase separation in biology; functional organization of a higher order. *Cell Communication and Signaling: CCS*, *14*(1), 1.

Moses, D., Yu, F., Ginell, G., Shamoon, N. M., Koenig, P. S., Holehouse, A. S., et al. (2020). *Probing the hidden sensitivity of intrinsically disordered proteins to their chemical environment*. bioRxiv. https://doi.org/10.1101/2020.08.17.252478.

Murade, C. U., & Shubeita, G. T. (2019). A molecular sensor reveals differences in macromolecular crowding between the cytoplasm and nucleoplasm. *ACS Sensors*, *4*(7), 1835–1843.

Neidhardt, F. C., Bloch, P. L., & Smith, D. F. (1974). Culture medium for enterobacteria. *Journal of Bacteriology*, *119*(3), 736–747.

Ohashi, T., Galiacy, S. D., Briscoe, G., & Erickson, H. P. (2007). An experimental study of GFP-based FRET, with application to intrinsically unstructured proteins. *Protein Science*, *16*(7), 1429–1438.

Ostrowska, N., Feig, M., & Trylska, J. (2019). Modeling crowded environment in molecular simulations. *Frontiers in Molecular Biosciences*, *6*, 86.

Periasamy, A., & Day, R. N. (1998). Visualizing protein interactions in living cells using digitized GFP imaging and FRET microscopy. In *Vol. 58. Methods in cell biology* (pp. 293–314). Elsevier.

Rivas, G., & Minton, A. P. (2016). Macromolecular crowding in vitro, in vivo, and in between. *Trends in Biochemical Sciences*, *41*(11), 970–981.

Rivas, G., & Minton, A. P. (2018). Toward an understanding of biochemical equilibria within living cells. *Biophysical Reviews, 10*(2), 241–253.

Rizzo, M. A., Springer, G. H., Granada, B., & Piston, D. W. (2004). An improved cyan fluorescent protein variant useful for FRET. *Nature Biotechnology, 22*(4), 445–449.

Sharaf, N. G., Barnes, C. O., Charlton, L. M., Young, G. B., & Pielak, G. J. (2010). A bioreactor for in-cell protein NMR. *Journal of Magnetic Resonance, 202*(2), 140–146.

Shinoda, H., Ma, Y., Nakashima, R., Sakurai, K., Matsuda, T., & Nagai, T. (2018). Acid-tolerant monomeric GFP from Olindias formosa. *Cell Chemical Biology, 25*(3), 330–338.

Soranno, A., Koenig, I., Borgia, M. B., Hofmann, H., Zosel, F., Nettels, D., et al. (2014). Single-molecule spectroscopy reveals polymer effects of disordered proteins in crowded environments. *Proceedings of the National Academy of Sciences of the United States of America, 111*(13), 4874–4879.

Spitzer, J., & Poolman, B. (2013). How crowded is the prokaryotic cytoplasm? *FEBS Letters, 587*(14), 2094–2098.

Tsien, R. Y. (1998). The green fluorescent protein. *Annual Review of Biochemistry, 67*(1), 509–544.

Uversky, V. N. (2009). Intrinsically disordered proteins and their environment: Effects of strong denaturants, temperature, pH, counter ions, membranes, binding partners, osmolytes, and macromolecular crowding. *The Protein Journal, 28*(7–8), 305–325.

van Rosmalen, M., Krom, M., & Merkx, M. (2017). Tuning the flexibility of glycine-serine linkers to allow rational design of multidomain proteins. *Biochemistry, 56*(50), 6565–6574.

Vinkenborg, J. L., Nicolson, T. J., Bellomo, E. A., Koay, M. S., Rutter, G. A., & Merkx, M. (2009). Imaging of intracellular free Zn2+ in real time using genetically-encoded FRET sensors. *Nature Methods, 6*(10), 737.

White, A. D., Nowinski, A. K., Huang, W., Keefe, A. J., Sun, F., & Jiang, S. (2012). Decoding nonspecific interactions from nature. *Chemical Science, 3*(12), 3488–3494.

Xia, Z., & Liu, Y. (2001). Reliable and global measurement of fluorescence resonance energy transfer using fluorescence microscopes. *Biophysical Journal, 81*(4), 2395–2402.

Zegarra, F. C., Homouz, D., Gasic, A. G., Babel, L., Kovermann, M., Wittung-Stafshede, P., et al. (2019). Crowding-induced elongated conformation of urea-unfolded apoazurin: Investigating the role of crowder shape in silico. *The Journal of Physical Chemistry B, 123*(17), 3607–3617.

Zhou, H.-X., Rivas, G., & Minton, A. P. (2008). Macromolecular crowding and confinement: Biochemical, biophysical, and potential physiological consequences. *Annual Review of Biophysics, 37*, 375–397.

Zimmerman, S. B., & Trach, S. O. (1991). Estimation of macromolecule concentrations and excluded volume effects for the cytoplasm of *Escherichia coli*. *Journal of Molecular Biology, 222*(3), 599–620.

Zosel, F., Soranno, A., Buholzer, K. J., Nettels, D., & Schuler, B. (2020). Depletion interactions modulate the binding between disordered proteins in crowded environments. *Proceedings of the National Academy of Sciences of the United States of America, 117*(24), 13480–13489.

CHAPTER FOUR

protaTETHER: A method for the incorporation of linkers in biomacromolecules

Anna I. Wurz, Collin T. O'Bryant, and Robert M. Hughes*
Department of Chemistry, East Carolina University, Greenville, NC, United States
*Corresponding author: e-mail address: hughesr16@ecu.edu

Contents

Abstract

The impacts of linkers on dynamics, expression, and activity of biomacromolecules are often overlooked. This may be due, in part, to the lack of facile methods for incorporation and analysis of linkers that vary iteratively in both length and sequence composition. The protaTETHER method addresses this gap by enabling the incorporation of focused linker libraries at potentially any region in a protein sequence. In this chapter,

Methods in Enzymology, Volume 647
ISSN 0076-6879
https://doi.org/10.1016/bs.mie.2020.10.001

we describe the generation and incorporation of linkers in a PKAc-GFP fusion protein and provide methods for the application and evaluation of the protaTETHER process.

1. Introduction

Protein linkers are frequently used in the construction of multi-domain biomacromolecules. However, the contributions of linkers to macromolecular function are often assumed to be negligible and, as a result, are often underestimated. In particular, protein fusions are a primary target for linker implementation, as countless studies use fluorescent protein tags to track the localization and activity of the protein of interest (POI) (Chen, Zaro, & Shen, 2013; Snapp, 2005). In addition, natural multi-domain proteins are usually joined by endogenous linker peptides. These linker peptides serve to connect the individual functioning domains and contribute to many attributes of proteins, such as maintaining proper structure, activity, and flexibility (Chen et al., 2013). Alternately, during expression of protein fusions, the complexes can become unstable, leading to unfolding, degradation or aggregation, and loss of function. The presence of an optimized linker sequence can therefore improve protein expression and activity, whereas the presence of a non-optimized linker can have the opposite effect (Reddy Chichili, Kumar, & Sivaraman, 2013). Moreover, recently developed applications of complex protein fusions, particularly in the field of optogenetics, require the fusion of a light-sensitive protein domains with both fluorescent proteins and POIs (Redchuk, Omelina, Chernov, & Verkhusha, 2017). Multi-component fusions such as these benefit from careful consideration of linker impacts on structure and function. When choosing the most compatible linker for a protein fusion, there are many factors to be considered, including linker length, linker sequence, and, consequently, linker structure (Chen et al., 2013). In this chapter, we introduce the protaTETHER cloning method, which introduces small linker libraries into target regions of protein sequences (variable linker regions or VLRs) with a restriction enzyme free cloning method (Norris & Hughes, 2018). A primary advantage of this approach is that it enables any region of a protein or protein fusion to be targeted for linker incorporation, promoting both the incorporation and identification of optimal linker length and sequence. Here we describe application of the protaTETHER method to a protein kinase A catalytic subunit (PKAc)-GFP fusion protein, providing a general template for linker design, cloning, and selection that can be applied to any protein or protein fusion.

2. Applications of linkers in biomacromolecules and protein fusions

Linkers play critical roles in protein structure, function, and dynamics. In this section we highlight various aspects of linkers in protein fusions, including the design of FRET pairs and the role of endogenous linkers in protein function (Fig. 1). These topics underscore the need for library-based methods for linker incorporation and evaluation.

2.1 FRET pairs

Developing a functional FRET sensor requires that the linker(s) connecting the fluorescent proteins promote maximum FRET signal strength and sensitivity (Shimozono & Miyawaki, 2008). Therefore, linker design and optimization are requisite for developing FRET biosensors with the desired response (Li et al., 2016). Numerous studies have focused on the role of linkers in FRET biosensors, including optimization of linker length and sequence (van Rosmalen, Krom, & Merkx, 2017), comparison of flexible vs rigid linker effects in L-lysine biosensors (Steffen et al., 2016), and linker effects in ERK biosensors (Vandame et al., 2014) and glutamate biosensors (Deuschle et al., 2005; Hires, Zhu, & Tsien, 2008), in addition to a linker-based approach to a general framework for kinase biosensors (Komatsu et al., 2011). Notably, a bacterial colony-based screening method for optimized linkers in FRET pairs has also been described (Ibraheem, Yap, Ding, & Campbell, 2011).

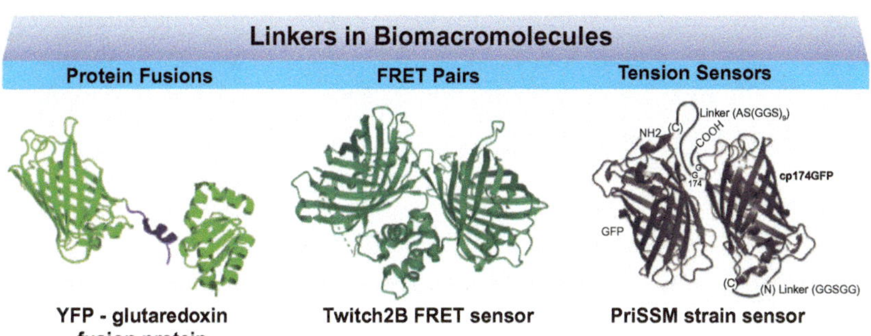

Fig. 1 Linkers play a central role in protein function. This includes protein fusions, FRET pairs, molecular tension sensors, and multidomain proteins. Protein structures created from RCS PDB files 6dav (YFP-glutaredoxin) and 6gel (Twitch2B). PriSSM strain sensor adapted from Iwai & Uyeda (Copyright 2008, National Academy of Sciences).

2.2 Force tension sensors

A second area in which linker design is paramount for function, also closely related with FRET biosensors, is that of tension sensors (Yang et al., 2015). These are important tools for biophysicists to study how mechanical forces impact protein dynamics and cellular behavior (Gates, LaCroix, Rothenberg, & Hoffman, 2019). Recent examples of linker design applied to force tension sensors include the examination of linker effects in vinculin-based tension sensors (LaCroix, Lynch, Berginski, & Hoffman, 2018; LaCroix, Rothenberg, Berginski, Urs, & Hoffman, 2015), and in actin-myosin, actinin, spectrin, and collagen-based tension sensors (Guo, Sachs, & Meng, 2014; Iwai & Uyeda, 2008; Yang et al., 2015).

2.3 Protein dynamics

Naturally occurring linkers, flexible regions, and intrinsically disordered regions in protein structures are an area of growing interest, particularly for their role in fundamental aspects of protein function, such as molecular recognition and enzyme allostery (Feller, Dehareng, & Da Lage, 2011; Schuurman-Wolters, de Boer, Pietrzyk, & Poolman, 2018; Wriggers, Chakravarty, & Jennings, 2005; Zhou, 2017). Furthermore, engineering flexibility or rigidity into a protein scaffold remains an important approach to investigate the contributions of flexible regions to protein structure. For example, in a recently published study, flexible glycines introduced onto the surface of an adenylate kinase resulted in enhanced enzyme activity, demonstrating the importance of flexibility in both protein structure/function relationships and protein evolution (Saavedra, Wrabl, Anderson, Li, & Hilser, 2018). In addition, intrinsically disordered regions in protein structures, experiencing recent interest for their previously unappreciated roles in protein function (Chen, Xie, & Wu, 2014; Maiti et al., 2019; Vuzman & Levy, 2012), may be an area in which experimental linker design and optimization proves fruitful (Li, Cao, Lai, & Liu, 2018).

2.4 Library-based methods for linker generation

Library-based methods for linker selection and optimization are important for a number of reasons: (1) linker effects in protein fusions can be unpredictable and/or counterintuitive (Sokolovski, Bhattacherjee, Kessler, Levy, & Horovitz, 2015), (2) linkers are often (but not always) required for fusion protein function (Zong, Maksimov, & Link, 2016), and (3) linkers can promote more efficient protein function without disrupting the primary sequence, and thus fundamental properties, of the protein of interest (Klement et al., 2015; Quehl, Schuurmann, Hollender, & Jose, 2017; Zhao, Du, & Liu, 2019). As a result, library-based methods for linker cloning and selection continue to be of interest in the scientific community. Recently published linker cloning methods include iFLinkC (Grawe, Ranglack, Weyrich, & Stein, 2020), P-Link (Belsare et al., 2014), and protaTETHER (Norris & Hughes, 2018). In the following sections of this chapter, we will describe the protaTETHER method and detail its application for the insertion of small linker libraries in a PKAc–GFP protein fusion.

3. Generation and incorporation of linkers with protaTETHER

3.1 The protaTETHER cloning process

Our cloning strategy combines a recently reported restriction-enzyme free cloning method (Fast Cloning (Li et al., 2011)) with the design of PCR primers that contain both homology to the target vector region and reiterative linker sequences, thereby enabling multiple annealing sites during PCR amplification and the generation of linker libraries (Fig. 2). The cloning strategy is designed to circumvent purification steps that might impact the quality and/or sequence representation present in the PCR-generated linker library by using restriction enzyme-free cloning. This is an effective means of skirting tedious purification steps while limiting PCR product loss (van den Ent & Lowe, 2006). In the first step of the process, the target vector is amplified utilizing forward and reverse primers adjacent to the region targeted for linker insertion (variable linker region or VLR; Fig. 2A). The vector amplicon is then digested with *Dpn*I to removed template DNA, and further purified by gel extraction. In the second step (Fig. 2B), linker libraries are created from forward and reverse primers containing redundant linker sequences that provide multiple annealing sites for forward primers, which results in a diversity of PCR amplicons. These PCR products are carried forward without further purification. In the third and final step (Fig. 2C), the vector PCR product is combined with the variable linker library and subsequently transformed into ultracompetent *E. coli*. Transformants are then isolated and characterized by Sanger sequencing.

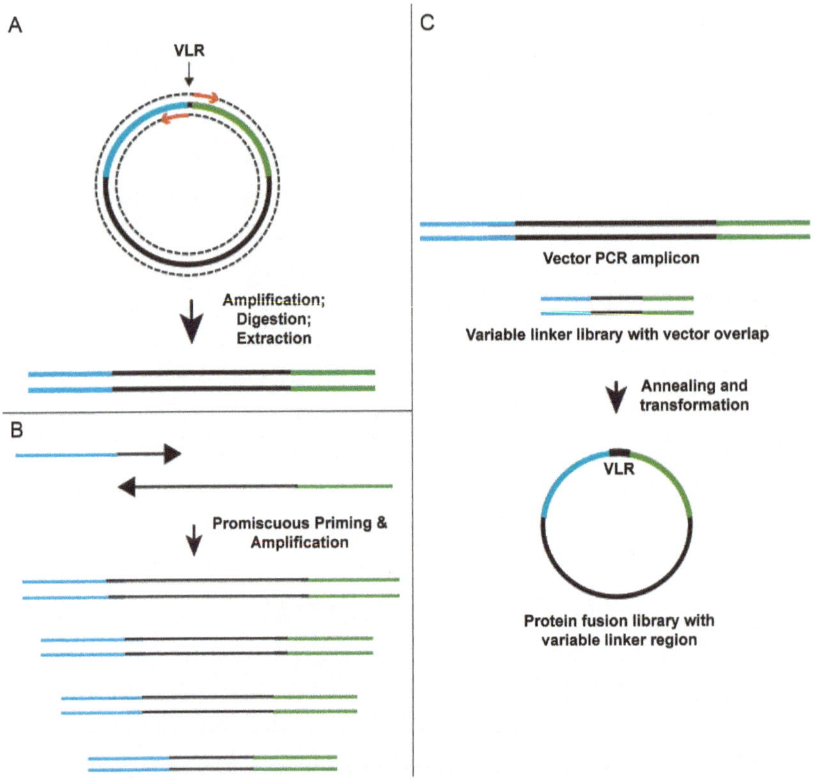

Fig. 2 Overview of protaTETHER cloning strategy. (A) Primers for vector amplification (red arrows) are designed to amplify region immediately adjacent to variable linker region (VLR). Post-amplification, the linear vector sequence is digested with *DpnI* and then gel extracted. PKAcs gene indicated by blue lines; GFP gene indicated by green lines. (B) Oligomer sequences are designed to generate multiple amplicons via reiterative sequence design. The short forward primer contains 17 bp of overlap with the end of the PKAcs gene in the target vector (blue) in addition to 1 $\times$ or 2 $\times$ linker repeat codons, while the longer reverse primer contains 17 bp of overlap with the start of the GFP in the target vector (green) in addition to 4–8 $\times$ complimentary linker repeat codons. (C) Both vector (gel purified) and linker library (crude) amplicons are combined, briefly annealed at room temperature, and transformed into ultracompetent *E. coli*. The resulting plasmids are screened to identify constructs containing variable linker inserts.

3.2 Equipment

1. PCR thermocycler (Eppendorf Mastercycler Pro).
2. Incubator shaker (New Brunswick Scientific Excella E24).
3. Gel rig and power supply for DNA electrophoresis.

4. Plate reader (Molecular Devices M2e with Microvolume Microplate adapter).
5. Refrigerated benchtop centrifuge (Eppendorf 5424R).
6. Heat block set to 42 °C.

3.3 Reagents

1. DNA polymerase (New England Biosciences Q5 Master Mix, or other high fidelity DNA polymerase).
2. *Dpn*I restriction enzyme and buffer (New England Biosciences).
3. Gel isolation and mini-prep column kit (ThermoFisher Scientific or other manufacturer).
4. Plasmid DNA template (PKAc-GFP ((Norris & Hughes, 2018) in plasmid phCMV (Genlantis)).
5. Forward and Reverse primer stocks (IDT, Inc.)
6. XL10-Gold Ultracompetent *E. coli* with beta mercaptoethanol (βME) (Stratagene, Inc.)
7. SOC Outgrowth Medium (New England Biosciences).
8. LB-Kanamycin liquid media and agar plates.
9. Protein ladder.
10. 6 × DNA Loading Dye.
11. Distilled deionized water (ddH_2O).
12. 1% agarose gel with Ethidium Bromide (or alternative contrast reagent).

4. Protocol

The plasmid template should be amplified adjacent to the region intended for linker insertion (the VLR or variable linker region, Fig. 2C). To achieve this, 17–24 bp forward and reverse primers should be designed to anneal to these specific sites with an annealing temperature around 60 °C (Li et al., 2011). The forward primer should anneal to the right side of the VLR, and the reverse primer should anneal to the left side of the VLR. In our example, 17 bp primers were designed for annealing to the N-terminal sequence of GFP (forward primer) and the C-terminal sequence of PKAc (reverse primer) (Table 1). All DNA concentrations were determined using a Molecular Devices plate reader equipped with a Microvolume Microplate adapter.

Table 1 Primer sequences for vector amplification and GGS library generation.

Primer Name	Sequence	Description
GFPNterm_Fwd	ATGGCTAGCAAAGGAGA	PKA-GFP vector primer F; homology with GFP N-term
PKACterm_Rev	AAACTCAGTAAACTCCT	PKA-GFP vector primer R; homology with PKA C-term
GGS1X_Fwd	*AGGAGTTTACTGAGTTT* GGCGGCAGC	Variable insert F primer; GGS; *17 bp homology with C-term of PKAcs*
GGS1X_Rev	*TCTCCTTTGCTAGCCAT* GCTGCCGCC	Variable insert R primer; GGS; *17 bp homology with N-term of GFP*
GGS2X_Fwd	*AGGAGTTTACTGAGTTT* $(GGCGGCAGC)_2$	Variable insert F primer; GGS x 2; *17 bp homology with C-term of PKAcs*
GGS2X_Rev	*TCTCCTTTGCTAGCCAT* $(GCTGCCGCC)_2$	Variable insert R primer; GGS x 2; *17 bp homology with N-term of GFP*
GGS3X_Fwd	*AGGAGTTTACTGAGTTT* $(GGCGGCAGC)_3$	Variable insert F primer; GGS x 3; *17 bp homology with C-term of PKAcs*
GGS3X_Rev	*TCTCCTTTGCTAGCCAT* $(GCTGCCGCC)_3$	Variable insert R primer; GGS x 3; *17 bp homology with N-term of GFP*
GGS4X_Rev	*TCTCCTTTGCTAGCCAT* $(GCTGCCGCC)_4$	Variable insert R primer; GGS x 4; *17 bp homology with N-term of GFP*

4.1 Target vector amplification: PKAc-GFP example

4.1.1. Design and order the vector amplifying primers. In our case, primers were sourced from Integrated DNA Technologies, Inc.

4.1.2. Dilute primers with distilled deionized H_2O to a concentration of 20 μM.

4.1.3. Dilute vector template with distilled deionized H_2O to 50 ng/μL.

4.1.4. Set up the vector PCR reaction by combining 2 μL of each forward and reverse primer with 2 μL of vector template. Then, add 19 μL of distilled deionized water and 25 μL of Q5 Master Mix to bring the final volume to 50 μL.

4.1.5. Use a gradient PCR thermocycler to amplify the target vector following manufacturer's instructions for the Q5 polymerase. In our case, annealing temperatures of 52 °C, 55 °C, and 58 °C were tested over 35 amplification cycles. Extension time was set to 4 min (30 s/kB of vector).

4.1.6. Confirm that PCR product of correct length has been generated by running 5 μL of the PCR reaction on a 1% agarose gel containing Ethidium Bromide contrast reagent. If amplification is not successful, try repeating with different template concentrations and different annealing temperatures. If this fails, try redesigning the PCR primers to optimize the melting temperature.

4.1.7. If amplification is successful, digest the vector by adding 1 μL of *Dpn*I directly to the PCR reaction. Incubate for 1 h at 37 °C.

4.1.8. Load the entire PCR reaction into a single well of a 1% agarose gel, isolate the band with the correct size by electrophoresis, excise the band with a sterile razor blade, then isolate using a DNA gel extraction kit.

4.1.9. Measure the concentration of the extracted DNA using a Molecular Devices M2e with Microvolume Microplate adapter or equivalent. Final concentration of the isolated template should be 5–20 ng/μL.

Tip: It is very important that the PCR reaction be thoroughly digested with DpnI, followed by gel isolation, to reduce PCR template contamination. Otherwise, background will be excessively high in subsequent bacterial transformation steps, complicating assessment of linker cloning efficiency.

4.2 Linker library amplification and transformation: GGS repeats

To generate a linker library for the protaTETHER method, forward and reverse primers need to be designed that pair a short forward primer with homology for the target vector (adjacent to the beginning of the VLR) with a longer, sequence redundant reverse primer with homology for target vector (adjacent to the end of the VLR). As a model system for primer design and library generation, we selected the frequently utilized GGS amino acid motif (Chen et al., 2013). For the forward primer, we designed sequences that overlapped with either one, two, or three GGS repeat sequences while designing reverse primers encoding one, two, three, or four GGS repeat sequences (Table 1). For our initial screen of primer pairs, we paired both GGS1X, GGS2X, and GGS3X forward and reverse primer pairs (**1, 2,** and **3**; Table 2) in addition to combinations of GGS1 × forward primer with

Table 2 Results of GGS linker incorporation via protaTETHER.

	Primer pair	Result	Colonies sequenced	Colonies with inserts
1	GGS1X_Fwd/ GGS1X_Rev	No inserts	12	0
2	GGS2X_Fwd/ GGS2X_Rev	No inserts	12	0
3	GGS3X_Fwd/ GGS3X_Rev	2X (**4**) and 3X (**2**) GGS	12	6
4	GGS1X_Fwd/ GGS3X_Rev	2X (**2**) and 3X (**4**) GGS	12	6
5	GGS2X_Fwd/ GGS3X_Rev	3X (**4**) GGS	12	4
6	GGS1X_Fwd/ GGS4X_Rev	2X (**1**) and 4X (**5**) GGS	12	6
7	GGS2X_Fwd/ GGS4X_Rev	1X (**1**), 2X (**2**), 3X (**2**), and 4X (**4**) GGS	12	9

GGS3X and GGS4X reverse primers (**4** and **6**; Table 2) and GGS2× forward primer with GGS3X and GGS4X reverse primers (**5** and **7**; Table 2). We then co-transformed the resulting PCR amplicons and the target vector containing PKAcs and GFP components into ultracompetent *E. coli*. PCR amplicons resulting from the pairing of 1× forward and reverse primers and amplicons resulting from the pairing of 2× forward and reverse primers failed to produce plasmids with linker inserts (Table 2; bold numbers indicate frequency of linker occurrence). By contrast, amplicons derived from 3× forward and reverse primers gave PKA-GFP fusions with 2× and 3× GGS repeat sequences. Transformation of PCR amplicons resulting from combination of a 1× GGS forward primer with 3× or 4× GGS reverse primers gave similar results, producing only a limited subset of linkers. However, transformation of the amplicons resulting from combination of the 2× GGS forward primer (GGS2X_Fwd; Table 2) with a 4× reverse primer (GGS4×_Rev; Table 2) provided the best coverage of all possible linker sequences (1×, 2×, 3×, and 4×), generating linkers ranging from 3 amino acids (GGS) to 12 amino acids (GGSGGSGGSGGS; Table 2, *line* 7). Importantly, 9 out of 12 colonies sequenced in this trial contained VLR inserts, resulting in a higher frequency of successful linker inserts than previously tested primer combinations. The superior combination of the 2 × forward primer with the 4 × reverse primer likely results from a combination of stronger annealing achieved by 2 × (T_m = 70 °C; salt-adjusted) vs 1 × (T_m = 34 °C; salt-adjusted) GGS overlap by the 2 × forward primer, while the greater sequence coverage results from additional sequence space encoded by the longer GGS4X reverse primer. Furthermore, the ability of the GGS2X_Fwd/GGS4X_Rev primer pair to produce the truncated GGS1X insert likely stems from annealing of the outermost GGS repeat sequence of the forward primer with the outermost GGS repeat sequence of the GGS4X reverse primer.

4.2.1. Design and order the linker library primers. In our case, primers were sourced from Integrated DNA Technologies, Inc.

4.2.2. Dilute primers with distilled deionized H_2O to a concentration of 20 μM.

4.2.3. Working on ice, set up the linker library PCR by combining 12.5 μL of both F and R primer stocks, followed by the addition of 25 μL of Q5 Master Mix (NEB).

4.2.4. Use a gradient PCR thermocycler to amplify the linker library, following manufacturer's instructions for the Q5 polymerase. In this example, an annealing temperature of 72 °C was used over 35 amplification cycles. Extension time was set to 30 s.

4.2.5. Confirm successful library amplification by running 5 μL of the PCR reaction on a 1% agarose gel containing Ethidium Bromide contrast reagent. If amplification is not successful (i.e., bright low MW bands are not visible on the gel), repeat with a lower annealing temperature (see Fig. 3).

4.2.6. Combine 2 μL of the linker library PCR (no purification necessary) with 2 μL of the purified target vector (20 ng/μL) in a PCR tube. Mix by pipetting and allow to sit at room temperature (20 min total) while proceeding to the next step.

4.2.7. Thaw an aliquot of ultracompetent cells on ice. Following manufacturer's instructions, aliquot 50 μL of competent cells into a 15 mL conical tube and add 2 μL of beta mercaptoethanol. Place on ice for 10 min, mixing by gently swirling every 2 min. Following this initial incubation, add 2 μL of the library + target vector cloning mixture to the competent cells and mix by gently swirling. Place mixture on ice for 30 min.

4.2.8. Heat shock the transformation mixture for exactly 30 s at 42 °C. Place the transformation on ice and incubate for 2 min. Following

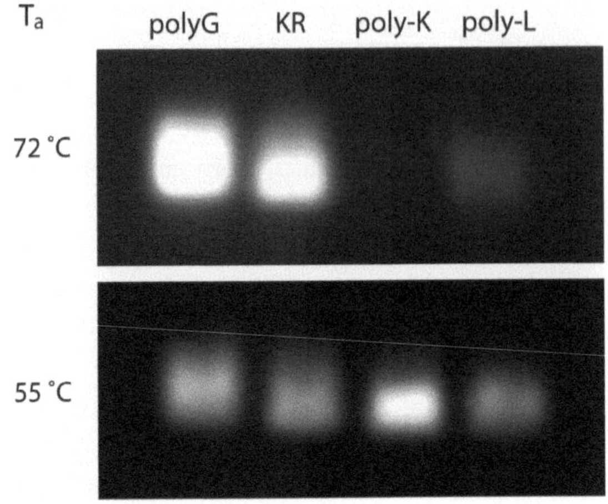

Fig. 3 Impact of annealing temperature on linker library amplification. Linker sequences with high melting temperatures (polyG and KR) amplify efficiently at a Tm of 72 °C, whereas linker sequences with lower melting temperatures (poly-K and poly-L) require a lower Tm (55 °C) for efficient amplification.

this, add 300 μL of SOC media. Place the resulting solution in a 37 °C shaking incubator, incubate for 1 h with shaking (260 rpm). Plate the entire mixture on a LB-Kanamycin plate (or divide into 50, 100, and 150 μL volumes on three separate plates to avoid potential colony overcrowding) and incubate overnight at 37 °C.

4.2.9. Pick colonies (10–12 recommended for initial library characterization), grow 3–5 μL mini prep cultures in LB-Kan, then isolate DNA to characterize linker insertion by Sanger sequencing.

Tip: (1) A reverse sequencing primer 50–100 bp downstream of the VLR is necessary for characterizing success of linker cloning experiment by sequencing. In our case, a reverse sequencing primer with homology near the N-terminal GFP sequence was used to characterize linker insertion. (2) Design primers for the intended expression host by using a codon optimization tool to optimize the repeating subunit of each linker. Use this optimized subunit to create the remaining primer sequence(s).

4.3 Primer design considerations: GGSD repeats

Based upon the results achieved with the GGS linker cloning experiment, we designed primers for generating a GGSD linker library. In this case, we expanded linker sequence space by creating a longer reverse primer (with 8 × GGSD sequence repeats), while still using a short forward primer (with a 1.5 × GGSD repeat sequence, mimicking the 2 × GGS repeat sequence of

Table 3 Primer sequences for GGSD linker library.

Primer name	Sequence	Description
GGSD1.5X_Fwd	AGGAGTTTACTGAGTTT GGCGGCAGCGACGGCGGC	Variable insert F primer; GGSD; *homology with C-term of PKA*
GGSD8X_Rev	TCTCCTTTGCTAGCCAT $(GTCGCTGCCGCC)_8$	Variable insert R primer; GGSD x 8; *17 bp homology with N-term of GFP*

Table 4 Results from GGSD library cloning experiment.

Primer pair	Result	Colonies sequenced	Colonies with inserts
GGSD1.5X_Fwd/ GGSD8X_Rev	0X (**6**), 1X (**0**), 2X (**0**), 3X (**1**), 4X (**1**), 5X (**1**), 6X (**0**), 7X (**1**), 8X (**2**), 9X (**1**), 10X (**1**)	14	8

the GGS library forward primer) (Table 3). This sequence enabled the generation and capture of longer linkers—even beyond the 8 × repeat sequence encoded in the reverse primer (Table 4; bold numbers indicate frequency of linker occurrence). The results in Table 4 were obtained using a procedure similar to the one in Section 4.2, with the minor modifications described below.

4.3.1. Design and order the linker library primers. In our case, primers were sourced from Integrated DNA Technologies, Inc. As the reverse primer ordered for this experiment was significantly longer than those previously designed, it was ordered on a smaller scale (4 nmol Ultramer DNA).

4.3.2. Dilute primers with distilled deionized H_2O to a concentration of 20 μM.

4.3.3. Working on ice, set up the linker library PCR by combining 12.5 μL of both F and R primer stocks, followed by the addition of 25 μL of Q5 Master Mix (NEB).

4.3.4. Use a gradient PCR thermocycler to amplify the linker library, following manufacturer's instructions for the Q5 polymerase. In this example, an annealing temperature of 72 °C was used over 35 amplification cycles. Extension time was set to 30 s.

4.3.5. Combine 2 μL of the linker library PCR (no purification necessary) with 2 μL of the amplified target vector in a PCR tube. Mix by pipetting and allow to sit at room temperature (20 min total) while proceeding to the next step.

4.3.6. Thaw an aliquot of ultracompetent cells on ice. Following manufacturer's instructions, aliquot 50 μL of competent cells into a 15 mL conical tube and add 2 μL of beta mercaptoethanol. Place on ice for 10 min, mixing by gently swirling every 2 min. Following this initial incubation, add 2 μL of the library + target vector cloning mixture to the competent cells and mix by gently swirling. Allow this mixture to sit on ice for 30 min.

4.3.7. Heat shock the transformation mixture for exactly 30 s at 42 °C. Following this, add 300 μL of SOC media. Place the resulting solution in a 37 °C shaking incubator, incubate for 1 h with shaking (260 rpm). Plate the entire mixture on a LB-Kan plate (or divide into 50, 100, and 150 μL volumes on three separate plates to avoid potential colony overcrowding) and incubate overnight at 37 °C.

4.3.8. Pick colonies (10–12 recommended for initial library characterization), grow 3–5 μL mini prep cultures in LB-Kan, then isolate DNA to characterize linker insertion by Sanger sequencing.

Tip: To improve library diversity, shorter reverse primers could be designed, containing 4X GGSD and 6X GGSD repeat sequences. Carrying out these cloning reactions in tandem with the 8X GGSD repeat sequence (all using the same 1.5X forward primer) would fill in the missing repeat sequences from the linker library (1X, 2X, and 6X GGSD) more efficiently than sequencing more colonies from the 8X GGSD repeat reaction.

5. Quantification of protaTETHER results

Upon reaching the desired linker library size and diversity, the next steps are to quantify expression and activity of the linker proteins. In this section we describe protocols for assessing expression and activity of a library of PKAc-VLR-GFP fusions.

5.1 Equipment

1. Humidified tissue culture incubator (37 °C/5% CO_2).
2. Biosafety hood or cabinet.
3. SDS-PAGE gel rig and power supply.
4. Western blot transfer apparatus.
5. Benchtop centrifuge (Eppendorf 5424R).
6. Hemocytometer.
7. Tissue culture microscope.
8. Cell media aspirator.

9. Plate reader (Molecular Devices M2e).
10. Confocal microscope equipped with GFP and DAPI filters (Olympus IX2-DSU).

5.2 Reagents and consumables

1. HeLa cell culture media (D-MEM/10% FBS/1% Pen-Strep).
2. Trypsin-EDTA (0.05%).
3. M-PER protein extraction reagent (ThermoScientific).
4. Six well tissue culture plates.
5. 1.5 mL microcentrifuge tubes.
6. 96-well clear bottom plate.
7. 35 mm glass-bottom dishes (Mattek).
8. HALT protease and phosphatase inhibitor (ThermoFisher).
9. Detergent compatible Bradford reagent (Pierce).
10. Protein ladder.
11. Primary antibodies: αGFP (Santa Cruz sc-8334); α-phosphoPKA substrate (Cell Signaling 9624); anti-GAPDH (ThermoFisher PIMA 515738).
12. Secondary antibodies: anti-mouse HRP; anti-rabbit HRP.
13. Lipofectamine 3000 reagent (ThermoFisher).
14. OptiMEM (Gibco).
15. Dulbecco's PBS (DPBS) with calcium and magnesium.
16. 6 × Laemmli sample buffer (LSB).
17. Bovine Serum Albumin (BSA) standard.
18. TBS with 0.1%Tween (TBST).
19. Distilled deionized H_2O (ddH_2O).
20. Cell fixative (4% PFA (Electron Microscopy Sciences) in DPBS).
21. HiFi DNA Assembly kit (New England Biosciences).
22. NEB-5α high efficiency chemically competent *E. coli* (New England Biosciences).
23. Imaging station (Azure c300).
24. Chemiluminescent substrate.
25. Restore western blot stripping buffer (ThermoFisher).

5.3 Expression of PKAc-VLR-GFP protein fusions

In this section we describe evaluation of protein fusion expression. This is critical, as not all members of the linker library will express at the same level. Depending upon the desired experimental outcome, linkers for a given protein may not be suitable if their presence negatively impacts protein expression. In the case of the PKAc-VLR-GFP fusions, we transfected HeLa cells with our expression plasmids and utilized anti-GFP western blots to assess protein expression (Fig. 4). Here we describe the protocol for transfection, protein harvest, and quantification via western blot.

5.3.1. HeLa cells cultured under standard conditions in a humidified incubator (37 °C; 5% CO_2; DMEM with 10% FBS and 1% Pen-Strep) should be thoroughly trypsinized (Trypsin-EDTA). After quenching trypsin, count cells with a hemocytometer. Plate cells in 6 well tissue culture plates at 2×10^6 cells per well.

5.3.2. The following day, check cell confluency on a tissue culture microscope. If cells are at the desired confluency (80% or higher), proceed with transfection. If cells are not at the desired confluency, wait an additional day before transfecting.

5.3.3. Transfect cells with Lipofectamine 3 K reagent. In our case, we used 1 μg of plasmid DNA per well, combined with 3.75 μg of Lipofectamine 3000 (L3K) reagent, and 5 μL of P3000 reagent. Solutions should be mixed separately (P3000 + DNA; L3K) in 125 μL of OptiMEM prior to combining for a 250 μL total volume. Allow solutions to sit at room temperature for 10–15 min. Add solution dropwise to cells, then return cells to incubator. Wait 16–24 h before assessing protein fusion expression.

5.3.4. Remove cells from incubator and aspirate media. Wash gently with 1 mL of DPBS, then remove wash solution by aspiration. Add 300 μL of M-PER plus 1X HALT protease and phosphatase inhibitor solution per well. Agitate on a shaker (200 rpm) for 10 min at room temperature prior to processing.

5.3.5. Pipette liquid in well up and down several times with a P1000 pipette. Carefully remove solution with pipette and place in a 1.5 mL microcentrifuge tube. Using a refrigerated benchtop centrifuge pre-chilled to 4 °C, clarify the lysates by spinning at 14000 rpm for 15 min.

5.3.6. Remove 100 μL of the clarified lysate and add to a clean tube. Keep the original tube on ice for use in the next step (5.3.7). Add 20 μL of 6 × LSB to the 100 μL sample. Heat the sample at 65 °C for 10 min.

5.3.7. Using the sample from the original tube, measure the total protein concentration with the Bradford assay reagent of your choice. We used 5 μL of the sample plus 200 μL of Bradford reagent per well of a 96-well plate (Costar, clear-bottom, black) and measured the

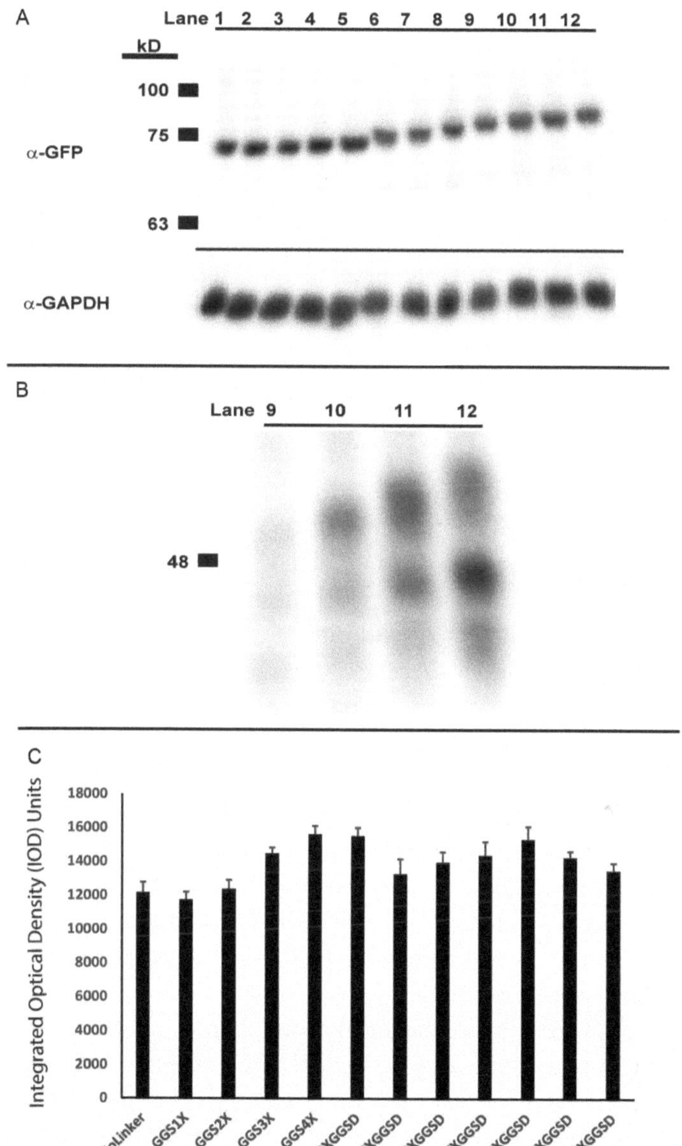

Fig. 4 (A) Western Blot (α-GFP) of PKA-VLR-GFP transfected HeLa cell lysates. VLR (Lanes 1–12) are as follows: No linker; GGS1X; GGS2X; GGS3X; GGS4X; GGSD3X; GGSD4X; GGSD5X; GGSD7X; GGSD8X; GGSD9X; GGSD10X. GAPDH was used as a loading control. (B) Immunopositive α-GFP degradation bands are visible in lanes 9–12 (GGSD7X; GGSD8X; GGSD9X; GGSD10X). (C) An average of three densitometry measurements per sample were compiled in FIJI (ImageJ) and normalized to GAPDH expression.

absorbance at 595 nm with a M2e Molecular Devices plate reader using the SoftMax Pro Bradford assay protocol. Measurement calibration was performed with a bovine serum albumin (BSA) standard curve (5 μL of standard solution plus 200 μL Bradford reagent per well).

5.3.8. After using the Bradford assay to calculate total protein content, load 20 μg of total protein per lane per sample on an SDS-PAGE gel. At the conclusion of electrophoresis, transfer the gel overnight at 20 V to a PVDF membrane in a western blot transfer apparatus. Block the membrane in 5% TBS-Tween for 1 h at room temperature, then incubate the blot with primary antibody (α-GFP; 1:1000 dilution in TBST-5% BSA) overnight at 4 °C on a rotating platform. Wash the blot in TBST (3 × 5 min). Add the appropriate secondary antibody (Invitrogen anti-mouse 2°; 1:10000 in TBST-5%BSA). Incubate on a rocker for 1 h at room temperature. Wash the blot in TBST (3 × 5 min). Rinse the blot in TBS, followed by a rinse in ddH_2O. The blot can then be developed with a chemiluminescent substrate (for an inexpensive alternative to commercial solutions, see Mruk et al. (Mruk & Cheng, 2011)) and imaged (Azure c300).

5.3.9. Using the Restore western blot stripping reagent, remove the antibodies and repeat the blot with an anti-GAPDH housekeeping antibody and the appropriate secondary antibody.

5.3.10. Import the blot image into ImageJ. Using the rectangular selection tool, select the area surrounding the band of interest in the first lane in the blot. Select additional lanes by pressing Ctrl +1 to generate new rectangles, which can be repositioned over adjacent lanes. Once all lanes have been selected, go to the Gel Analysis menu and select Plot Lanes. Use the line tool to connect the baselines underneath the curves. Obtain integrated optical density values by clicking the wand tool in the area underneath the curve for each lane plotted. These values will automatically be recorded in a separate Results window. These values should be exported to Excel for analysis.

5.3.11. Normalize the resulting data to the band intensities obtained for the GAPDH housekeeping blot.

5.4 Activity of PKAc-VLR-GFP protein fusions

Enzyme activity in a protein fusion can be directly impacted by linker length and flexibility. In our case, we were interested in assessing the impacts of linker length and flexibility on the activity of the catalytic subunit of the

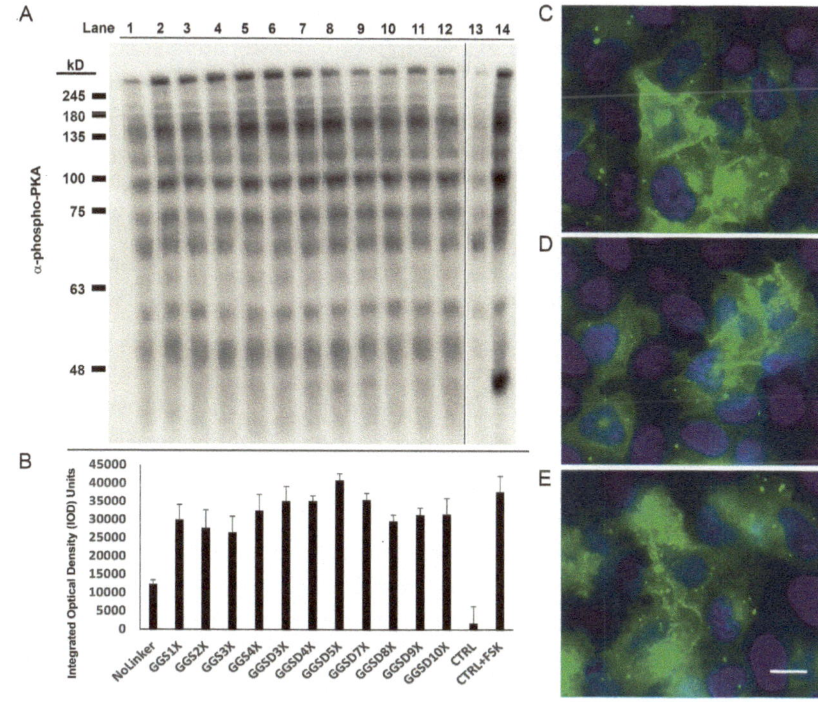

Fig. 5 (A) Western Blot (α-PKA phosphosubstrate) of PKA-VLR-GFP transfected HeLa cell lysates. VLR (Lanes 1–14) are as follows: No linker; GGS1X; GGS2X; GGS3X; GGS4X; GGSD3X; GGSD4X; GGSD5X; GGSD7X; GGSD8X; GGSD9X; GGSD10X; no transfect control; Forskolin-treated control. GAPDH was used as a loading control. (B) Band intensities were quantified by densitometry analysis in ImageJ. Identical rectangular selection areas (ranging from the dark bands above 245 kD MW marker down to the second faint band below the 48 kD MW marker) were used to measure the total optical density of each lane. Error bars represent the average of three replicates for each condition. (C–E) No linker, GGSD, and GGS iterations of PKA-VLR-GFP constructs are membrane localized. Images shown are (C) PKA-GGSD3X-GFP (D) PKA-GGS4X-GFP (E) PKA-GFP. Confocal microscopy images (Olympus IX2-DSU Tandem Spinning Disk Confocal; 60× objective) were acquired using eGFP and DAPI filter cubes. Channel overlays created with FIJI (ImageJ). Scale bar = 10 μm.

cAMP-dependent protein kinase A (PKAc). While there are numerous methods for determining PKAc activity, we used a western blotting method that employs an antibody for serines and threonines phosphorylated in the context of the generic PKAc recognition sequence (RRXS*/T*) (Fig. 5). This method enables semi-quantitative assessment of linker-dependent PKAc activity trends.

5.4.1. Follow steps 5.3.1–5.3.7 from the preceding section.

5.4.2. After using the Bradford assay to calculate total protein content, load 20 μg of total protein per lane per sample on an SDS-PAGE gel. At the conclusion of electrophoresis, transfer the gel overnight at 20 V to a PVDF membrane in a western blot transfer apparatus. Block the membrane in 5% TBS-Tween for 1 h at room temperature, then incubate the blot in primary antibody (α-phospho PKA substrate antibody; 1:1000 dilution in TBST-5% BSA) overnight at 4 °C on a rotating platform. Wash the blot in TBST (3 × 5 min). Add the appropriate secondary antibody (Invitrogen anti-rabbit 2°; 1:3000 in TBST-5%BSA). Incubate on a rocker for 1 h at room temperature. Wash the blot in TBST (3 × 5 min). Rinse the blot in TBS, followed by a short rinse in ddH_2O. The blot can then be developed with a chemiluminescent substrate and imaged (Azure c300).

5.4.3. Using the Restore western blot stripping reagent, remove the antibodies and repeat the blot with an anti-GAPDH housekeeping antibody and the appropriate secondary antibody.

5.4.4. Import the blot image into ImageJ. To assess total phosphorylation, using the rectangular selection tool, select the entire first lane in the blot. Select additional lanes by pressing Ctrl +1 to generate new rectangles, which can be repositioned over adjacent lanes. Once all lanes have been selected, go to the Gel Analysis menu and select Plot Lanes. Use the line tool to connect the baselines underneath the curves. Obtain integrated optical density values by clicking the wand tool in the area underneath the curve for each lane plotted. These values will automatically be recorded in a separate Results window. These values should be exported to Excel for analysis.

5.4.5. Normalize the resulting data to the band intensities obtained for the GAPDH housekeeping blot.

Tip: This blotting step can be repeated with a more specific phosphoPKA substrate antibody, such as an anti-phospho VASP antibody. Note: an alternative to using a housekeeping protein such as GAPDH is to quantify total lysate loading with a conventionally stained SDS-PAGE gel.

5.5 Subcellular localization of PKAc-VLR-GFP protein fusions

5.5.1. HeLa cells cultured under standard conditions in a humidified incubator (37 °C; 5% CO_2; DMEM with 10% FBS and 1% Pen-Strep) should be thoroughly trypsinized (Trypsin-EDTA). After quenching trypsin, count cells with a hemocytometer. Plate cells in 35 mm glass-bottom Mattek dishes (No. 1.5, uncoated) at 2×10^6 cells per well.

5.5.2. The following day, check cell confluency on a tissue culture microscope. If cells are at the desired confluency (80% or higher), proceed with transfection. If cells are not at the desired confluency, wait an additional day before transfecting.

5.5.3. Transfect cells with Lipofectamine 3 K reagent. In our case, we used 1 μg of plasmid DNA per well, combined with 3.75 μg of Lipofectamine 3000 (L3K) reagent, and 5 μL of P3000 reagent. Solutions should be mixed separately (P3000 + DNA; L3K) in 125 μL of OptiMEM prior to combining for a 250 μL total volume. Allow solutions to sit at room temperature for 10–15 min. Add solution dropwise to cells, then return cells to incubator. Wait 16–24 h before assessing protein fusion expression.

5.5.4. Add solution dropwise to cells, then return cells to incubator. Wait 16–24 h before assessing protein fusion localization.

5.5.5. Remove cell culture media by aspiration. Gently wash cells 1 × in DPBS. Remove by aspiration. Add fixative (4% PFA/DPBS) and incubate cells at room temperature for 15 min. Remove fixative, then wash cells 1 × with DPBS. Remove wash and then add 1–2 mL DPBS. If desired, add nuclear stain (Hoescht 33342; 5 μg/mL working concentration) for 5 min in PBS, then wash 1 × 5 min with DPBS prior to imaging.

5.5.6. Using a confocal microscope equipped with GFP and DAPI filters, assess protein fusion localization. In the case of the PKA-VLR-GFP construct, membrane localization was anticipated due to the presence of an N-terminal membrane myristoylation domain. This pattern was observed for the majority of the linkers incorporated into the PKA-GFP fusion (Fig. 5C–E). However, the linker sequence can also significantly disrupt anticipated subcellular localization patterns as observed in Norris et al. (Norris & Hughes, 2018).

Tip: Cells can also be imaged live to ensure that there are no localization artifacts due to cellular fixation. Additional PKAc-dependent activity assays can be performed on the fixed cells, including immunostaining for pCREB as described in Norris et al. (Norris & Hughes, 2018).

6. Summary

Design and incorporation of linkers in protein sequences is an important experimental consideration that is often overlooked. The protaTETHER method addresses this issue by facilitating the cloning and characterization of small linker libraries into biomacromolecules. This method may have utility in the optimization of biosensors, optogenetic protein fusions, and in experiments that characterize dynamic and functional aspects of endogenous linker regions in multi-domain proteins.

Acknowledgments

We acknowledge the East Carolina University Division of Research, Economic Development and Engagement, the Harriot College of Arts and Sciences, and the Department of Chemistry for financial and facility support of this project.

References

Belsare, K. D., Joëlle Ruff, A., Martinez, R., Shivange, A. V., Mundhada, H., Holtmann, D., et al. (2014). P-LinK: A method for generating multicomponent cytochrome P450 fusions with variable linker length. *Biotechniques*, *57*, 13–20. https://doi.org/10.2144/000114187.

Chen, J., Xie, Z.-R., & Wu, Y. (2014). A multiscale model for simulating binding kinetics of proteins with flexible linkers. *Proteins*, *82*, 2512–2522. https://doi.org/10.1002/prot.24614.

Chen, X., Zaro, J. L., & Shen, W. C. (2013). Fusion protein linkers: Property, design and functionality. *Advanced Drug Delivery Reviews*, *65*, 1357–1369. https://doi.org/10.1016/j.addr.2012.09.039.

Deuschle, K., Okumoto, S., Fehr, M., Looger, L. L., Kozhukh, L., & Frommer, W. B. (2005). Construction and optimization of a family of genetically encoded metabolite sensors by semirational protein engineering. *Protein Science*, *14*, 2304–2314. https://doi.org/10.1110/ps.051508105.

Feller, G., Dehareng, D., & Da Lage, J.-L. (2011). How to remain nonfolded and pliable: The linkers in modular alpha-amylases as a case study. *The FEBS Journal*, *278*, 2333–2340. https://doi.org/10.1111/j.1742-4658.2011.08154.x.

Gates, E. M., LaCroix, A. S., Rothenberg, K. E., & Hoffman, B. D. (2019). Improving quality, reproducibility, and usability of FRET-based tension sensors. *Cytometry. Part A*, *95*, 201–213. https://doi.org/10.1002/cyto.a.23688.

Grawe, A., Ranglack, J., Weyrich, A., & Stein, V. (2020). iFLinkC: An iterative functional linker cloning strategy for the combinatorial assembly and recombination of linker peptides with functional domains. *Nucleic Acids Research*, *48*, e24. https://doi.org/10.1093/nar/gkz1210.

Guo, J., Sachs, F., & Meng, F. (2014). Fluorescence-based force/tension sensors: A novel tool to visualize mechanical forces in structural proteins in live cells. *Antioxidants & Redox Signaling*, *20*, 986–999. https://doi.org/10.1089/ars.2013.5708.

Hires, S. A., Zhu, Y., & Tsien, R. Y. (2008). Optical measurement of synaptic glutamate spillover and reuptake by linker optimized glutamate-sensitive fluorescent reporters. *Proceedings of the National Academy of Sciences of the United States of America*, *105*, 4411–4416. https://doi.org/10.1073/pnas.0712008105.

Ibraheem, A., Yap, H., Ding, Y., & Campbell, R. E. (2011). A bacteria colony-based screen for optimal linker combinations in genetically encoded biosensors. *BMC Biotechnology*, *11*, 105. https://doi.org/10.1186/1472-6750-11-105.

Iwai, S., & Uyeda, T. Q. P. (2008). Visualizing myosin-actin interaction with a genetically-encoded fluorescent strain sensor. *Proceedings of the National Academy of Sciences of the United States of America*, *105*, 16,882–16,887. https://doi.org/10.1073/pnas.0805513105.

Klement, M., Liu, C., Loo, B. L. W., Choo, A. B.-H., Ow, D. S.-W., & Lee, D.-Y. (2015). Effect of linker flexibility and length on the functionality of a cytotoxic engineered antibody fragment. *Journal of Biotechnology*, *199*, 90–97. https://doi.org/10.1016/j.jbiotec.2015.02.008.

Komatsu, N., Aoki, K., Yamada, M., Yukinaga, H., Fujita, Y., Kamioka, Y., et al. (2011). Development of an optimized backbone of FRET biosensors for kinases and GTPases. *Molecular Biology of the Cell*, *22*, 4647–4656. https://doi.org/10.1091/mbc.E11-01-0072.

LaCroix, A. S., Lynch, A. D., Berginski, M. E., & Hoffman, B. D. (2018). Tunable molecular tension sensors reveal extension-based control of vinculin loading. *eLife*, *7*, e33927. https://doi.org/10.7554/eLife.33927.

LaCroix, A. S., Rothenberg, K. E., Berginski, M. E., Urs, A. N., & Hoffman, B. D. (2015). Construction, imaging, and analysis of FRET-based tension sensors in living cells. *Methods in Cell Biology*, *125*, 161–186. https://doi.org/10.1016/bs.mcb.2014.10.033.

Li, M., Cao, H., Lai, L., & Liu, Z. (2018). Disordered linkers in multidomain allosteric proteins: Entropic effect to favor the open state or enhanced local concentration to favor the closed state? *Protein Science*, *27*, 1600–1610. https://doi.org/10.1002/pro.3475.

Li, G., Huang, Z., Zhang, C., Dong, B. J., Guo, R. H., Yue, H. W., et al. (2016). Construction of a linker library with widely controllable flexibility for fusion protein design. *Applied Microbiology and Biotechnology*, *100*, 215–225. https://doi.org/10.1007/s00253-015-6985-3.

Li, C., Wen, A., Shen, B., Lu, J., Huang, Y., & Chang, Y. (2011). FastCloning: A highly simplified, purification-free, sequence- and ligation-independent PCR cloning method. *BMC Biotechnology*, *11*, 92. https://doi.org/10.1186/1472-6750-11-92.

Maiti, S., Acharya, B., Boorla, V. S., Manna, B., Ghosh, A., & De, S. (2019). Dynamic studies on intrinsically disordered regions of two paralogous transcription factors reveal rigid segments with important biological functions. *Journal of Molecular Biology*, *431*, 1353–1369. https://doi.org/10.1016/j.jmb.2019.02.021.

Mruk, D. D., & Cheng, C. Y. (2011). Enhanced chemiluminescence (ECL) for routine immunoblotting: An inexpensive alternative to commercially available kits. *Spermatogenesis*, *1*, 121–122. https://doi.org/10.4161/spmg.1.2.16606.

Norris, J. L., & Hughes, R. M. (2018). protaTETHER—A method for the incorporation of variable linkers in protein fusions reveals impacts of linker flexibility in a PKAc-GFP fusion protein. *FEBS Open Bio*, *8*, 1029–1042. https://doi.org/10.1002/2211-5463.12414.

Quehl, P., Schuurmann, J., Hollender, J., & Jose, J. (2017). Improving the activity of surface displayed cytochrome P450 enzymes by optimizing the outer membrane linker. *Biochimica et Biophysica Acta—Biomembranes*, *1859*, 104–116. https://doi.org/10.1016/j.bbamem.2016.10.022.

Redchuk, T. A., Omelina, E. S., Chernov, K. G., & Verkhusha, V. V. (2017). Near-infrared optogenetic pair for protein regulation and spectral multiplexing. *Nature Chemical Biology*, *13*, 633–639. https://doi.org/10.1038/nchembio.2343.

Reddy Chichili, V. P., Kumar, V., & Sivaraman, J. (2013). Linkers in the structural biology of protein-protein interactions. *Protein Science*, *22*, 153–167. https://doi.org/10.1002/pro.2206.

Saavedra, H. G., Wrabl, J. O., Anderson, J. A., Li, J., & Hilser, V. J. (2018). Dynamic allostery can drive cold adaptation in enzymes. *Nature*, *558*, 324–328. https://doi.org/10.1038/s41586-018-0183-2.

Schuurman-Wolters, G. K., de Boer, M., Pietrzyk, M. K., & Poolman, B. (2018). Protein linkers provide limits on the domain interactions in the ABC importer GlnPQ and determine the rate of transport. *Journal of Molecular Biology*, *430*, 1249–1262. https://doi.org/10.1016/j.jmb.2018.02.014.

Shimozono, S., & Miyawaki, A. (2008). Engineering FRET constructs using CFP and YFP. *Methods in Cell Biology, 85*, 381–393. https://doi.org/10.1016/S0091-679X(08)85016-9.

Snapp, E. (2005). Design and use of fluorescent fusion proteins in cell biology. *Current Protocols in Cell Biology, 27*(1), 21.4.1–21.4.13. Chapter 21:Unit 21.4. https://doi.org/10.1002/0471143030.cb2104s27.Design.

Sokolovski, M., Bhattacherjee, A., Kessler, N., Levy, Y., & Horovitz, A. (2015). Thermodynamic protein destabilization by GFP tagging: A case of interdomain allostery. *Biophysical Journal, 109*, 1157–1162. https://doi.org/10.1016/j.bpj.2015.04.032.

Steffen, V., Otten, J., Engelmann, S., Radek, A., Limberg, M., Koenig, B. W., et al. (2016). A toolbox of genetically encoded FRET-based biosensors for rapid L-lysine analysis. *Sensors (Basel), 16*(10), 1604. https://doi.org/10.3390/s16101604.

van den Ent, F., & Lowe, J. (2006). RF cloning: A restriction-free method for inserting target genes into plasmids. *Journal of Biochemical and Biophysical Methods, 67*, 67–74. https://doi.org/10.1016/j.jbbm.2005.12.008.

van Rosmalen, M., Krom, M., & Merkx, M. (2017). Tuning the flexibility of glycine-serine linkers to allow rational design of multidomain proteins. *Biochemistry, 56*, 6565–6574. https://doi.org/10.1021/acs.biochem.7b00902.

Vandame, P., Spriet, C., Riquet, F., Trinel, D., Cailliau-Maggio, K., & Bodart, J.-F. (2014). Optimization of ERK activity biosensors for both ratiometric and lifetime FRET measurements. *Sensors (Basel), 14*, 1140–1154. https://doi.org/10.3390/s140101140.

Vuzman, D., & Levy, Y. (2012). Intrinsically disordered regions as affinity tuners in protein-DNA interactions. *Molecular BioSystems, 8*, 47–57. https://doi.org/10.1039/c1mb05273j.

Wriggers, W., Chakravarty, S., & Jennings, P. A. (2005). Control of protein functional dynamics by peptide linkers. *Biopolymers, 80*, 736–746. https://doi.org/10.1002/bip.20291.

Yang, C., Zhang, X., Guo, Y., Meng, F., Sachs, F., & Guo, J. (2015). Mechanical dynamics in live cells and fluorescence-based force/tension sensors. *Biochimica et Biophysica Acta, 1853*, 1889–1904. https://doi.org/10.1016/j.bbamcr.2015.05.001.

Zhao, W., Du, G., & Liu, S. (2019). An efficient thermostabilization strategy based on self-assembling amphipathic peptides for fusion tags. *Enzyme and Microbial Technology, 121*, 68–77. https://doi.org/10.1016/j.enzmictec.2018.11.004.

Zhou, H.-X. (2017). Gating motions and stationary gating properties of ionotropic glutamate receptors: Computation meets electrophysiology. *Accounts of Chemical Research, 50*, 814–822. https://doi.org/10.1021/acs.accounts.6b00598.

Zong, C., Maksimov, M. O., & Link, A. J. (2016). Construction of lasso peptide fusion proteins. *ACS Chemical Biology, 11*, 61–68. https://doi.org/10.1021/acschembio.5b00745.

CHAPTER FIVE

Optimizing the linker length for fusing an alcohol dehydrogenase with a cyclohexanone monooxygenase

Alejandro Gran-Scheuch[a,b,†], Friso Aalbers[a,c,†], Yannick Woudstra[a], Loreto Parra[d], and Marco W. Fraaije[a,*]

[a]Molecular Enzymology Group, University of Groningen, Groningen, The Netherlands
[b]Department of Chemical and Bioprocesses Engineering, Pontificia Universidad Católica de Chile, Santiago, Chile
[c]Department of Biology and Biotechnology "L. Spallanzani", University of Pavia, Pavia, Italy
[d]Institute for Biological and Medical Engineering, Pontificia Universidad Católica de Chile, Santiago, Chile
*Corresponding author: e-mail address: m.w.fraaije@rug.nl

Contents

† These authors contributed equally.

Methods in Enzymology, Volume 647
ISSN 0076-6879
https://doi.org/10.1016/bs.mie.2020.09.008

Abstract

The use of enzymes in organic synthesis is highly appealing due their remarkably high chemo-, regio- and enantioselectivity. Nevertheless, for biosynthetic routes to be industrially useful, the enzymes must fulfill several requirements. Particularly, in case of cofactor-dependent enzymes self-sufficient systems are highly valuable. This can be achieved by fusing enzymes with complementary cofactor dependency. Such bifunctional enzymes are also relatively easy to handle, may enhance stability, and promote product intermediate channeling. However, usually the characteristics of the linker, fusing the target enzymes, are not thoroughly evaluated. A poor linker design can lead to detrimental effects on expression levels, enzyme stability and/or enzyme performance. In this chapter, the effect of the length of a glycine-rich linker was explored for the case study of ε-caprolactone synthesis through an alcohol dehydrogenase-cyclohexanone monooxygenase fusion system. The procedure includes cloning of linker variants, expression analysis, determination of thermostability and effect on activity and conversion levels of 15 variants of different linker sizes. The protocols can also be used for the creation of other protein-protein fusions.

1. Introduction

Biocatalysis refers to the use of enzymes to catalyze the synthesis of chemicals. For this, the field has exploited the accumulated knowledge of catalytic mechanisms, kinetic parameters and structure-function relationships of enzymes. This fundamental knowledge enables the development of improved biocatalysts for the chemo-, regio- and enantioselective synthesis of (fine) chemicals (Adams, Brown, Diaz-Rodriguez, Lloyd, & Roiban, 2019; Woodley, 2020). The use of biocatalysts in the synthesis of industrially relevant chemicals, such as lactones, is well described (Delgove, Elford, Bernaerts, & De Wildeman, 2018; Messiha et al., 2018; Woodley, 2020). These compounds are highly attractive as platform chemicals for the pharmaceutical and chemical industries, because they are often used as precursors for valuable chemicals, such as floral scents, or for polymer synthesis (Delgove et al., 2018; Magali, de Revel, & Marchand, 2017; Messiha et al., 2018; Zhu et al., 2020). For example, ε-caprolactone, is a valued 6-carbon lactone that can be used as a precursor for the synthesis of caprolactam (Ritz, Fuchs, Kieczka, & Moran, 2000). This product is further polymerized to form nylon-6, a popular polyamide polymer that is used in the manufacturing industries because of its high strength and elasticity. In addition to the production of polyamides, the most significant chemical feature of ε-caprolactone is its capacity to polymerize through a straightforward ring-opening reaction (Dahlhoff, Niederer, & Hoelderich, 2001; Künkel et al., 2000). The obtained product is polycaprolactone, a biodegradable thermo polyester used as additive in resins, coating, colorant materials and polyurethanes (Köpnick, Schmidt, Brügging, Rüter, & Kaminsky, 2000; Labet & Thielemans, 2009; Mallin, Wulf, & Bornscheuer, 2013). Nowadays, ε-caprolactone has a high global demand and is produced on a scale of over 10,000 tons per year (Thaore, Chadwick, & Shah, 2018). However, the classical synthesis of the lactone is not eco-friendly, as this compound is generally obtained through a Baeyer-Villiger oxidation using peracetic acid and cyclohexanone in an anhydrous solvent such as acetone (Labet & Thielemans, 2009; Ritz et al., 2000). Otherwise, it can be obtained by letting an excess of cyclohexanone react with air at 25–50 °C in the presence of acetaldehyde and metal-oxidant catalysts. As a byproduct, acetic acid is obtained, which must be removed by distillation (Ritz et al., 2000). A biocatalytic process would offer an attractive greener alternative for the synthesis of ε-caprolactone.

Broadly, enzymes have evolved for millions of years to catalyze specific reactions with a high efficiency. The use of enzymes may reduce energy costs by performing catalysis at milder conditions than classical chemical methods, such as lower temperature and pressure (Fraaije et al., 2005; Messiha et al., 2018; Savoca, Tonoli, Atobatele, & Verderio, 2018). Enzymes can be employed as (i) isolated enzymes, (ii) immobilized enzymes, (iii) cell free extracts or (iv) within whole cells (Wachtmeister & Rother, 2016; Woodley, 2008). The latter is the most widely used method for large scale reactions, since (1) it is the cheapest way to formulate enzymes, (2) it provides greater enzyme stability, and (3) it satisfies biochemical requirements such as the presence of cofactors and cosubstrates. However, the use of whole cells also implies more complex mass balance calculations and transfer concerns, such as inadequate supply of oxygen or substrate/product transfer in/out the cells (Baldwin & Woodley, 2006; Doig et al., 2002; Kadisch, Willrodt, Hillen, Bühler, & Schmid, 2017; Milker et al., 2017; Sieben et al., 2016). Evidently, each of these approaches has its own pros and cons, and the application will depend on the specific reaction to be achieved.

Baeyer-Villiger monooxygenases (BVMO, E.C. 1.14.13) have been described to produce ε-caprolactone using cyclohexanone as a substrate (Mallin et al., 2013; Mthethwa et al., 2017; Parra, Acevedo, & Reetz, 2015; Staudt, Bornscheuer, Menyes, Hummel, & Gröger, 2013). In particular, type I BVMOs from the class B flavin-dependent monooxygenases contain flavin adenine dinucleotide (FAD) as a prosthetic group, which is required for catalysis. In the catalytic cycle, the FAD is reduced by NADPH (Van Berkel, Kamerbeek, & Fraaije, 2006). Subsequently, the reduced enzyme reacts with dioxygen to form a reactive peroxyflavin enzyme intermediate which reacts with a ketone to form an ester or lactone (Romero, Gomez Castellanos, Gadda, Fraaije, & Mattevi, 2017). Release of the product and $NADP^+$ completes the catalytic cycle. The most studied BVMO for the synthesis of ε-caprolactone is a cyclohexanone monooxygenase (AcCHMO) isolated from the soil bacterium *Acinetobacter calcoaceticus* NCIMB 9871 (Donoghue, Norris, & Trudgill, 1976). Although this enzyme shows a remarkably broad substrate scope, it exhibits very poor stability. Specifically, its apparent melting temperature (T_M^{app}) is around 36 °C and it has a low tolerance toward cosolvents (Secundo et al., 2011). Recently, a more robust CHMO homolog was described and crystallized (Romero, Gómez Castellanos, Mattevi, & Fraaije, 2016). This variant, TmCHMO, was discovered in the actinobacterium *Thermocrispum municipale* DSM 44069. Even though this BVMO displays high sequence identity of 57% with AcCHMO, TmCHMO showed greater robustness in presence of cosolvents and a significantly higher T_M^{app}. A clear disadvantage for CHMO and BVMOs as biocatalysts is their NADPH dependence. This cosubstrate is relatively expensive and therefore problematic to be used for biosynthetic routes. To remedy this, CHMO has been explored as fusion enzyme with different NADPH regenerating enzymes, such as phosphite dehydrogenase, formate dehydrogenase, glucose dehydrogenase or alcohol dehydrogenase (ADH) (Mourelle-Insua, Aalbers, Lavandera, Gotor-Fernández, & Fraaije, 2019; Torres Pazmiño et al., 2009). Interestingly, some alcohol dehydrogenases (ADH, E.C.1.1.1.X) are able to oxidize cyclohexanol to cyclohexanone while reducing $NAD(P)^+$. The combination of such an ADH with a BVMO can catalyze a cascade reaction, in which the ADH regenerates NADPH while producing the substrate for the BVMO. For the particular reaction of this study, ADH catalyzes the oxidation of cyclohexanol (forming NADPH) which is converted into ε-caprolactone by CHMO (consuming NADPH) (Mallin et al., 2013; Sattler et al., 2014; Schmidt et al., 2015; Staudt et al., 2013).

For the biocatalytic cascade reaction above, it is attractive to create a fused bifunctional ADH-BVMO fusion. For such fusion, it is important to introduce a proper linker. The linker is expected to allow enough flexibility and proper folding of the fused enzymes. The close proximity of two active-sites may also increase the conversion rates by enhancing the channeling of the reaction intermediates (Wheeldon et al., 2016). Although this substrate channeling effect is questioned in some studies (Ellis et al., 2019; Kuzmak, Carmali, von Lieres, Russell, & Kondrat, 2019; Rabe, Müller, Skoupi, & Niemeyer, 2017), and likely only applies when the concentration of the intermediate remains low, there are a number of studies that found higher conversions for a cascade reaction with fused enzymes compared to the equimolar combination of separate enzymes (Aalbers & Fraaije, 2019).

The ADH-BVMO cascade system has been previously studied for the optimization of the biocatalytic conversion of cyclohexanol into ε-caprolactone (Fig. 1) (Kohl, Srinivasamurthy, Böttcher, Kabisch, & Bornscheuer, 2018; Schmidt et al., 2015). However, the reaction design is not trivial and some critical parameters are worth mentioning. For example, a poor substrate/cosubstrate set up can adversely affect the conversion rates. Particularly, AcCHMO suffers from inhibition by cyclohexanol, cyclohexanone and

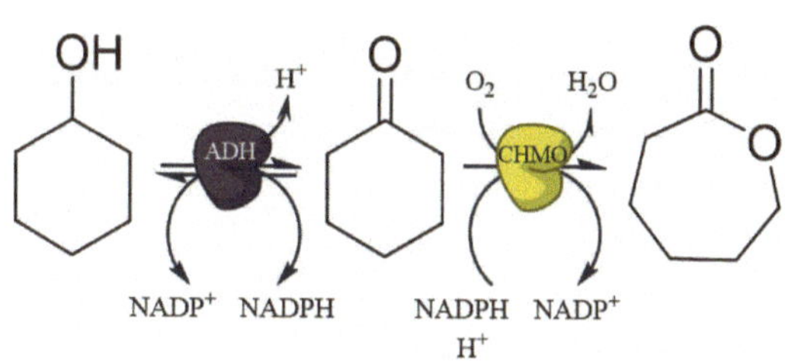

Fig. 1 TbADH-TmCHMO cascade reaction. A schematic representation of the cascade reaction is shown. The cascade starts with the reversible oxidation of cyclohexanol to cyclohexanone and $NADP^+$ reduction by TbADH. Subsequently, the TmCHMO partner accepts the cyclohexanone and performs a Baeyer-Villiger oxidation. For this reaction, the hydride donor is consumed and lactone and water are produced. Then, the oxidized $NADP^+$ can be re-occupied for the first reaction.

caprolactone (Scherkus et al., 2017). For TmCHMO, an activity reduction of 50% was observed at 66 mM ε-caprolactone, and 75% at 2 mM cyclohexanol (Aalbers & Fraaije, 2017). Therefore, a good cascade strategy must be found that balances optimal kinetic conditions for both enzymes.

In a previously published report, three different ADHs were explored as TmCHMO fusion partners: (i) TbADH from *Thermoanaerobacter brockii*, (ii) PfADH from *Pyrococcus furiosus* and (iii) MiADH from *Mesotoga infera* (Aalbers & Fraaije, 2017). To avoid product inhibition, a lipase was employed as additional biocatalyst, to decrease product concentration by catalyzing the ring opening of ε-caprolactone. The best performing fusion (TbADH-TmCHMO) achieved full conversion of 200 mM cyclohexanol with a turnover number (TON) of 13,000, while using separate enzymes only 41% conversion was obtained (TON = 5600).

Even though the TbADH-TmCHMO fusion gave promising results for the cascade reaction, the study did not try to optimize the linker. In that sense, both length and physicochemical characteristics of the linker can be tuned. Key aspects to consider are flexibility/rigidity and hydrophilicity/hydrophobicity of the linker sequence. Jeon et al. investigated the use of two different linkers in a similar two-step whole-cell biocatalytic process—conversion of long-chain sec-alcohols into esters—(Jeon, Baek, Bornscheuer, & Park, 2015). Fusions were evaluated consisting of an ADH from *Micrococcus luteus* NCTC2665 and BVMOs of *Pseudomonas putida* KT2440 or *Rhodococcus jostii* RHA1. The use of a glycine-rich linker had a better performance on the conversions than a rigid α-helix linker, and the fusion also outperformed the same combination of separate enzymes. Moreover, various studies have described that linkers with nonpolar residues, such as the glycine-rich linker, are beneficial by enhancing the flexibility between both partners (Argos, 1990; Chen, Zaro, & Shen, 2013). This would provide freedom for correct folding and conformational changes. In a critical study on flexible linkers, that also included computational modeling work, linker length and composition turned out to be highly important (Van Rosmalen, Krom, & Merkx, 2017). Shorter linkers and sequences containing more glycines increased FRET efficiencies between the ECFP and EYFP domains of the fusion. Other studies have described that the size of the linker can critically perturb the biocatalytic properties of fusions, such as enzyme activity, stability or coupling efficiency (Corrado, Knaus, & Mutti, 2018; Heine et al., 2017; Tischler et al., 2018). Therefore, in addition to the physicochemical composition of the linker, the length itself may have a key role on the biocatalytic features of the fusion.

Thus, it is attractive to evaluate whether this is also the case for other fusions, such as the ADH-CHMO fusion, and to study the effects of linker length on its performance of its cascade reaction.

In this chapter, we describe an experimental procedure for exploring the effect of the length of a glycine-rich linker on the biocatalytic properties of a ADH-BVMO fusion. Specifically, the TbADH-TmCHMO fusion for the production of ε-caprolactone from cyclohexanol was studied. The glycine-rich linker SSGGSGGSGGSAGTA was selected based on previous work. Fusion variants were designed with different linker size using the P-LinK methodology (Belsare et al., 2014). The fusion enzymes were prepared with a linker size of 1–15 amino acids. The complete procedure is as follows: (i) verifying the expression levels of the fusion, (ii) spectrophotometrically analyze the oxidation state of TmCHMO, (iii) evaluate the effect of the linker length on the thermostability, (iv) activity and (v) conversion levels.

2. General method and statistical analysis

Common security measures for a biosafety level-1 laboratory must be followed. Personal protective elements as glasses, lab coat and chemically resistant gloves are required for the researcher when performing the experiments. For the statistical analysis, the determination of thermostability, activity and conversions level were performed in triplicates or duplicates. The results were analyzed using GraphPad Prism v6.05 for Windows (GraphPad Software, La Jolla, CA, United States). Statistical difference was evaluated using ordinary one-way ANOVA, using significance $P < 0.05$. Multiple comparisons against the control was performed using Dunnett's multiple comparison test. Nevertheless, the user can use other suitable software according to his preferences.

3. Molecular design of linker variants

The most straightforward way to obtain a fusion enzyme is by ordering a synthetic DNA construct for expression of the respective protein. On the other hand, there are various ways to create a construct to express a fusion enzyme using the two respective genes. A couple of well-described and convenient cloning approaches are: (1) traditional restriction/ligation by adding a unique restriction site at the reverse primer of the first gene (for the first enzyme) and the forward primer of the second gene (for the second enzyme of the fusion), such that they can be ligated. It is worth mentioning that the stop codon of the first gene must be removed. (2) Golden Gate cloning, similar to restriction/ligation, except that it employs a restriction site (*Bsa*I) that is not recognized after ligation, which greatly improves efficiency for ligation, and reduces steps compared to restriction/ligation. A disadvantage is that a Golden gate vector is required. Lastly, using (3) Gibson assembly/In-Fusion® cloning by designing primers with 15–30 base pairs (bp) overlapping regions, the amplified fragments can be assembled through an isothermal reaction with three enzymes: DNA polymerase, 5′ exonuclease, and DNA ligase. In each of these methods there is some room to build a linker region, by adding codons that code for a linker to the reverse primer of the first gene and to the forward primer of the second gene. Alternatively, with each method an existing linker can also be amplified as fragment, and ligated together with the two genes.

Once a construct coding for a fusion enzyme is obtained, different linker sequences can be designed and introduced. In this section, we describe the use of the P-LinK method that has been described to change the linker length between two domains of a cytochrome P450 monooxygenase (Belsare et al., 2014). This method is analogous to QuikChange with partially overlapping primers, with as difference that the primers are partly phosphorothioated, which enables efficient chemical ligation of the vectors after amplification (referred to as PLIC reaction from Phosphorothioated Ligation Independent Cloning). This is an advantage in terms of efficiency and number of steps needed to complete the construct compared to, for example, Gibson cloning. There are other methods to introduce different linkers, of which one recent example is detailed in chapter "Synthetic protein switches: Combinatorial linker engineering with iFLinkC" by Gräwe et al. of this edition of *Methods in Enzymology*, as well as in a recent publication (Gräwe, Ranglack, Weyrich, & Stein, 2020).

3.1 Equipment

- PCR Thermocycler machine (T100 thermal Cycler, Bio-Rad).
- Shaker (Innova 44, New Brunswick Scientific).
- Thin-walled PCR tubes (DNAse free and/or sterile).
- Electrophoresis chamber (Bio-Rad).
- Thermomixer or water bath.
- Speedvac equipment.
- UV transilluminator (M20, Appligene).

3.2 Reagents

- PCR oligonucleotides containing 12 phosphorothioates (PTO) (see note 1) at the 5′-end (see Section 3.2: steps 1 and 2).
- Pfu polymerase (as master mix, or with separate components: dNTPs mix, polymerase buffer).
- PCR cleanup kit (Qiagen).
- Plasmid MiniPrep kit (Qiagen).
- Sterile dH_2O
- Agarose gel (1%, w/v) (from agarose powder).
- *Dpn*I enzyme (NEB).
- Sterile 80% glycerol
- Iodine solution (100 mM I_2 in ethanol).
- Sterile LB medium (for 1 L, 10 g tryptone, 10 g NaCl and 5 g yeast extract). For LB agar, add agar-agar at 2% (w/v) final concentration (20 g for 1 L).
- *Escherichia coli* NEB 10-β chemo competent cells, genotype: Δ*(ara-leu) 7697 araD139 fhuA* Δ*lacX74 galK16 galE15 e14*-ϕ*80*d*lacZ*Δ*M15 recA1 relA1 endA1 nupG rpsL* (Str^R) *rph spoT1* Δ*(mrr-hsdRMS-mcrBC)*.

3.3 Procedure

1. Design oligonucleotides as follows: the protocol has been optimized for having (at least) 12 phosphorothioates (PTOs) at the 5′ end of the reverse primer and forward primers. For every five linker lengths (1–5 amino acids), the same reverse (RV) primer is used and five different forward (FW) primers. See the example in Table 1.
2. Solubilize PTO oligonucleotide stocks to 100 μM in 10 mM Tris HCl pH 7.0 or autoclaved milli-Q water. Then, dilute to 20 μM for working stocks. Store the 100 μM stocks at −20 °C to preserve.
3. Set up PCR reactions with either PCR master mix, following the manufacturer's instruction, or using separate components, as follows in Table 2. Include a reaction without any primers added, to serve as control for the DpnI digestion.
4. Transfer the tubes to a thermocycler and start the following PCR program (Table 3). The lid must be at 105 °C to prevent condensation.
5. Analyze the PCR products by loading them in a 1% agarose gel (20–50 mL) with 1.5 μL of RotiSafe as gel stainer. Mix 3 μL of PCR sample with 0.5 μL of loading dye (6 ×), and add it to the well, then examine the gel under UV light in a UV transilluminator (Fig. 2A) (see note 2).

Table 1 List of primers for the linker variant design.

	PTO region (12)[a]	Linker region	Gene-overlapping region
RV1	GGCCAGAATAACAAC	–	TGGTTTGATCAGATC
T_m[b]	47 °C	–	45 °C
FW1	GTTGTTATTCTGGCC	TCG	ATGAGCACCACCCAGACCCCG
FW2	GTTGTTATTCTGGCC	TCGAGT	ATGAGCACCACCCAGACCCCG
FW3	GTTGTTATTCTGGCC	TCGAGTGGT	ATGAGCACCACCCAGACCCCG
FW4	GTTGTTATTCTGGCC	TCGAGTGGTGGC	ATGAGCACCACCCAGACCCCG
FW5	GTTGTTATTCTGGCC	TCGAGTGGTGGCTCT	ATGAGCACCACCCAGACCCCG
T_m[b]	47 °C	10–56 °C	69 °C
RV2	AGAGCCACCACT[c]	–	CGAGGCCAGAATAACAACTGGTTTG
T_m	53 °C	–	68 °C
FW6	AGTGGTGGCTCT	GGT	ATGAGCACCACCCAGACCCCGGACC
FW7	AGTGGTGGCTCT	GGTGGG	ATGAGCACCACCCAGACCCCGGACC
FW8	AGTGGTGGCTCT	GGTGGGAGC	ATGAGCACCACCCAGACCCCGGACC
FW9	AGTGGTGGCTCT	GGTGGGAGCGGT	ATGAGCACCACCCAGACCCCGGACC
FW10	AGTGGTGGCTCT	GGTGGGAGCGGTGGC	ATGAGCACCACCCAGACCCCGGACC
FW11	AGTGGTGGCTCT	GGTGGGAGCGGTGGCTCA	ATGAGCACCACCCAGACCCCGGACC
T_m	53 °C	10–74 °C	80 °C

RV3	TGAGCCACCGCT[a]	–	CCCACCAGAGCCACCACTCGAGGCCAGAAT AACAACTGGTTTG
T_m	59°C	–	82°C
FW12	AGCGGTGGCTCA	GCT	ATGAGCACCACCCAGACCCCGGACC
FW13	AGCGGTGGCTCA	GCTGGT	ATGAGCACCACCCAGACCCCGGACC
FW14	AGCGGTGGCTCA	GCTGGTACC	ATGAGCACCACCCAGACCCCGGACC
FW15	AGCGGTGGCTCA	GCTGGTACCGCG	ATGAGCACCACCCAGACCCCGGACC
FW16	AGCGGTGGCTCA	GCTGGTACCGCGGGC	ATGAGCACCACCCAGACCCCGGACC
T_m	59°C	10–71°C	82°C

[a]The PTO region of the FW primers and RV primer are always the reverse complement of one another. Ensure that this hybridization area has a melting temperature higher than 42°C to be stable during heat shock.
[b]Melting temperature of region of primer as calculated by NEB Tm calculator (https://tmcalculator.neb.com/). The linker region shows here codes for: SSGGS; the first part of our glycine-rich linker (SSGGSGGSGGSAGTAG).
[c]The PTO region for both RV6–11 and RV12–16 are actually in the linker region, hence the name in the middle column is now "linker extension region." Note that the gene-overlapping region is slightly extended in this set of primers (underlined part is the same as in the 1–5 primers).

Table 2 Polymerase chain reaction setup.

Volume	Components	Final concentration
5 μL	10× HF polymerase buffer	1×
1 μL	50× dNTP mixture	0.2 mM[a]
1 μL	FW primer	0.4 μM
1 μL	RV primer	0.4 μM
0.5 μL	Polymerase (e.g. Phusion)	5.0 U
1 μL	Template DNA	0.4 ng μL^{-1}
x μL	Sterile milli-Q/dH_2O up to 50 μL	
50 μL	Final volume	

[a]0.2 mM of each dATP, dCTP, dGTP, and dTTP.

Table 3 Thermocycling conditions.

Step	Time	Temperature
1	2 min	98 °C
2	30 s	98 °C
3	30 s	68 °C (−1 °C/cycle)
4	7 min	72 °C
		Repeat steps 2–4 for 12 cycles
5	30 s	98 °C
6	30 s	57 °C
7	7 min	72 °C
		Repeat steps 5–7 for 15 cycles
8	11 min	72 °C

6. Add 10 U of DpnI and CutSmart buffer (1× final concentration) to each reaction that gave product, and incubate them at 37 °C for at least 1 h or overnight (16 h).
7. Purify the PCR products with a PCR cleanup kit, and in the final step elute with 22 μL dH_2O, to ensure a high concentration of DNA
8. Determine DNA concentration with a NanoDrop spectrophotometer. A concentration of approximately 0.02–0.04 pmol μL^{-1} (=μM) is desired. To translate that value to ng μL^{-1}, the following online tool can be used: https://nebiocalculator.neb.com/#!/dsdnaamt (see note 3).

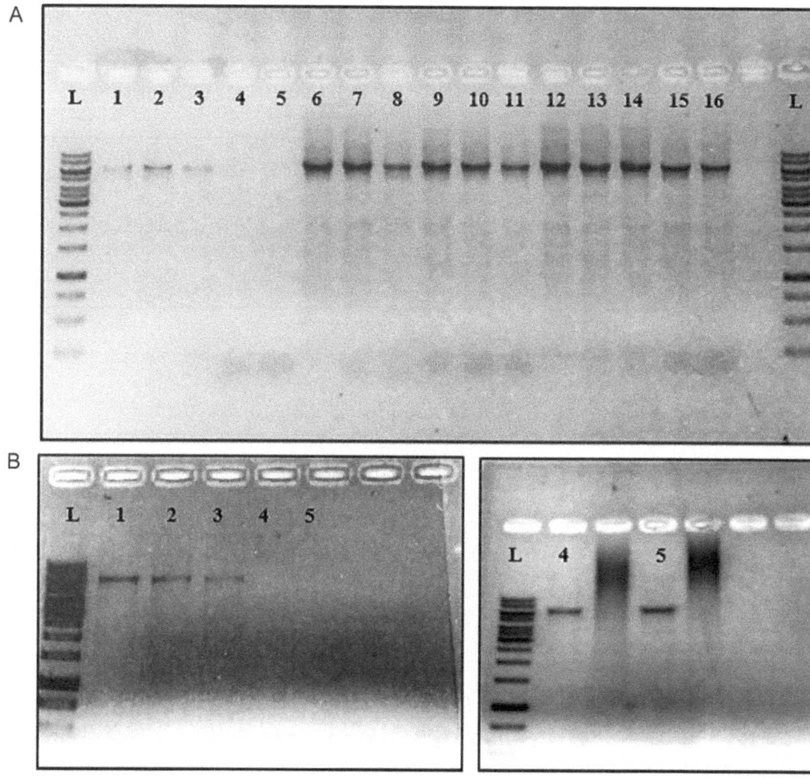

Fig. 2 Agarose gel analysis of the PCR products. The numbers indicate the linker length of the PCR product. Lanes marked with L are loaded with the ladder (Page Ruler).

9. To increase the DNA concentration, one can apply the samples to a speedvac, to reach the 0.02–0.04 pmol μL^{-1} concentration range. Optionally, an additional PCR reaction can be performed using the corresponding purified PCR product as template.
10. Place the purified PCR products and thin-wall PCR tubes on ice. Transfer 4 μL of each PCR product (at 0.02–0.04 pmol μL^{-1}) to a thin-wall PCR tube, as well as one for the control, and keep on ice.
11. To anneal the PCR product and form the circularized vector, a PLIC reaction is carried out. This reaction consists of two components: (1) the PCR product and (2) iodine cleavage solution. This iodine solution can be prepared as shown in Table 4. Add 2 μL of iodine cleavage mixture to each thin-wall PCR tube that contains 4 μL of PCR reaction, for a total volume of 6 μL. The solution should be yellowish. Keep the solutions on ice (see note 4).

Table 4 Iodine solution reaction setup.

Volume	Component	Notes
5 µL	Tris HCl buffer (500 mM; pH 9)	Since only a small volume is needed, prepare a low amount and store at 4 °C
3 µL	Iodine-EtOH solution (100 mM)	Keep the solution protected from light (in a nontransparent tube, or wrapped in aluminum foil), and keep on ice or at 4 °C. It is best to prepare this solution fresh. In our case, we mixed 0.254 g of iodine with 10 mL of 99% ethanol
2 µL	dH_2O	
10 µL	Total volume	This solution is kept on ice or at 4 °C. This volume can be used for five PLIC reactions

12. Transfer the tubes into PCR thermocycler with a preheated lid (80 °C), and start the program with an incubation of 10 min at 70 °C (PTO cleaving step). Then, for the annealing/hybridization step, incubate the tubes at 25 °C for 5 min. After this step is done, the tubes can be transferred into ice, or stored at 4 °C. The solutions are ready to be used to transform chemically competent *E. coli* cells

13. Set a thermomixer or water bath to 42 °C. Pre-warm 1 mL SOC or LB media to 37 °C in the thermo block, then transform the samples (PLIC reaction with each PCR product) and include the PLIC reaction control (without iodine) and *Dpn*I control.

14. Thaw 18 tubes of 50–100 µL chemically competent *E. coli* NEB 10-β cells on ice, and label them accordingly (1–18). Add the 2.5 µL to the, respectively, labeled tube and mix gently by flicking the tube, and directly place back on ice.

15. Incubate on ice for 20–30 min.

16. Perform a heat shock by incubating the tubes for 30 s at 42 °C in the thermomixer or water bath. Place them back into the ice after the 30 s.

17. Add 900 µL pre-warmed SOC or LB media to each tube, and incubate them at 37 °C for 60 min while shaking. This can be done in a large shaking incubator with a tube rack, or in a thermomixer at 700 r.p.m. This is necessary to ensure proper aeration.

18. Centrifuge the tubes at 5000 *g* for 1 min, remove 700–800 µL of the supernatant, and resuspend the remaining 100–200 µL with a pipette using sterile tips, and transfer to an LB agar plate with the appropriate antibiotic (50 $\mu g\,mL^{-1}$ ampicillin). Spread the cell suspension with sterile glass beads, an inoculation loop, a bend glass pipette, or a glass/metal spreader. If one of these tools was heated with a flame to sterilize, ensure that it is cooled down prior to touching the cell suspension.

19. Incubate the agar plates at 37 °C overnight.
20. If everything went well, the control plates should be empty or nearly empty, while the plates from the PLIC reaction should have greater than 100 colonies. Typically, the more colonies, the higher the chance the correct construct will be obtained.
21. Pick a colony from each plate, except the controls, to inoculate 5 mL LB with 50 μg mL^{-1} ampicillin, to grow overnight at 30 °C (>18 h) or 37 °C (12–18 h).
22. Transfer ~2 mL from the overnight culture to a 2 mL eppendorf tube, and spin down the cells at 5000 g for 1 min. Use a MiniPrep kit to purify the plasmid from this pellet, according to the kit's instructions. Store the remaining 3 mL of the culture at 4 °C.
23. Send the plasmid for sequencing, to ensure the right construct is obtained. After sequence confirmation, use the corresponding stored culture to make a −70/−80 °C glycerol stock by mixing 400 μL culture with 400 μL of sterile 80% glycerol, and store at −70/−80 °C (see note 5). Alternatively, the corresponding plasmid can be used to transform chemically competent cells, and these can be used to make a −70/−80 °C stock (see note 6).

3.4 Notes

1. A notable disadvantage of this method is the need for phosphorothioated primers, which are quite costly or require in-house primer synthesis. There are other methods to introduce different linkers, of which one recent example is detailed in this edition of *Methods in Enzymology*, as well as in a recent publication (Gräwe et al., 2020). Moreover, the method detailed above could also be carried out with Gibson assembly instead of PLIC reaction, using regular primers, with some adaptations.
2. Since not all PCR reactions yielded detectable products, the missing reactions were repeated (using the same program as above). For step 3, the annealing temperature parameter was modified to 62 °C (−1 °C/cycle), while for step 6, it was changed to 50 °C. This yielded product for linker lengths 1–3 (Fig. 2B). To obtain PCR product for 4 and 5, the program from Table 5 was used. In addition, one can add DMSO (3–10%) to a PCR reaction. In our case, we added two PCR reactions that included 5 μL DMSO (10%, v/v).

Table 5 Alternative thermocycling conditions.

Step	Time	Temperature
1	2 min	98 °C
2	30 s	98 °C
3	30 s	70 °C (−0.5 °C/cycle)
4	7 min	72 °C
		Repeat step 2–4 for 24 cycles
5	30 s	98 °C
6	30 s	58 °C
7	7 min	72 °C
		Repeat steps 5–7 for 15 cycles
8	11 min	72 °C

3. For this chapter, the PCR products were around 6800 bp, since the vector is 6789 bp. Hence, 0.02–0.04 pmol μL^{-1} corresponds to a DNA concentration of 84–168 ng μL^{-1}.
4. Optionally, an additional control can be added where 4 μL of PCR reaction is mixed with 2 μL of dH_2O, to have an indication of whether the PLIC reaction worked.
5. For this scenario case, variants with the 1–3 and 6–15 linker length were obtained, while the constructs with 4 and 5 amino acid were not.
6. *E. coli* NEB 10-β is a derivate strain from DH10B, commonly preferred for high quality plasmid preparation, mostly due the deficiency in endonuclease I (*endA1*) and recombinase (*recA1*).

4. Purification of His_{x6} tagged proteins

This section describes the expression and purification of the obtained fused variants and the single components TmCHMO and TbADH after the cloning and sequence verification of the plasmid of each construct. The expression plasmids have a pBAD backbone that included a N-terminal hexahistidine tag to facilitate the further purification. The pBAD expression system is tightly controlled using L-arabinose and within the plasmids contains the gene for ampicillin resistance. For an optimal yield, some aspects of protein expression should be considered, such as the evaluation of the optical density of the culture before induction, concentration of the inducer, time and temperature of induction, oxygenation levels, different types of plasmid backbones and host strains (Qiagen, 2003; Spriestersbach, Kubicek, Schaefer, Block, & Maertens, 2015). Then, after cell lysis the enzymes are obtained through immobilized metal affinity chromatography (IMAC) using nickel-sepharose HP resin. To increase the speed of the purification process, the resin is bed in a gravity flow column. For a more precise purification, other systems can be employed, such as fast protein liquid chromatography (FPLC). Using FPLC, the purity of the sample can be even higher, but it would be more time consuming. Finally, the level of expression and purity of the proteins can be evaluated by sodium dodecyl sulfate-polyacrylamide gel electrophoresis (SDS-PAGE). The fusions were well expressed in the host, and yields in the range of 300 $mg L^{-1}$ were obtained. Additionally, electrophoresis showed the expected size for the variants according to the molecular size estimation.

4.1 Equipment

- Shaker (Innova 44, New Brunswick Scientific).
- Cooling centrifuge (Centrifuge 5804R, Eppendorf).
- Sonicator (Sonics, Vibra Cell).
- Orbital rotator (Nutating Mixer, VWR International).
- Electrophoresis system (Power Pac HC and Mini-Protean Tetra system, Bio-Rad).
- Nano drop spectrophotometer (NanoDrop 1000, Thermo Scientific).

4.2 Buffers, strain and reagents

- Sterile LB medium (for 1 L, add 10 g tryptone, 10 g NaCl and 5 g yeast extract). For LB agar, add agar-agar at 2% (w/v) final concentration.
- Sterile TB medium (for 1 L, add 12 g tryptone, 24 g yeast extract, 5 g glycerol and after autoclaving 100 mL sterile 10 × TB salts). For 1 L of 10 × TB salts dissolve 23.1 g KH_2PO_4 and 125.4 g K_2HPO_4.
- Lysis buffer (50 mM Tris HCl pH 8.0, 500 mM NaCl, 30 μM FAD, 1 mM PMSF, 10 mM $MgCl_2$, 0.5 $mg mL^{-1}$ DNAse I and 1 $mg mL^{-1}$ lysozyme).
- Buffer A (50 mM Tris HCl pH 8.0, 500 mM NaCl and 5 mM imidazole).
- Buffer B (50 mM Tris HCl pH 8.0, 500 mM NaCl and 400 mM imidazole).
- Buffer C (50 mM Tris HCl pH 8.0 and 150 mM NaCl).

- Nickel-Sepharose resin (GE Healthcare).
- Ampicillin stock at 50 mg mL^{-1}.

4.3 Procedure

1. *E. coli* NEB 10-β glycerol stocks containing the plasmids with the cloned genes were taken from −80 °C freezer and kept on a cooling tube rack. In parallel, prepare sterile culture tubes with 5 mL LB supplemented with 50 μg mL^{-1} ampicillin (see note 1).
2. Inoculate and grow the pre-cultures in a shaker at 37 °C with constant agitation (135 r.p.m.) over night.
3. The next morning, in a 250 mL flask inoculate at 1:100 in 50 mL TB medium supplemented with 50 μg mL^{-1} ampicillin.
4. Incubate the flasks at 37 °C until desired OD_{600nm} and induce at optimal expressions conditions (see note 2).
5. Centrifuge for 15 min at 3000 *g* at 4 °C and discard the supernatants.
6. Gently suspend the pellet in 10 mL lysis buffer. Optionally, to facilitate the resuspension of the pellet in the lysis buffer, can be used syringe with a sharp needle.
7. Transfer the suspension into a cooled sonication vessel or into a plastic or glass container that is placed inside a beaker with water and ice to effectively cool the suspension while it is subjected to sonication.
8. Sonicate for 10 min at 70% amplitude, 5 sec on and 5 sec off.
9. To remove the insoluble fraction (cell debris, insoluble proteins) centrifuge the obtained cell extracts (CE) for 60 min at 15,000 *g* at 4 °C.
10. Filter the cell free extracts (CFE) using 0.45 μm filter(s).
11. Wash 3 mL of the nickel-sepharose resin in a gravity flow column with 3–5 column volumes of milli-Q grade water.
12. Equilibrate the resin with at least five column volumes of buffer A.
13. Transfer the filtered CFEs and the 3 mL pre-equilibrated resin in suitable tubes and incubate them for 60 min in an orbital rotator at 4 °C.
14. Transfer each sample into a gravity flow column and discard the flow-through (eluate).
15. Wash with 10–15 column volumes of buffer A (see note 3).
16. Elute with 5 mL of buffer B (see note 4).
17. To remove the imidazole of the protein mixture, use a commercial size exclusion chromatography column, such as Econo-Pac® 10DG desalting prepacked gravity flow columns. Another suitable desalting column can also be used (if you will use another desalting column,

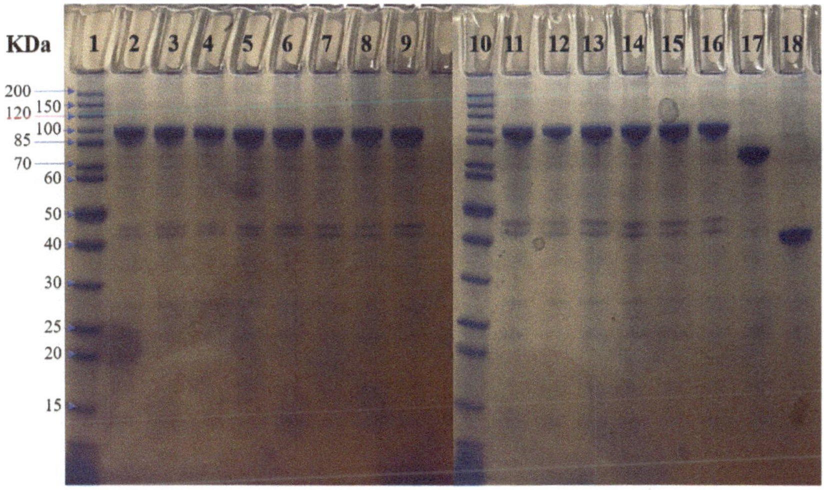

Fig. 3 SDS-PAGE of linker variants and single enzymes. After purification, 1 μg of protein was loaded per well in an SDS-PAGE. Each well was named according to the size of the amino acid linker. (1) Protein ladder, (2) 1 amino acid (aa), (3) 2 aa, (4) 3 aa, (5) 6 aa, (6) 7 aa, (7) 8 aa, (8) 9 aa, (9) 10 aa, (10) protein ladder, (11) 11 aa, (12) 12 aa, (13) 13 aa, (14) 13 aa, (15) 14 aa, (16) 15 aa, (17) TmCHMO and (18) TbADH. The expected size of the fusions is ~102 KDa.

follow the manufacturer instructions and skip until step 20). First, wash the column with 20 mL of buffer C.

18. Subsequently, after buffer C reaches the top frit, load up to 3.3 mL of the eluted sample obtained in step 16 and discard the first 3.3 mL.
19. Add 4 mL of buffer C and collect the eluted 4 mL.
20. Determine the amount of protein using a nano drop spectrophotometer.
21. To determine the level of purity, analyze the desalted samples through SDS-PAGE. Load 1–2 μg of purified protein per well (Fig. 3).
22. Finally, flash freeze the protein samples with liquid nitrogen and store them at −80 °C until use. Try to prepare small volumes aliquots (equal to or less than 500 μL).

4.4 Notes

1. During the inoculation and induction, the media (LB and TB), glasses, pipette tips and flasks must be autoclaved. During the protein purification process, keep the samples on ice the whole time.
2. For the pBAD expression system, first test small culture volumes in different expression conditions. We preferred to test 5 mL culture at 0.002–0.2% (w/v) L-arabinose and incubate at 17, 24, 30 or 37 °C for

16, 24, 48 or 72 h. For this method, cells were induced at OD_{600nm} 0.8 with 0.02% L-arabinose for 48 h at 17 °C at 135 r.p.m. For more information about high level expression and purification of histidine tagged proteins see the handbook *The QIAexpressionist* (Qiagen, 2003).

3. Optionally, follow the absorbance at 280 nm and continue the washing with the binding buffer until the absorbance reach a steady baseline.
4. Flavin-dependent proteins, such as TmCHMO due to their bound prosthetic cofactor, have a characteristic UV–visible spectrum with high absorbance about 440 nm. Consequently, these enzymes display a distinctive yellow color. Therefore, during purification, if there are relatively high amounts of protein, it is possible to easily see the protein mobility within the resin.

5. UV–visible spectral analysis of TmCHMO

As described above, TmCHMO contains FAD as prosthetic group. The bound flavin cofactor provides a useful probe to study and characterize the enzyme based on the UV–visible absorbance features of FAD (Chapman & Reid, 1999). FAD can exist in three different redox states: (1) the two-electrons reduced flavin hydroquinone, (2) the one-electron reduced flavin semiquinone and (3) the fully oxidized flavin (Massey, 2000). Each redox state has a specific UV–visible absorbance spectrum, hence the redox state of the protein can be easily determined by spectrophotometric analysis (Macheroux, 1999). Moreover, FAD has various protonation states, which alters the spectrophotometric properties. For example, the flavin semiquinone state absorbs at relatively long wavelengths (>550 nm), while the anionic flavin semiquinone displays a strong absorbance at 380 nm (Chapman & Reid, 1999; Massey, 2000). Particularly, oxidized TmCHMO exhibits two absorbance maxima, at 376 and 440 nm (Romero et al., 2016). TmCHMO in the oxidized state has a molar extinction coefficient of 14.0 mM^{-1} cm^{-1} at 440 nm, similar to other described flavoproteins. Another useful spectral feature for flavin-containing proteins is the *A*280:*A*440 absorbance ratio. The absorbance at 280 nm can be used to determine the amount of total protein, due to absorbance of aromatic amino acids. By measuring the *A*280:*A*440 it is possible to estimate the amount of holo protein (=FAD-containing enzyme) in a sample.

For this section, the absorbance spectra of the purified enzyme variants were recorded and the UV–visible spectra were analyzed. For all the fused enzymes, the typical spectral characteristics of oxidized FAD were observed

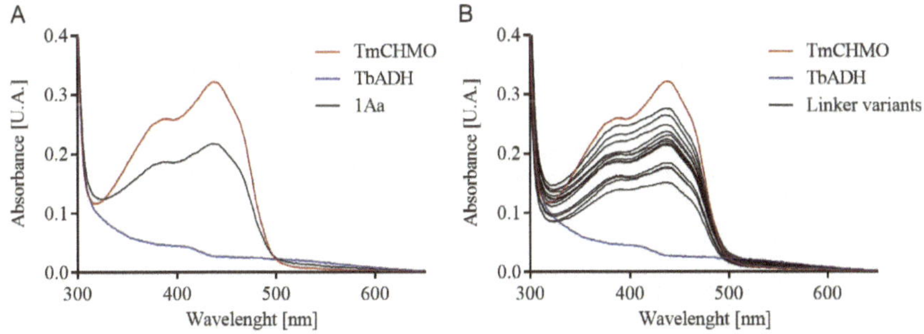

Fig. 4 Spectrophotometric determination of fused variants. For each variant, the UV–visible spectrum (300–600 nm) was recorded. For all purified FAD-containing proteins, the fully oxidized redox state and the distinctive peaks at 376 and 440 nm were observed. The full spectra of (A) TmCHMO, TbADH and one linker variant is shown and (B) for all the evaluated enzymes.

(Fig. 4). Furthermore, all tested linkers allowed binding of the FAD cofactor because the *A*280:*A*440 nm values ranged between 11.4 and 12.5 (Table 6).

5.1 Equipment

- UV–visible spectrophotometer (see note 1) (Synergy H1 Microplate Reader, BioTek).

5.2 Buffer

- Buffer A (50 mM Tris HCl pH 8.0 and 150 mM NaCl).

5.3 Procedure

1. Take the protein samples from the −80 °C freezer to 0 °C (transfer them in a bucket or holder on ice). Keep the samples on cold until completely thawed, depending on the volume of the aliquot it should thaw in about 15 min.
2. Turn on the spectrophotometer. Leave it for at least 20 min which will allow the lamp to heat up and to measure more reliably. Depending on the spectrophotometer this time can vary. For optimal instrument performance follow the instructions of the manufacturer.
3. Measure buffer A as a blank (see note 2).
4. Prepare a first measurement and verify that the absorbance at 440 nm is equal to or less than 0.5 [A.U.]. Dilute with buffer A if necessary
5. When the UV–visible spectrum has the requested absorbance, record the absorbance at 600 nm, this value is later subtracted as protein background.

Table 6 Spectrophotometric UV–visible evaluation of fused variants.

Variant	**Yield (mg L^{-1})**	***A280:A440***
TmCHMO	500	11.7
TbADH	400	–
1 Aa	300	12.5
2 Aa	330	12.0
3 Aa	360	12.4
6 Aa	300	11.4
7 Aa	330	12.1
8 Aa	280	12.2
9 Aa	260	12.6
10 Aa	290	11.4
11 Aa	350	10.9
12 Aa	270	13.0
13 Aa	350	11.7
13 Aa*	350	11.9
14 Aa	350	12.2
15 Aa	250	12.3

The concentration of each variants was spectrophotometrically d-etermined using the molar extinction coefficient of TmCHMO ($\varepsilon_{440nm} = 14.0\,mM^{-1}\,cm^{-1}$). Additionally, the *A280:A440* ratios were calculated for all the constructs.

5.4 Notes

1. For this section, a microplate reader with absorbance correction and 96-well plates were used. For each measurement, 300 μL protein samples were analyzed.
2. If a spectrophotometer and reusable cuvettes are used, cuvettes must be properly washed after each measurement. Optionally, wash the cuvette with a 20% ethanol solution, then wash with milli-Q grade water and finally dry it.

6. Apparent melting temperature determination

Assessing the manner in which enzymes respond to temperature variation can give useful biophysical information. In general, thermostability analysis can be approached thermodynamically or kinetically (Bommarius & Paye, 2013; Polizzi, Bommarius, Broering, & Chaparro-Riggers, 2007; Sanchez-Ruiz, 2010). Various methods have been extensively described to determine the temperature effect on purified protein samples, mainly by monitoring the perturbations on the secondary or tertiary structure by circular dichroism, differential scanning calorimetry, absorbance or fluorescence (Faham et al., 2004; Galka, Baturin, Manley, Kehler, & O'neil, 2008; Grinberg, Gevondyan, Grinberg, & Grinberg, 2001; Hughes, Rees, Heathcote, & Jones, 2006; Lau & Bowie, 1997; Mancusso, Karpowich, Czyzewski, & Wang, 2011). A faster, cheaper and easier alternative, is to evaluate the apparent melting temperature (T_M^{app}) by measuring fluorescence changes within a temperature gradient. The T_M^{app} parameter is defined as the midpoint of the protein unfolding transition (Cummings, Farnum, & Nelen, 2006). The ThermoFluor technique is a well-described method which requires a small amount of sample (Pantoliano et al., 2001). This technique permits to determine the protein T_M^{app} by measuring the fluorescence intensity of a solvatochromic dye, such as ANS (1-anilino-8-naphthanlenesulfonate) or SYPRO orange (Geerlof et al., 2006; Mezzasalma et al., 2007). The fundament of the method lies in the difference of the quantum yield of the dye when it is in water compared with its fluorescence in a nonpolar environment, such as when bound to hydrophobic moieties of unfolded proteins. During a temperature ramp, the sample is excited at a specific wavelength and the emission is recorded. At increasing temperatures, the protein begins to unfold exposing its hydrophobic patches. The dye binds to the unfolded protein, increasing its fluorescence intensity. A similar technique, denominated ThermoFAD, is based on the same concept (Forneris, Orru, Bonivento, Chiarelli, & Mattevi, 2009). Yet, in ThermoFAD a solvatochromic dye is not required as it takes advantage of the intrinsic fluorescence of the flavin cofactor. The fluorescence of flavin cofactors is typically quenched when bound in a protein, resulting in a fluorescence increase when the protein unfolds. During the temperature ramp, the sample is excited at a wavelength range between 450 and 490 nm and the fluorescence is measured using an emission filter of 515–530 nm. The data obtained is a sigmoidal curve of the

Table 7 Apparent melting temperature determination.

Variant	TmCHMO (°C)[a, b]	TbADH (°C)[b]
TmCHMO	53.0	–
TbADH	–	93.5
1 Aa	51.5	90.5
2 Aa	52.0	90.0
3 Aa	52.0	93.0
6 Aa	52.0	91.5
7 Aa	52.0	91.0
8 Aa	52.0	91.0
9 Aa	52.0	91.0
10 Aa	52.0	91.0
11 Aa	52.0	90.5
12 Aa	51.5	91.5
13 Aa	52.0	91.5
13 Aa*	52.0	91.0
14 Aa	52.0	91.0
15 Aa	52.0	91.0

Using a temperature ramp, thermostability for all constructs was measured using both ThermoFAD[a] and ThermoFluor[a, b] techniques. Each experiment was performed in duplicate.

fluorescence intensity over time. The apparent melting temperature is taken at the temperature at which the highest slope of the curve is observed. For this section, both techniques (ThermoFluor and ThermoFAD) were used and the results were summarized in Table 7.

6.1 Equipment

- Real-time PCR detection system (CFX96 C1000 Touch Thermal Cycler, Bio-Rad).
- 96-Well qPCR Plates (Bio-Rad).
- Adhesive qPCR Plate Sealing Film (Bio-Rad).
- Bench centrifuge with plate adaptor (Centrifuge X-4001, Salmenkipp).

6.2 Buffer

- Buffer A (50 mM Tris HCl pH 8.0 and 150 mM NaCl).
- Fluorescent dye as convenient, SYPRO orange was used for this section (see note 2).

6.3 Procedure

1. Take the protein samples from the −80 °C freezer to 0 °C (transfer them in a bucket or holder on ice). Keep the samples on cold until completely thawed, depending on the volume of the aliquot it should thaw in about 15 min.
2. The concentration of the protein sample must be at least 1 $mg\,mL^{-1}$. Dilute with buffer A if is necessary. Perform the experiments in triplicate.
3. On the real-time PCR detection system (qPCR instrument), configure the temperature ramp program. The program has a gradient from 20 to 99 °C that measures the fluorescence intensity every 0.5 °C after a 10 s delay for signal. Depending on the technique and dye, configure a suitable excitation and emission filter (see note 2).
4. Pipette 20 μL of protein sample into a well of the specified qPCR plate. Seal the plate with a sealing film (notes 3 and 4).
5. Transfer the plate to the instrument and start the program.
6. Depending on the instrument in use, after the temperature gradient the slope of the sigmoidal curve can be plotted to determine the apparent melting temperature.

6.4 Notes

1. Although there are some spectral changes among flavin-containing proteins, in general members of this class of enzymes show a fluorescence excitation maximum at 370–380 nm and 440–450 nm. While for the fluorescence emission, the maximum intensity is at 535 nm.
2. For the ThermoFluor technique, SYPRO orange dye was used. The real-time PCR detection system was configured to excite between 450 and 490 nm and detect fluorescence intensity using an emission filter in the 560–580 nm range. If another solvatochromic dye is used, follow the instruction of the manufacturer.
3. Discard the bubbles of the plate by centrifugation. Use a delicate task wiper to clean fingerprints and impurities from the seal and to fix the plate on the instrument.

4. Both ThermoFAD and ThermoFluor techniques are extremely convenient: using 96-well qPCR plates it is possible to analyze up to 96 samples at the same time in less than 1 h. Furthermore, depending on the objective of the study, the effect of the pH, miscible cosolvents, storage conditions, cosubstrates or presence of different additives can be easily analyzed.

7. Evaluation of the linker size on the enzyme activity

For biocatalytic purposes, it is highly valuable to determine the effect of the linker length on enzyme activity. The activity of TmCHMO and TbADH can be readily measured by spectrophotometrically following the absorbance of NADPH at 340 nm ($\varepsilon_{340nm} = 6.22\,mM^{-1}\,cm^{-1}$). For ADHs, the reaction is monitored by the measuring the formation of the reduced nicotinamide cosubstrate (Bogin, Peretz, & Burstein, 1997). While for TmCHMO, the oxidation rate is followed (Romero et al., 2016). However, determining the activity for fused enzymes that share their coenzyme is not trivial. In this case, one enzyme can disrupt the measurement for the other. Specifically, the reversible reaction of TbADH could affect the kinetic analysis for TmCHMO by NADPH-mediated reduction of the ketone. Nevertheless, for this scenario, the broad substrate acceptance of TmCHMO can be exploited and activity can be analyzed using an alternative substrate. TmCHMO can oxidize S-containing substrates, such as thioanisole (phenyl methyl sulfide). The use of this thioether as BVMO test substrate prevents interfering ADH activity. For this section, the oxidation of thioanisole (TmCHMO activity) or cyclohexanol (TbADH activity) was spectrophotometrically monitored.

7.1 Equipment

- UV–visible spectrophotometer with orbital rotor (see note 1) (Synergy H1 Microplate Reader, BioTek).
- Multi pipette suitable for volumes of 300 μL.
- 96-Well plates.

7.2 Buffer and reagents

- Buffer A (50 mM Tris HCl pH 8.0 and 150 mM NaCl).
- $NADP^+$ stock solution at 300 μM prepared in buffer A.
- NADPH stock solution at 300 μM prepared in buffer A.

- Cyclohexanol stock solution at 40 mM prepared in buffer A.
- Protein sample at 2.0 μM prepared in buffer A (note 2).
- Thioanisole (phenyl methyl sulfide) stock solution at 4.0 mM prepared in buffer A and 10% (v/v) 1,4-dioxane as cosolvent.

7.3 Procedure

1. Take the protein samples from the −80 °C freezer to 0 °C (transfer them in a bucket or holder on ice). Keep the samples on cold until completely thawed, depending on the volume of the aliquot it should thaw in about 15 min.
2. Turn on the spectrophotometer, this allows the lamps to heat up and measure in a more reliably way. Leave the spectrophotometer on for at least 20 min, depending on the equipment this time can vary, for a better instrument performance follow the instructions of the manufacturer.
3. In the spectrophotometer, configure a kinetic program measuring absorbance at 340 nm for 120 s with constant orbital shaking, include a pathlength correction. Due the sensor delays between measurements, prepare two rows for each kinetic analysis.
4. For the kinetic analysis:
   a. For dehydrogenase activity, pipette 75 μL of cyclohexanol solution and 75 μL of protein sample into each well.
   b. For sulfide oxidation measurement, pipette 75 μL of thioanisole solution and 75 μL of protein sample into each well.
5. Gently suspend the solution by pipetting, avoid the formation of bubbles.
6. Using a multi pipette, transfer 150 μL of the $NADP^+$ solution for the TbADH kinetic analysis. While for TmCHMO, transfer 150 μL of the NADPH solution (note 3). Quickly transfer the plate into the microplate reader and start the measurement.
7. After the kinetic measurement, continue with the next two rows.
8. To perform the analyze, plot the corrected absorbance values over time in seconds. First, determine the slope of the initial linear range on [A.U.] s^{-1} units.
9. Then, translate the obtained slope to the rate of NADPH per seconds using the molar extinction coefficient of NADPH ($\varepsilon_{340nm} = 0.00622\,\mu M^{-1}\,cm^{-1}$). The obtained units are in ($\mu M\,s^{-1}$).
10. Finally, correct the obtained data for the enzyme concentration in order to calculate the k_{obs} (s^{-1}) (Fig. 5).

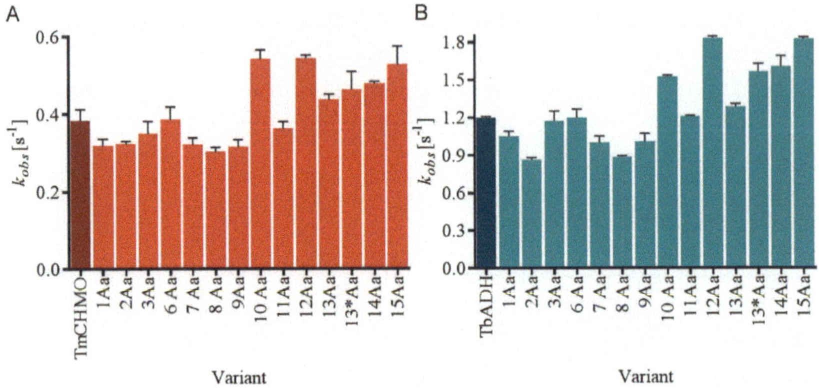

Fig. 5 Effect of linker length on enzyme activity. For each variant, alcohol dehydrogenase and sulfoxidation activity was evaluated. (A) For alcohol dehydrogenase activity, the $NADP^+$ reduction rates were measured at 10 mM cyclohexanol and 150 μM $NADP^+$. (B) For sulfoxidation activity, NADPH oxidation rates were studied using 1.0 mM thioanisole and 150 μM NADPH. As a control, the single non-fused enzymes were also evaluated. Each experiment was performed in triplicate.

7.4 Notes

1. A microplate reader (with absorbance correction) and 96-well plates were used for this section. For each reaction, the final volume was 300 μL. If a spectrophotometer and reusable cuvettes are used, the cuvettes must be properly washed at each measurement. Optionally, wash the cuvette with 20% ethanol solution, then wash with milli-Q grade water and finally dry it.
2. For the 2.0 μM protein solution in buffer A, keep all the tubes on ice until the reaction starts. Perform the experiments in triplicate.
3. For kinetic analysis, measurements must be rapid to avoid loss of the absorbance change of the first few seconds. Thereby, prepare all the reactions near to the instrument. After transferring NADPH or $NADP^+$ solution to the wells, quickly mix using the multi pipette avoiding the formation of bubbles. Subsequently, transfer the plate in the spectrophotometer and start the measurement.

8. Evaluation of fused variants in small-scale bioconversions

As a final biocatalytic evaluation, the linker variants can be tested in small-scale conversions and the dimensionless turnover number (TON) can be determined. The TON is defined as the ratio of the obtained moles of product divided by the moles of used enzyme. For this calculation, the cyclohexanol conversion and the produced ketone and lactone are monitored. By varying the experimental conditions, such as the temperature or pH, optimal parameters for biocatalytic performance of the fusion enzymes can be determined (see note 1). Due the chemical nature of the substrate and products, the conversions were monitored using gas chromatography (GC). This technique resolves analytes based on their volatility and polarity (Yip, Chun, & Chan, 2013). For this section, compounds were analyzed and quantified using a GC coupled to a mass spectrometer (MS) with electron ionization and quadruple separation. GC is nowadays often coupled to MS as it results in a powerful analytical tool (Bothner et al., 2000; Mellon, 2003). The general principle of coupled GC–MS is the same for all other coupled MS techniques. Broadly, the GC instrument separates the components into fractions which are transferred to the MS module. Then, the ions are detected, resulting in a MS spectrum for each analyte. Finally, a total ion current of the continuous mass scanning is obtained (Lindon, Tranter, & Koppenaal, 2016). GC–MS is a mature and sensitive technique, which is often equipped with powerful data analysis software for the identification of molecular ions or distinctive fragmented ion patterns. Finally, the identification is automatically performed by comparing with spectral MS libraries, such as databases from the National Institute of Standards and Technology (NIST), Golm Metabolome Database (GMD), or the Human Metabolome Database (HMDB) (Lindon et al., 2016; Yip et al., 2013).

8.1 Equipment

- Shaker (Innova 44, New Brunswick Scientific).
- Vortex.
- Bench centrifuge (Heraeus Fresco 17 Centrifuge, Thermo Scientific).
- GC–MS (GC–MS QP2010 ultra with electron ionization and quadrupole separation, Shimadzu) (note 2).
- HP-5MS column (Agilent, 30 m × 0.25 mm × 0.25 μm).

8.2 Buffer and reagent

- Buffer A (50 mM Tris HCl pH 8.0 and 150 mM NaCl).
- Buffer B (50 mM Tris HCl pH 8.0, 40 μM FAD and 150 mM NaCl).
- Protein mixtures at 2.0 μM in buffer B.
- $NADP^+$ stock solution at 300 μM prepared in buffer A.

- NADPH stock solution at 300 μM prepared in buffer A.
- Cyclohexanol stock solution at 60 mM prepared in buffer A.
- Cyclohexanone stock solution at 60 mM prepared in buffer A.
- ε-Caprolactone standard solutions (1.0–15 mM).
- FAD stock solution at 1.0 mM in buffer A.
- Ethyl acetate with 0.025% (v/v) mesitylene (external standard).
- $MgSO_4$ anhydrous.

8.3 Procedure

1. Take the protein samples from the −80 °C freezer to 0 °C (transfer them in a bucket or holder on ice). Keep the samples on cold until completely thawed, depending on the volume of the aliquot it should thaw in about 15 min.
2. Transfer 125 μL of cyclohexanol solution in a 10 mL vial (see note 3) (for a TmCHMO control reaction, transfer 125 μL of cyclohexanone solution). Subsequently, transfer 125 μL of enzyme mixture.
3. Subsequently, transfer 250 μL of $NADP^+$ solution to the 10 mL vial and incubate them in a shaker at 25 or 37 °C for 24 h (for the TmCHMO control reaction, add 250 μL of NADPH solution).
4. After incubation, transfer the solution in a 2 mL tube and add 500 μL of ethyl acetate (1 volume reaction).
5. Vortex the tubes for 30 s, subsequently centrifuge at 16,000 *g* and transfer the organic layer to another tube.
6. Repeat step 4 and 5 two more times collecting the organic layer in a same tube.
7. Add a spatula of $MgSO_4$ (roughly 5 mg) in the organic solution and vortex it for 30 s to remove the residual water for further GC analysis.
8. Centrifuge the tubes at 16,000 *g* and transfer the organic supernatants in GC-vials (use suitable inserts if necessary).
9. Perform the same extraction procedure for ε-caprolactone standard solutions to prepare a calibration curve.
10. For the GC analysis, configure the temperature program as:
    a. Hold 5 min at 50 °C.
    b. Increase 5 °C min^{-1} for 2 min.
    c. Increase 2 °C min^{-1} for 10 min.
    d. Hold for 10 min.
11. To calculate the substrate conversions, integrate the peaks of the substrate, mesitylene and product(s) from the chromatogram. Then,

Table 8 Bioconversions analysis of self-sufficient bifunctional fusions.

Variant	Conversion (%)	TON ± SEM
TbADH + TmCHMO	39	11,700 ± 600
1 Aa	57	17,100 ± 150
2 Aa	65	19,500 ± 600
3 Aa	66	19,800 ± 900
6 Aa	62	18,600 ± 900
7 Aa	69	20,700 ± 1800
8 Aa	55	16,500 ± 600
9 Aa	49	14,700 ± 1200
10 Aa	58	17,400 ± 900
11 Aa	56	16,800 ± 300
12 Aa	46	13,800 ± 300
13 Aa	69	20,700 ± 1200
13 Aa*	70	21,000 ± 300
14 Aa	85	25,550 ± 900
15 Aa	48	14,400 ± 900

Conversion levels and TONs for linker variants and the single enzymes are shown. For each bioconversion 15 mM cyclohexanol and 0.5 μM of enzyme were used. Reactions were performed in buffer A at 37 °C and 135 r.p.m. for 24 h. Conversions were calculated based on the substrate depletion from the GC–MS analysis. The turnover number was calculated as the ratio of moles of product obtained divided by the moles of enzyme used in the reaction. Each experiment was performed in duplicate.

normalize the integrated values with the area of the external standard. Compare the results with a control reaction carried out without enzyme and quantify the percentage of conversion.

12. Finally, if cyclohexanone is observed in the chromatogram, calculate the production of ε-caprolactone using the calibration curve and determine the TON values (note 5) (Table 8).

8.4 Notes

1. To find optimal conditions for conversion, two enzyme concentrations (0.5 and 5 μM), two substrate concentrations (10–20 mM) and two different temperatures (24 and 37 °C) were tested. The condition of 0.5 μM

enzyme and 15 mM cyclohexanol at 37 °C for 24 h allowed to observe differences between the TON values when comparing the performance of the different fusion enzymes, avoiding full or very low substrate conversions.

2. If the mass spectrometer is not available for the identification, bioconversions can also be determined by regular GC analysis. In that case, the analytical standards of cyclohexanol, cyclohexanone and ε-caprolactone must be previously tested.
3. For bioconversions using monooxygenases as biocatalysts, the liquid: gas volume ratio must allow enough dioxygen for the reaction. For this methodology, 500 μL reaction volumes were prepared in 10 mL vial.
4. For the analysis and identification of the molecules, LabSolutions GC–MS software from Shimadzu with a MS-spectral library from NIST was used. Another suitable GC–MS and library may be used as preferred.
5. If the presence of cyclohexanone in the chromatogram is not observed, the conversion of the substrate can be used for the TON determination. Standards with cyclohexanol, cyclohexanone and ε-caprolactone must be evaluated in advance.

9. Summary and conclusions

This chapter explores the effect of the linker size on the biocatalytic properties of a self-sufficient multifunctional TbADH-TmCHMO fusion biocatalyst. Fourteen different linker variants of the previously reported glycine-rich sequence were obtained and evaluated. All the obtained variants exhibited a high expression level (250 and 360 mg L^{-1}) and were obtained as FAD-containing proteins. The latter is suggested by the obtained *A*280:*A*440 ratios. The protocol for enzyme purification includes an excess of FAD during cell lysis which may promote holo enzyme formation. It was previously reported that *E. coli* cells may not be able to support biosynthesis of sufficient amounts of the FAD cofactor (Milker, Fink, et al., 2017; Milker, Goncalves, Fink, & Rudroff, 2017). Regarding the thermostability, for TmCHMO—using the ThermoFAD method—it was observed that the length of the linker did not show a noticeable effect on its unfolding profile. On the other hand, using the ThermoFluor method—where the T_M^{app} of ADH is also monitored—the linker size had only a small effect compared with the not-fused TbADH. For all the constructs, the ADH fusion partner exhibited a somewhat lower T_M^{app}. In particular, for three variants the differences were up to 3 °C. Nevertheless, for TbADH this effect is not operationally significant due its relatively high thermostability ($T_M^{app} = 93.5$ °C). Regarding the effect on enzyme activity, the alcohol oxidation and sulfoxidation activities were similar for the nine shortest variants. The fusions with the length of 10, 12 and 15 amino acids, showed for both activities a slightly increased k_{obs}. The small-scale bioconversions resulted in highest turnover numbers of the fusions with 2, 3, 6, 7, 13 and 14 amino acid linkers (TONs between 20,000 and 25,000). While for the reaction with the single enzymes, the obtained TON was around 12,000. Clearly, the use of optimal linkers results in superior performance. The fusion with the 14 amino acids linker displayed the highest TON and seems to be the best candidate for biocatalytic purposes.

This contribution provides an easy and efficient protocol to generate a collection of fusion enzymes with different linkers. This allowed to determine the effect of the linker size on the properties of the self-sufficient TbADH-TmCHMO fusions for the synthesis of ε-caprolactone. The employed procedure can be easily adapted for generating a similar library of fusion enzymes from other enzymes. Also, the other protocols, such as the protocol to determine apparent melting temperatures of a collection of (flavo)proteins, can be of value in other enzyme engineering or biocatalytic studies. Such easy-to-use protocols will help to improve procedures for generating optimized biocatalysts.

Acknowledgments

This work received funding from the European Union (EU), project ROBOX (grant agreement no. 635734) under the EU's Horizon 2020 Program Research and Innovation actions H2020-LEIT BIO-2014-1, from the Institute for Biological and Medical Engineering from Pontificia Universidad Católica de Chile and CORFO's grant 14ENI2-26862. A PhD scholarship from CONICYT (Comisión Nacional de Investigación Científica y Tecnológica de Chile), from the graduated school of Pontificia Universidad Católica de Chile and from the Faculty of engineering of Pontificia Universidad Católica de Chile to AG-S are acknowledged. The views and opinions expressed in this article are only those of the authors and do not necessarily reflect those of the European Union Research Agency. The European Union is not liable for any use that may be made of the information contained herein.

References

Aalbers, F. S., & Fraaije, M. W. (2017). Coupled reactions by coupled enzymes: Alcohol to lactone cascade with alcohol dehydrogenase–cyclohexanone monooxygenase fusions. *Applied Microbiology and Biotechnology*, *101*(20), 7557–7565.

Aalbers, F. S., & Fraaije, M. W. (2019). Enzyme fusions in biocatalysis: Coupling reactions by pairing enzymes. *ChemBioChem*, *20*(1), 20–28.

Adams, J. P., Brown, M. J. B., Diaz-Rodriguez, A., Lloyd, R. C., & Roiban, G. (2019). Biocatalysis: A pharma perspective. *Advanced Synthesis & Catalysis*, *361*(11), 2421–2432.

Argos, P. (1990). An investigation of oligopeptides linking domains in protein tertiary structures and possible candidates for general gene fusion. *Journal of Molecular Biology*, *211*(4), 943–958.

Baldwin, C. V. F., & Woodley, J. M. (2006). On oxygen limitation in a whole cell biocatalytic Baeyer–Villiger oxidation process. *Biotechnology and Bioengineering*, *95*(3), 362–369.

Belsare, K. D., Ruff, A. J., Martinez, R., Shivange, A. V., Mundhada, H., Holtmann, D., et al. (2014). P-LinK: A method for generating multicomponent cytochrome P450 fusions with variable linker length. *BioTechniques*, *57*(1), 13–20.

Bogin, O., Peretz, M., & Burstein, Y. (1997). *Thermoanaerobacter brockii* alcohol dehydrogenase: Characterization of the active site metal and its ligand amino acids. *Protein Science*, *6*(2), 450–458.

Bommarius, A. S., & Paye, M. F. (2013). Stabilizing biocatalysts. *Chemical Society Reviews*, *42*(15), 6534–6565.

Bothner, B., Chavez, R., Wei, J., Strupp, C., Phung, Q., Schneemann, A., et al. (2000). Monitoring enzyme catalysis with mass spectrometry. *Journal of Biological Chemistry*, *275*(18), 13455–13459.

Chapman, S. K., & Reid, G. A. (1999). *Flavoprotein protocols. Vol. 131*. Springer Science & Business Media.

Chen, X., Zaro, J. L., & Shen, W. C. (2013). Fusion protein linkers: Property, design and functionality. *Advanced Drug Delivery Reviews*, *65*(10), 1357–1369.

Corrado, M. L., Knaus, T., & Mutti, F. G. (2018). A chimeric styrene monooxygenase with increased efficiency in asymmetric biocatalytic epoxidation. *ChemBioChem*, *19*(7), 679–686.

Cummings, M. D., Farnum, M. A., & Nelen, M. I. (2006). Universal screening methods and applications of ThermoFluor®. *Journal of Biomolecular Screening*, *11*(7), 854–863.

Dahlhoff, G., Niederer, J. P. M., & Hoelderich, W. F. (2001). ε-Caprolactam: New by-product free synthesis routes. *Catalysis Reviews*, *43*(4), 381–441.

Delgove, M. A. F., Elford, M. T., Bernaerts, K. V., & De Wildeman, S. M. A. (2018). Toward upscaled biocatalytic preparation of lactone building blocks for polymer applications. *Organic Process Research & Development*, *22*(7), 803–812.

Doig, S. D., Avenell, P. J., Bird, P. A., Gallati, P., Lander, K. S., Lye, G. J., et al. (2002). Reactor operation and scale-up of whole cell Baeyer-Villiger catalyzed lactone synthesis. *Biotechnology Progress*, *18*(5), 1039–1046.

Donoghue, N. A., Norris, D. B., & Trudgill, P. W. (1976). The purification and properties of cyclohexanone oxygenase from *Nocardia globerula* CL1 and *Acinetobacter* NCIB 9871. *European Journal of Biochemistry*, *63*(1), 175–192.

Ellis, G. A., Klein, W. P., Lasarte-Aragones, G., Thakur, M., Walper, S. A., & Medintz, I. L. (2019). Artificial multienzyme scaffolds: Pursuing in vitro substrate channeling with an overview of current progress. *ACS Catalysis*, *9*(12), 10812–10869.

Faham, S., Yang, D., Bare, E., Yohannan, S., Whitelegge, J. P., & Bowie, J. U. (2004). Side-chain contributions to membrane protein structure and stability. *Journal of Molecular Biology*, *335*(1), 297–305.

Forneris, F., Orru, R., Bonivento, D., Chiarelli, L. R., & Mattevi, A. (2009). ThermoFAD, a Thermofluor-adapted flavin ad hoc detection system for protein folding and ligand binding. *FEBS Journal*, *276*(10), 2833–2840.

Fraaije, M. W., Wu, J., Heuts, D. P. H. M., Van Hellemond, E. W., Spelberg, J. H. L., & Janssen, D. B. (2005). Discovery of a thermostable Baeyer-Villiger monooxygenase by genome mining. *Applied Microbiology and Biotechnology*, *66*(4), 393–400.

Galka, J. J., Baturin, S. J., Manley, D. M., Kehler, A. J., & O'neil, J. D. (2008). Stability of the glycerol facilitator in detergent solutions. *Biochemistry*, *47*(11), 3513–3524.

Geerlof, A., Brown, J., Coutard, B., Egloff, M.-P., Enguita, F. J., Fogg, M. J., et al. (2006). The impact of protein characterization in structural proteomics. *Acta Crystallographica Section D: Biological Crystallography*, *62*(10), 1125–1136.

Gräwe, A., Ranglack, J., Weyrich, A., & Stein, V. (2020). iFLinkC: An iterative functional linker cloning strategy for the combinatorial assembly and recombination of linker peptides with functional domains. *Nucleic Acids Research*, *48*(4), e24.

Grinberg, A. V., Gevondyan, N. M., Grinberg, N. V., & Grinberg, V. Y. (2001). The thermal unfolding and domain structure of Na^{+}/K^{+}-exchanging ATPase. A scanning calorimetry study. *European Journal of Biochemistry*, *268*(19), 5027–5036.

Heine, T., Tucker, K., Okonkwo, N., Assefa, B., Conrad, C., Scholtissek, A., et al. (2017). Engineering styrene monooxygenase for biocatalysis: Reductase-epoxidase fusion proteins. *Applied Biochemistry and Biotechnology*, *181*(4), 1590–1610.

Hughes, A. V., Rees, P., Heathcote, P., & Jones, M. R. (2006). Kinetic analysis of the thermal stability of the photosynthetic reaction center from *Rhodobacter sphaeroides*. *Biophysical Journal*, *90*(11), 4155–4166.

Jeon, E. Y., Baek, A. H., Bornscheuer, U. T., & Park, J. B. (2015). Enzyme fusion for whole-cell biotransformation of long-chain sec-alcohols into esters. *Applied Microbiology and Biotechnology*, *99*(15), 6267–6275.

Kadisch, M., Willrodt, C., Hillen, M., Bühler, B., & Schmid, A. (2017). Maximizing the stability of metabolic engineering-derived whole-cell biocatalysts. *Biotechnology Journal*, *12*(8), 1600170.

Kohl, A., Srinivasamurthy, V., Böttcher, D., Kabisch, J., & Bornscheuer, U. T. (2018). Co-expression of an alcohol dehydrogenase and a cyclohexanone monooxygenase for cascade reactions facilitates the regeneration of the NADPH cofactor. *Enzyme and Microbial Technology*, *108*(August 2017), 53–58.

Köpnick, H., Schmidt, M., Brügging, W., Rüter, J., & Kaminsky, W. (2000). Polyesters. In *Ullmann's encyclopedia of industrial chemistry*.

Künkel, A., Becker, J., Börger, L., Hamprecht, J., Koltzenburg, S., Loos, R., et al. (2000). Polymers, biodegradable. In *Ullmann's encyclopedia of industrial chemistry* (pp. 1–29).

Kuzmak, A., Carmali, S., von Lieres, E., Russell, A. J., & Kondrat, S. (2019). Can enzyme proximity accelerate cascade reactions? *Scientific Reports*, *9*(1), 1–7.

Labet, M., & Thielemans, W. (2009). Synthesis of polycaprolactone: A review. *Chemical Society Reviews*, *38*(12), 3484–3504.

Lau, F. W., & Bowie, J. U. (1997). A method for assessing the stability of a membrane protein. *Biochemistry*, *36*(19), 5884–5892.

Lindon, J. C., Tranter, G. E., & Koppenaal, D. (2016). *Encyclopedia of spectroscopy and spectrometry*. Academic Press.

Macheroux, P. (1999). UV-visible spectroscopy as a tool to study flavoproteins. In *Flavoprotein protocols* (pp. 1–7). Humana Press.

Magali, P., de Revel, G., & Marchand, S. (2017). First identification of threep-menthane lactones and their potentialprecursor, menthofuran, in red wines. *Food Chemistry*, *217*, 294–302.

Mallin, H., Wulf, H., & Bornscheuer, U. T. (2013). A self-sufficient Baeyer-Villiger biocatalysis system for the synthesis of ε-caprolactone from cyclohexanol. *Enzyme and Microbial Technology*, *53*(4), 283–287.

Mancusso, R., Karpowich, N. K., Czyzewski, B. K., & Wang, D. N. (2011). Simple screening method for improving membrane protein thermostability. *Methods*, *55*(4), 324–329.

Massey, V. (2000). The chemical and biological versatility of riboflavin. *Biochemical Society Transactions*, *28*(4), 283–296.

Mellon, F. A. (2003). *Mass spectrometry: Principles and instrumentation. 2003* (pp. 3739–3749).

Messiha, H. L., Ahmed, S. T., Karuppiah, V., Suardíaz, R., Ascue Avalos, G. A., Fey, N., et al. (2018). Biocatalytic routes to lactone monomers for polymer production. *Biochemistry*, *57*(13), 1997–2008.

Mezzasalma, T. M., Kranz, J. K., Chan, W., Struble, G. T., Schalk-Hihi, C., Deckman, I. C., et al. (2007). Enhancing recombinant protein quality and yield by protein stability profiling. *Journal of Biomolecular Screening*, *12*(3), 418–428.

Milker, S., Fink, M. J., Oberleitner, N., Ressmann, A. K., Mihovilovic, M. D., & Bornscheuer, U. T. (2017). Kinetic modeling of an enzymatic redox cascade in vivo reveals cofactor-caused bottlenecks. *ChemCatChem*, *9*, 3420–3427.

Milker, S., Goncalves, L. C. P., Fink, M. J., & Rudroff, F. (2017). *Escherichia coli* fails to efficiently maintain the activity of an important flavin monooxygenase in recombinant overexpression. *Frontiers in Microbiology*, *8*, 2201.

Mourelle-Insua, Á., Aalbers, F. S., Lavandera, I., Gotor-Fernández, V., & Fraaije, M. W. (2019). What to sacrifice? Fusions of cofactor regenerating enzymes with Baeyer-Villiger monooxygenases and alcohol dehydrogenases for self-sufficient redox biocatalysis. *Tetrahedron*, *75*(13), 1832–1839.

Mthethwa, K. S., Kassier, K., Engel, J., Kara, S., Smit, M. S., & Opperman, D. J. (2017). Fungal BVMOs as alternatives to cyclohexanone monooxygenase. *Enzyme and Microbial Technology*, *106*, 11–17.

Pantoliano, M. W., Petrella, E. C., Kwasnoski, J. D., Lobanov, V. S., Myslik, J., Graf, E., et al. (2001). High-density miniaturized thermal shift assays as a general strategy for drug discovery. *Journal of Biomolecular Screening*, *6*(6), 429–440.

Parra, L. P., Acevedo, J. P., & Reetz, M. T. (2015). Directed evolution of phenylacetone monooxygenase as an active catalyst for the Baeyer-Villiger conversion of cyclohexanone to caprolactone. *Biotechnology and Bioengineering*, *112*(7), 1354–1364.

Polizzi, K. M., Bommarius, A. S., Broering, J. M., & Chaparro-Riggers, J. F. (2007). Stability of biocatalysts. *Current Opinion in Chemical Biology*, *11*(2), 220–225.

Qiagen. (2003). *The QIA expressionist: A handbook for high-level expression and purification of 6xHis-tagged protein* (5th ed.). Qiagen.

Rabe, K. S., Müller, J., Skoupi, M., & Niemeyer, C. M. (2017). Cascades in compartments: En route to machine-assisted biotechnology. *Angewandte Chemie International Edition*, *56*(44), 13574–13589.

Ritz, J., Fuchs, H., Kieczka, H., & Moran, W. C. (2000). Caprolactam. In *Ullmann's encyclopedia of industrial chemistry*.

Romero, E., Gomez Castellanos, J. R., Gadda, G., Fraaije, M. W., & Mattevi, A. (2017). Same substrate, many reactions: Oxygen activation in flavoenzymes. *Chemical Reviews*, *118*(4), 1742–1769.

Romero, E., Gómez Castellanos, J. R., Mattevi, A., & Fraaije, M. W. (2016). Characterization and crystal structure of a robust cyclohexanone monooxygenase. *Angewandte Chemie International Edition*, *55*(51), 15852–15855.

Sanchez-Ruiz, J. M. (2010). Protein kinetic stability. *Biophysical Chemistry*, *148*(1–3), 1–15.

Sattler, J. H., Fuchs, M., Mutti, F. G., Grischek, B., Engel, P., Pfeffer, J., et al. (2014). Introducing an in situ capping strategy in systems biocatalysis to access 6-aminohexanoic acid. *Angewandte Chemie International Edition*, *53*(51), 14153–14157.

Savoca, M. P., Tonoli, E., Atobatele, A. G., & Verderio, E. A. M. (2018). Biocatalysis by transglutaminases: A review of biotechnological applications. *Micromachines*, *9*(11), 562.

Scherkus, C., Schmidt, S., Bornscheuer, U. T., Gröger, H., Kara, S., & Liese, A. (2017). Kinetic insights into ε-caprolactone synthesis: Improvement of an enzymatic cascade reaction. *Biotechnology and Bioengineering*, *114*(6), 1215–1221.

Schmidt, S., Scherkus, C., Muschiol, J., Menyes, U., Winkler, T., Hummel, W., et al. (2015). An enzyme cascade synthesis of ε-caprolactone and its oligomers. *Angewandte Chemie International Edition*, *54*(9), 2784–2787.

Secundo, F., Fialà, S., Fraaije, M. W., De Gonzalo, G., Meli, M., Zambianchi, F., et al. (2011). Effects of water miscible organic solvents on the activity and conformation of the Baeyer–Villiger monooxygenases from *Thermobifida fusca* and *Acinetobacter calcoaceticus*: A comparative study. *Biotechnology and Bioengineering*, *108*, 491–499.

Sieben, M., Steinhorn, G., Müller, C., Fuchs, S., Ann Chin, L., Regestein, L., et al. (2016). Testing plasmid stability of *Escherichia coli* using the continuously operated shaken BIOreactor system. *Biotechnology Progress*, *32*(6), 1418–1425.

Spriestersbach, A., Kubicek, J., Schaefer, F., Block, H., & Maertens, B. (2015). Purification of His-tagged proteins. In *Methods in enzymology* (Vol. 559, pp. 1–15). Elsevier.

Staudt, S., Bornscheuer, U. T., Menyes, U., Hummel, W., & Gröger, H. (2013). Direct biocatalytic one-pot-transformation of cyclohexanol with molecular oxygen into ε-caprolactone. *Enzyme and Microbial Technology*, *53*(4), 288–292.

Thaore, V., Chadwick, D., & Shah, N. (2018). Sustainable production of chemical intermediates for nylon manufacture: A techno-economic analysis for renewable production of caprolactone. *Chemical Engineering Research and Design*, *135*, 140–152.

Tischler, D., Schwabe, R., Siegel, L., Joffroy, K., Kaschabek, S. R., Scholtissek, A., et al. (2018). VpStyA1/VpStyA2B of *Variovorax paradoxus* EPS: An aryl alkyl sulfoxidase rather than a styrene epoxidizing monooxygenase. *Molecules*, *23*(4), 809.

Torres Pazmiño, D. E., Riebel, A., de Lange, J., Rudroff, F., Mihovilovic, M. D., & Fraaije, M. W. (2009). Efficient biooxidations catalyzed by a new generation of self-sufficient Baeyer–Villiger monooxygenases. *ChemBioChem*, *10*(16), 2595–2598.

Van Berkel, W. J. H., Kamerbeek, N. M., & Fraaije, M. W. (2006). Flavoprotein monooxygenases, a diverse class of oxidative biocatalysts. *Journal of Biotechnology*, *124*, 670–689.

Van Rosmalen, M., Krom, M., & Merkx, M. (2017). Tuning the flexibility of glycine-serine linkers to allow rational design of multidomain proteins. *Biochemistry*, *56*(50), 6565–6574.

Wachtmeister, J., & Rother, D. (2016). Recent advances in whole cell biocatalysis techniques bridging from investigative to industrial scale. *Current Opinion in Biotechnology*, *42*, 169–177.

Wheeldon, I., Minteer, S. D., Banta, S., Barton, S. C., Atanassov, P., & Sigman, M. (2016). Substrate channelling as an approach to cascade reactions. *Nature Chemistry*, *8*(4), 299.

Woodley, J. M. (2008). New opportunities for biocatalysis: Making pharmaceutical processes greener. *Trends in Biotechnology*, *26*(6), 321–327.

Woodley, J. M. (2020). New frontiers in biocatalysis for sustainable synthesis. *Current Opinion in Green and Sustainable Chemistry*, *21*, 22–26.

Yip, L. Y., Chun, E., & Chan, Y. (2013). Proteomic and metabolomic approaches to biomarker discovery. In *Proteomic and metabolomic approaches to biomarker discovery* Academic Press. Chapter 8.

Zhu, Y., Kang, S., Yan, H., Lv, H. P., Zhang, Y., & Lin, Z. (2020). Enantiomeric distributions of volatile lactones and terpenoids in white teas stored for different durations. *Food Chemistry*, *320*, 126632.

CHAPTER SIX

Predicting the effect of disordered linkers on effective concentrations and avidity with the "C_{eff} calculator" app

Magnus Kjaergaard[a,b,c,*], Juliana Glavina[d], and Lucia Beatriz Chemes[d,*]

[a]Department of Molecular Biology and Genetics, Aarhus University, Aarhus, Denmark
[b]The Danish Research Institute for Translational Neuroscience (DANDRITE), Aarhus, Denmark
[c]Center for Proteins in Memory (PROMEMO), Aarhus, Denmark
[d]Instituto de Investigaciones Biotecnológicas "Dr. Rodolfo A. Ugalde", IIB-UNSAM, IIBIO-CONICET, Universidad Nacional de San Martín, San Martín, Argentina
[*]Corresponding authors: e-mail address: magnus@mbg.au.dk; lchemes@iib.unsam.edu.ar

Contents

Abstract

Linkers are crucial to the functions of multidomain proteins as they couple functional units to encode regulation such as auto-inhibition, enzyme targeting or tuning of interaction strength. A linker changes reactions from bimolecular to unimolecular, and the equilibrium and kinetics is thus determined by the properties of the linker rather than concentrations. We present a theoretical workflow for estimating the functional consequences of tethering by a linker. We discuss how to: (1) Identify flexible linkers from

Methods in Enzymology, Volume 647
ISSN 0076-6879
https://doi.org/10.1016/bs.mie.2020.09.012

sequence. (2) Model the end-to-end distance distribution for a flexible linker using a worm-like chain. (3) Estimate the effective concentration of a ligand tethered by a flexible linker. (4) Calculate the decrease in binding affinity caused by auto-inhibition. (5) Calculate the expected avidity enhancement of a bivalent interaction from effective concentration. The worm-like chain modeling is available through a web application called the "C_{eff} calculator" (http://ceffapp.chemeslab.org), which will allow user-friendly prediction of experimentally inaccessible parameters.

1. Introduction

Proteins are often modular and consist of several domains strung together. By merging functional domains, multidomain proteins become more than the sum of their domains and may acquire emergent properties. This is particularly important in signaling, where enzymes and scaffolding proteins have evolved by assembly of different protein interaction modules that can change their activity or interaction status, giving rise to the complex biochemical switches that control the function of signaling networks (Scott & Pawson, 2009). In addition, small modular elements called "short linear motifs" or SLiMs also encode function within proteins and can cooperate with other motifs and domains in determining a protein's function. The functions of a multidomain protein are not only defined by which domains it contains, but also depend on how the domains are connected. Functional modules can either be forced together or isolated from each other depending on the architecture of the protein. The emergent properties of multidomain protein are thus often determined by the connections between domains/SLiMs, which typically take the form of linkers that may vary from rigid to completely flexible. In macromolecular complexes, adaptor or scaffolding proteins play similar roles to the linker although through non-covalent interactions.

When protein domains or SLiMs are physically connected, reactions can occur internally in the protein or protein complex. Common examples include auto-inhibition, when an inhibitor is tethered to its protein target, or auto-activation when an enzyme is its own substrate as in e.g., kinases and proteases. When reactions occur intermolecularly, the kinetics and equilibrium depend on the concentrations of the reactants. Instead, when a reaction occurs intramolecularly, the reaction is concentration independent and the kinetics and equilibrium depend on the connection between the reactants. By enforcing proximity, a linker can cause a weak intramolecular interaction to outcompete an intermolecular interaction that is nominally much stronger (Borcherds et al., 2017). Conversely, a linker can also ensure that a strong interaction is prevented by physical strain. Changes in linkers or scaffolding proteins likely regulate many multidomain proteins, although we are only starting to uncover such effects. A quantitative understanding of linkers is crucial for understanding cellular switches and also for the design and tuning of multidomain proteins in biotechnology including biosensors and multivalent drugs (van Dongen et al., 2007; van Rosmalen, Krom, & Merkx, 2017; Zhou, 2003a).

There is no unified definition of what constitutes a linker, and therefore even unquestionable linkers are rarely annotated as such. Systematic studies of linkers have mostly been confined to systems related by homology, where the linker could be defined as the region between folded domains (Szabo et al., 2019). From a functional point-of-view, the linker may include every region that connects two parts of the protein that functionally interact. In a three-domain protein, the middle domain may be considered part of the linker from the perspective of first and last domain. Case studies show that linkers can range in length from a handful to hundreds of amino acids, and in structure from rigid to highly flexible. The most common and generalisable type of linker is an intrinsically disordered region (IDR). Intrinsically disordered regions do not form fixed tertiary structures due to their hydrophilicity and often high net charge, and are thus advantageous as linkers in multidomain proteins as they allow almost free re-orientation of the domains relative to each other (Mittal, Holehouse, Cohan, & Pappu, 2018). This allows a tethered enzyme to search for substrates or a multivalent interaction to search for secondary binding sites.

Linkers can be understood quantitatively through the effective concentration they enforce. The effective concentration of a tethered ligand is defined as the concentration of free ligand that encounters the binding site as often as the tethered ligand. Intra-molecular binding events including enzyme catalysis and processivity (Dyla & Kjaergaard, 2020; Speltz & Zalatan, 2020; Szabo et al., 2019) and avidity of multivalent interactions (Bertagna, Toptygin, Brand, & Barrick, 2008; Zhou, 2001) are thus controlled by effective concentrations. Intramolecular binding can be approximated by replacing the ligand concentration with a constant effective concentration defined by the linker architecture. An intramolecular interaction will thus be saturated when the effective concentration largely exceeds the K_D of the interaction. Multivalent interactions occur when two molecules bind through at least two structurally independent sites. In the following, we will consider the simplest case of a bivalent interaction.

Avidity mainly arises from decreased dissociation rates, as all binding sites must be released simultaneously for full dissociation to occur. Therefore, multivalent complexes are often much more stable than their component monomeric interactions. After the first binding event of a bivalent molecule, subsequent binding events are controlled by the effective concentration and the monovalent K_D analogous to intramolecular binding. The effective concentration thus controls the population of singly bound protein from which the full dissociation must occur, and high effective concentrations favor rebinding over dissociation enhancing enzyme processivity (Szabo et al., 2019). Therefore the stability of multivalent complexes depends on the effective concentration enforced by the linker, and in turn scales with linker length (Evers, van Dongen, Faesen, Meijer, & Merkx, 2006; Mack et al., 2012; Sørensen, Jendroszek, & Kjaergaard, 2019; van Rosmalen et al., 2017).

Effective concentrations can be measured directly by competition experiments, where a linked system is titrated with free ligand (Krishnamurthy, Semetey, Bracher, Shen, & Whitesides, 2007; Sørensen & Kjaergaard, 2019). The free ligand will displace half of the tethered ligand when its free concentration equals the effective concentration of the linked ligand. Such direct measurement has only been accomplished in designed model systems. In most natural systems, direct measurement is impractical due to the high concentrations of free ligands needed, and the lack of an observable signal from displacement of two identical interactions. Effective concentration can also be measured by determining the binding affinity of the untethered ligands and comparing these values to the global affinity of the tethered system (Borcherds et al., 2017; Zhou, 2001). However, this requires quantitative measurements with both the untethered and tethered system in vitro, which is not always possible. Alternatively, effective concentrations are typically modeled based on theoretical models of the linker (Zhou, 2001, 2004), which will be the main focus of this protocol. This allows estimation of the effective concentration when it cannot be measured directly, and comparison of experimental and theoretical values, allowing to discriminate whether the linker has a simple tethering role, or whether other mechanisms are at play. Finally, the use of these models can also allow one to choose the optimal linker for a biotechnological application such as optimization of avidity in a multivalent protein drug. Theoretical models of intrinsically disordered regions are often homopolymer approximations, which means that the chain is assumed to be featureless and with uniform compaction. This assumption is not guaranteed to be a good approximation of your favorite IDR but reproduces many features of IDRs such as hydrodynamic volume and end-to-end distances surprisingly well. Homopolymer models are thus a useful baseline assumption that can then be adapted to account for sequence-specific interactions. The most important sequence features that predict IDR compactions are net charge, patterning of opposite charges and residues that can undergo π-interactions. The first step towards calculating effective concentrations is to determine the length of the linker and the spacing between binding sites in your system.

2. Defining linker length and binding site spacing

The database for experimentally validated disordered proteins (www.disprot.org) contained 187 regions annotated as "flexible linkers/spacers" and 39 regions annotated with "tethering" functions at the time of writing (June 2020). However, linkers are rarely annotated and the first step towards investigating a linker is to identify it. A "linker" is defined from the perspective of what is being linked, and may as such contain folded domains with other functions. Here, however we will only consider classical linkers, which is understood as an interdomain region that does not have other functions than tethering. Often significant insights on the linker regions can be predicted from sequence alone using bioinformatics tools. As the length is one of the most important determinants of a linker, it is crucial to define the boundaries of the flexible segment.

2.1 Identifying disordered linker domains

The recommended method of mapping linker boundaries depends on whether the linked entity is a folded domain or a SLiM. For SLiMs without well-defined structure, the linker can usually be assumed to start at after the last residue of the motif. For folded domains of known 3D structure, the linker is usually defined as starting at the first or last residue that is not resolved in the structure. If the precise structure of the domain is not known, folded domains can be mapped based on sequence homology using web servers such as CDD (http://www.ncbi.nlm.nih.gov/Structure/cdd/cdd.shtml) or PFAM (https://pfam.xfam.org/) (El-Gebali et al., 2019; Marchler-Bauer et al., 2015) (Fig. 1). Disordered regions are occasionally annotated as e.g., conserved "low complexity regions", so a conserved region does not always represent a folded domain. Regions interspersed between conserved domains may either be linkers or an unknown folded domain. Disordered linkers can usually be identified from sequence alone due to their compositional bias, which in practice is assessed by computational disorder

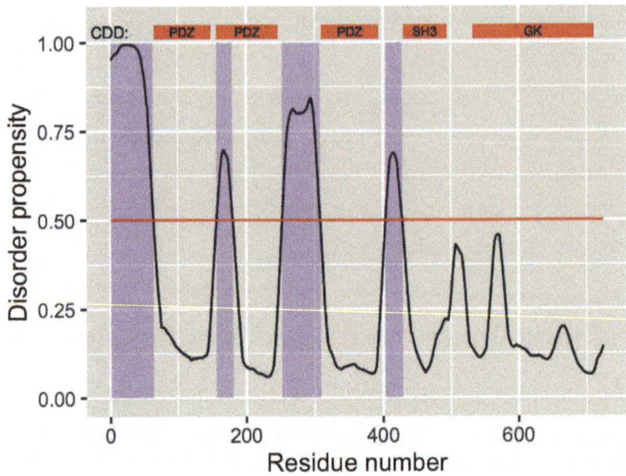

Fig. 1 Sequence based prediction of interdomain linkers. Conserved domains are predicted for rat PSD95 (UniProt: P31016) using CDD (red bars). Disorder propensities are predicted for the same sequence using a MFDp2, which suggest a propensity for disorder in the regions joining the conserved domains. Regions predicted to be disordered are shown in blue shading. The boundaries of disordered linkers can be established from ends of the conserved domains and the sharp rise in disorder propensity.

predictors. Many such predictors have been developed over the years using different criteria and data to define disordered segments. Most predictors agree on which regions have a high relative disorder propensity, but have different sensitivity and specificity. A practical solution is to use a meta-predictor such as e.g. MobiDB (https://mobidb.bio.unipd.it/), which collates the consensus of different predictors, or a single predictor that performs well in recent benchmarks (Nielsen & Mulder, 2019) such as MFDp2 (http://biomine.ece.ualberta.ca/MFDp2/) (Mizianty, Peng, & Kurgan, 2013). Disordered segments can also be recognized by comparing related sequences as disordered regions experience a different selection pressure than folded proteins. Typically individual residues are not conserved, but the overall parameters of the linker such as length, disorder propensity and physical parameters will likely be conserved (Brown, Johnson, Dunker, & Daughdrill, 2011). Therefore, a sequence alignment of homologous proteins can aid pin-pointing the boundary between folded and disordered segments as the sequence conservation will usually drop at the beginning of the linker. In total, the process outlined above should lead to definition of the first and last residues of the linkers and thus its length, which will be the key input of the modeling below.

2.2 Measuring the distance between binding sites (r_o)

Intramolecular and multivalent interactions are also modulated by the distance separating the attachment sites of the linker. For intramolecular binding, the relevant distance is that between the linker attachment site and the binding site (Fig. 3A). In a bivalent interaction, the relevant distance is the distance between binding sites. This distance is named r_o in the model (Eq. 3), and should be determined before detailed modeling of linkers. The most accurate way to do this is by measuring this distance in a structure of the complex, if such a structure is available from the Protein Data Bank (https://www.rcsb.org/). Many programs allow visualization and analysis of protein structures including Chimera (https://www.cgl.ucsf.edu/chimera/) and PyMol (https://pymol.org/edu/?q=educational/), which contain tools to measure distances and are freely available. Depending on the system, the appropriate distance may be from the N-terminus of one part to the C-terminus of the other or vice versa. If the complex structure is known for a related system, sometimes the distance can be defined from ancillary structural knowledge of the system (an example of this is provided in Section 4.1). If no structure of related complexes is available, modeling will be less precise but will not necessarily be futile. For long linkers, the effective concentration is less sensitive to the distance between binding sites. E.g., for a 100-residue linker, changing the r_o from 20 to 40 Å will result in less than 2-fold difference in the effective concentration. Thus, for complexes of unknown structure with long linkers, an approximate distance may simply be estimated from the size of the domains.

3. Motivation for simplified linker models

A disordered linker confines the attached domains to a small volume with an elevated effective concentration. Functional interpretation of linker properties often depends on estimating this volume. The reach of a linker thus defines both which interactions are sterically possible and the effective concentrations that control intramolecular binding events and avidity of multivalent interactions. To predict these effects a reasonable approximation of the linker geometry is needed. Theoretical models developed for homopolymers agree well with experiments in disordered proteins (Schimmel & Flory, 1967; Zhou, 2001, 2004) even though they do not take into account factors such as sequence composition, intramolecular interactions or charge attraction/repulsion effects. Therefore, homopolymer models are good first approximations to understanding systems joined by flexible linkers. Disordered linkers are best understood by considering their end-to-end distance, which defines their reach and the effective concentration they enforce on the tethered domains or motifs. As the linker is structurally heterogeneous, the end-to-end distance is best represented by a probability distribution $p(r)$ (Fig. 2A). By taking into account this distribution and the distance between binding sites, it is possible to calculate the effective concentration (C_{eff}) enforced by a linker. In Section 3.1, we explain the Worm Like Chain (WLC) model that allows C_{eff} calculations and in Section 3.2 we present the "C_{eff} calculator," a ready-made tool for calculating $p(r)$ and C_{eff}. Readers who do not wish to go into the detail of the model can skip to protocol Section 3.2. In Section 3.3, we cover an "advanced" feature involving adapting the model to account for different linker sequence compositions.

3.1 Polymer models for disordered linkers (WLC)

Flexible linkers can be approximated by a simplified model that describes the linker as a flexible polymer (Schimmel & Flory, 1967). The two main models used to describe homopolymers are the Gaussian Chain (GC) and the Worm Like Chain (WLC). In the following, we will use the WLC

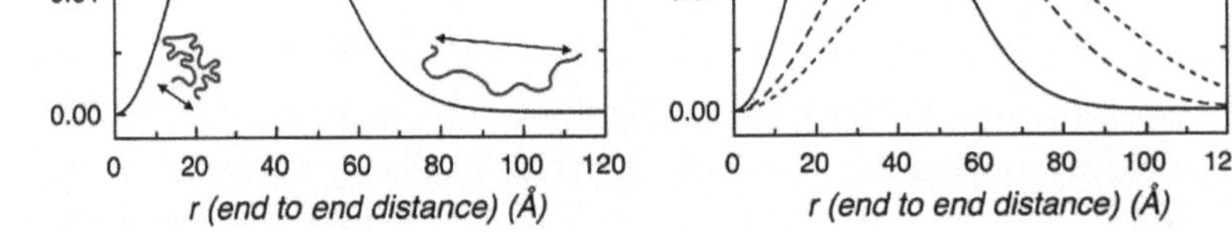

Fig. 2 Probability distribution of the end-to-end distances of a flexible linker. (A) The $p(r)$ plot shows the distribution of end-to-end distances for a 70-residue linker with $L_p = 3$ Å. The end-to-end distance with highest probability is 32.8 Å (peak of the distribution). Representative conformations are depicted in blue, and double arrows mark the end-to-end distance of each conformation. (B) Effect of changing the stiffness of a linker. For the same length, the linker becomes more extended as the stiffness increases (higher L_p). The values depicted represent highly flexible linkers ($L_p = 3$ Å) as those seen in glycine rich chains, intermediate flexibility ($L_p = 6$ Å) and low flexibility ($L_p = 9$ Å) as seen in proline-rich linkers (>20% proline). The peak of the distribution shifts from 32.8 Å ($L_p = 3$ Å) to 57.7 Å ($L_p = 9$ Å).

model, which describes the behavior of short protein linkers better and works equally well as the GC model for long protein linkers. The WLC model (Zhou, 2001, 2004) describes a linker as a semi-flexible polymer represented as a continuous cylinder with a fixed but randomly directed radius of curvature. The polymer conformation is defined by two variables: the contour length (L_c) and the persistence length (L_p) (Zhou, 2004, Bertagna et al., 2008). The contour length is the length of a fully extended chain and is calculated as $L_c = N_{res} * b_o$, where N_{res} is the number of amino acids in the linker and b_o is the average size of an amino acid, taken to be 3.8 Å (Zhou, 2001). The persistence length L_p describes the stiffness of the chain, and it represents the distance across which the direction of the chain becomes uncorrelated. Experimental measurements have shown that most protein linkers can be well described by L_p values ranging between 3 Å and 6 Å (van Rosmalen et al., 2017; Zhou, 2001, 2004).

3.1.1 Defining the end-to-end distance distribution of the WLC model

The WLC model provides a mathematical expression for the probability density function *p*(*r*), which describes the distribution of end-to-end distances of a polymer (Rubinstein & Colby, 2003). This distribution depends only on the persistence length (L_p) and the contour length (L_c) and can be written as:

$$p(r) = 4\pi r^2 \left(\frac{3}{4\pi L_p L_c}\right)^{\frac{3}{2}} exp\left(\frac{-3r^2}{4L_p L_c}\right) \zeta\left(r, L_p, L_c\right) \qquad (1)$$

Where the last term in Eq. (1) is defined as:

$$\zeta\left(r, L_p, L_c\right) = 1 - \left\{\frac{5L_p}{4L_c} - \frac{2r^2}{L_c^2} + \frac{33r^4}{80L_p L_c^3} + \frac{79L_p^2}{160L_c^2} + \frac{329L_p r^2}{120L_c^3} - \frac{6799r^4}{1600L_c^4} + \frac{3441r^6}{2800L_p L_c^5} - \frac{1089r^8}{12800L_p^2 L_c^6}\right\} \qquad (2)$$

Fig. 2 shows the shape of the end-to-end distance distribution. For a linker of defined length, the end-to-end distances show a bell-shaped distribution that peaks at intermediate extension and is unlikely to be either very compact or fully extended (Fig. 2A). Increasing the stiffness and thus L_p of a linker of fixed length leads to a more extended ensemble (Fig. 2B) (Bertagna et al., 2008; Zhou, 2001). In protein linkers, this variation can be brought about by changes in the sequence composition from more flexible to more rigid amino acids (GS linkers versus proline-rich linkers) (van Rosmalen et al., 2017; Walther et al., 2007).

3.1.2 Calculating effective concentration from the WLC model

The effective concentration describes the local concentration of a tethered ligand around its binding site. Once the probability density function $p(r)$ for a linker is known, the effective concentration enforced by the linker restrained to a distance r_o between the attachment sites is given by:

$$C_{eff} = \frac{p(r_0)}{4\pi r_0^2}\frac{10^{27}\text{Å}^3 l^{-1}}{N_A} \tag{3}$$

where N_A is Avogadro's constant, and C_{eff} represents the molar concentration of one linker end at a distance of r_0 from the site of tethering.

Fig. 3 shows how C_{eff} depends on the linker length for a tethered system, where r_o represents the distance restraining the two ends of the linker when the complex is formed (Fig. 3A). Fig. 3B shows the shape for the C_{eff} curve corresponding to the example shown in Fig. 3A. The dependence of the effective concentration on linker length always follows the same shape: When the linker is short, the linker rarely attains sufficient extension for the second motif to reach the secondary binding site, and hence C_{eff} is low (left part of the graph in Fig. 3B). The optimal linker length is achieved when the average dimensions of the linker match the separation between the binding sites (r_o), which yields a maximum C_{eff} value at the binding site and in the curve in Fig. 3B. For longer linkers, C_{eff} decays more slowly as the linker explores a larger volume around the secondary motif corresponding to the right side of Fig. 3B. The example binding curve in Fig. 3B shows an example based on the MdmX auto-inhibitory interaction described in Section 4.1, where a maximal C_{eff} of 2.24 mM is attained with a 69-residue linker. The maximal value of C_{eff} that can be attained decays as the separation between binding sites, increases (Table 1), with largest C_{eff} values (10–100 mM) obtained at small separations (10–20 Å) and lower C_{eff} values (low mM range) obtained for separations of 40–60 Å. However, even for larger distances between binding sites, tethering can lead to affinity enhancement if the C_{eff} enforced by the linker is larger than the dissociation constant of the weakest motif. Many tethered systems enforce C_{eff} values in the mM range, allowing strong affinity enhancement (Bertagna et al., 2008; Zhou, 2004).

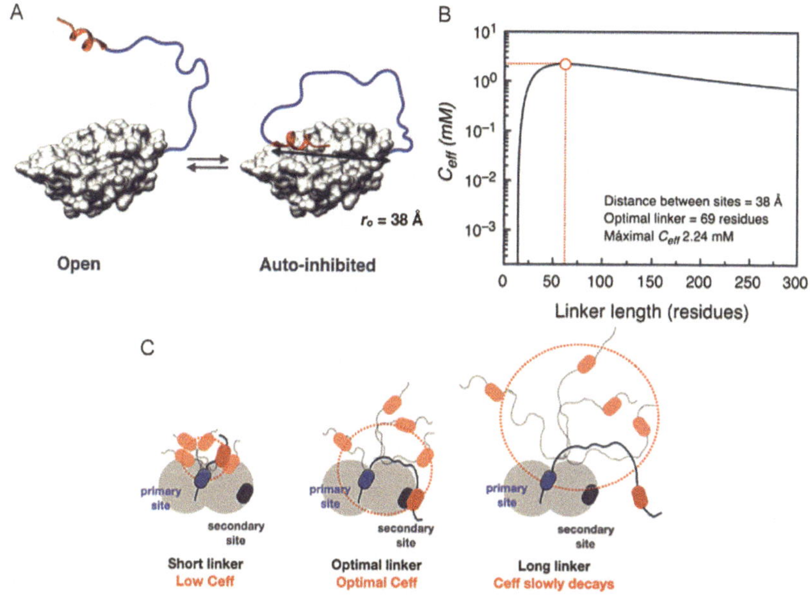

Fig. 3 Understanding the C_{eff} curve for a flexible linker. (A) As a representative example, we show a system where a protein domain is auto-inhibited by intramolecular binding of a helical motif. The distance between contact sites (r_o) is determined in the auto-inhibited complex structure by measuring the distance between the last residue of the folded domain (beginning of the linker) and the site where the binding motif contacts the folded domain (end of the linker). In this case $r_o = 38$ Å (PDB:3DAB). (B) The C_{eff} curve for the example shown in A is depicted. The maximal C_{eff} value (2.24 mM) is reached with a 69-residue linker. (C) Representative diagram showing how C_{eff} depends on linker length for a tethered bimolecular interaction. A folded domain is shown in gray, and a disordered protein harboring two binding motifs is shown as black line with the motifs depicted in blue and red, respectively. Once the first motif is bound the reaction is intramolecular. For short linkers, the second motif rarely reaches the secondary binding site and C_{eff} is low. For an optimal linker, the linker extension matches the separation between binding sites, providing the maximal concentration of the second motif at the secondary binding site. For a longer linker, C_{eff} slowly decays as the linker explores a larger conformational space away from the secondary binding site.

3.2 Calculating *p*(*r*) and C_{eff} using the online application "C_{eff} calculator"

The most convenient way to perform WLC modeling of linkers is to implement the equations in a web application that calculates and graphs the desired plots and values for user defined values of linker length and r_o. To facilitate this task we present a web app developed by our group and called "C_{eff} Calculator" (http://ceffapp.chemeslab.org), that allows the visualization of p(r) and C_{eff} plots for different systems, and can be accessed through

Table 1 Values of C_{eff} and optimal linker length for different separations between binding sites.

	$L_p = 4$ Å	
Distance between sites (Å)	**Maximal C_{eff} (mM)**	**Optimal linker length (res)[a]**
10	216	2
20	17.1	12
30	4.73	28
40	1.95	51
50	0.99	81
60	0.57	117
100	0.12	327

[a]The optimal linker length was chosen as the integer number of residues yielding the highest C_{eff} value.

a graphical user interface without any requirement for coding (Fig. 4). The "C_{eff} Calculator" was developed in *R* using the Shiny package (Chang et al., 2020). The app contains two main tabs named "C_{eff} Plot" and "*P*(*r*) Plot" that can be accessed from the Navigation panel (Fig. 4). Both tabs require the input of system specific parameters (linker length: N_{res}, and binding site separation: r_o) and basic WLC parameters (L_p and b_o). The value of b_o (3.8 Å) is fixed for amino acid chains and should not be modified except when using the app to model other polymers. The standard value of L_p is 4 Å, and it should not be changed except for advanced applications (see Section 3.3).

The "C_{eff} Plot" tab (Fig. 4A) calculates the effective concentration as a function of linker length for a given spacing between linker attachment sites (r_o) and also provides the C_{eff} value for a specific linker length. The specific parameters required are the "Distance between contact sites" (r_o) and the specific linker length. The Plot panel automatically adjusts the plot axes and displays the C_{eff} plot, which can be downloaded using the camera icon located above the plot. The tabulated values of the plot can be downloaded using the "Download C_{eff} Plot values" button as a tab separated file that can be opened as a spreadsheet and plotted independently. Below the Plot panel, the output panel displays the desired C_{eff} parameters: (1) As many users will focus on a specific system with a defined linker length, the C_{eff} calculator takes a linker length parameter and calculates the effective concentration for this linker. (2) Furthermore, the program calculates the linker length at which optimal affinity enhancement is reached, a parameter often desired

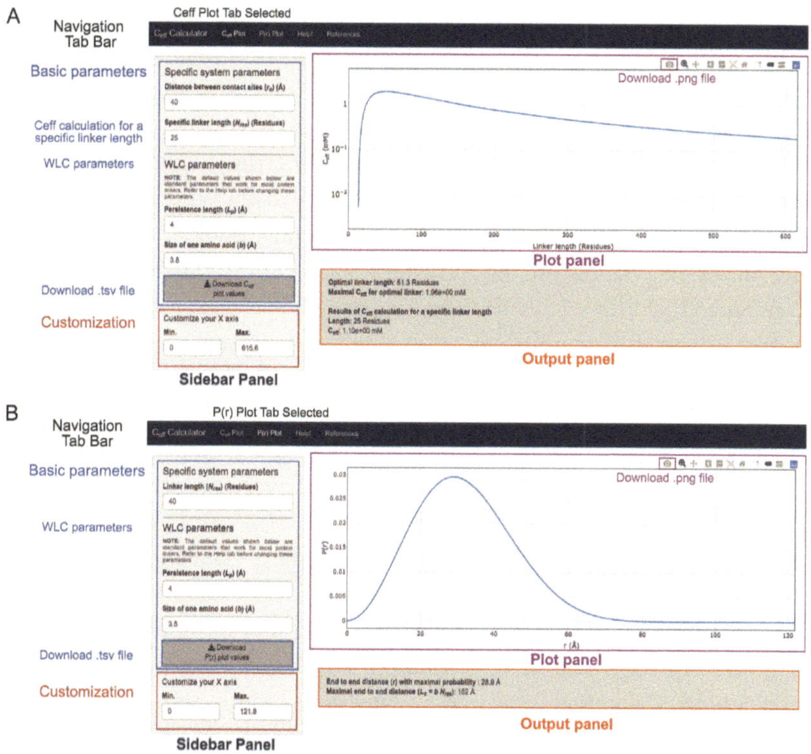

Fig. 4 The "C_{eff} *Calculator*" online application. The navigation tab bar at the top of the web page allows switching between different tabs. Only the C_{eff} Plot Tab and the *P*(*r*) Plot Tab are described. (A) The C_{eff} Plot Tab has three panels. The Sidebar panel allows the input of: (1) Basic parameters (blue square) are divided in two sections: The upper section allows input of specific system parameters: distance between contact sites, r_o and input of a specific linker length that allows calculation of the C_{eff} value for a specific linker length. The lower section allows input of the WLC parameters: persistence length, L_p, and the size of one amino acid, *b*, and allows download of the C_{eff} plot values for plotting with the desired software. (2) Customization (red square) allows setting custom limits for the *x* axis. The plot panel (purple square) shows the C_{eff} plot according to the specific system parameters and allows download of the plot as a png file. The Output panel (orange square) reports (1) the optimal linker length and maximal C_{eff} for the system and (2) the C_{eff} value for a specific linker length. (B) The *P*(*r*) Plot Tab also has three panels. The Sidebar panel allows the input of: (1) Basic parameters (blue square): linker length (N_{res}) and the lower section allows input of the WLC parameters and downloading of the *p*(*r*) plot values. (2) Customization (red square) allows setting limits for the *x*-axis. The plot panel (purple square) shows the *p*(*r*) plot and allows downloading the plot as a png file. The Output panel (orange square) reports (1) the end-to-end distance (*r*) with maximal probability and (2) the maximal end-to-end distance, calculated as: $L_c = b_o\ N_{res}$, for the system. The C_{eff} Calculator was developed in *R* using the Shiny package (Chang, Cheng, Allaire, Xie, & McPherson, 2020).

in biotechnological applications involving linker design and in evaluating whether natural systems are evolutionarily optimized for tight binding.

The "*P*(*r*) Plot" tab shows the distance probability distribution for a linker of defined length (Fig. 4B). The plot axes are automatically scaled by the app, but can be selected by the user using the customization panel. Below the plot, the output panel reports "End-to-end distance with maximal probability" and the "Maximal end-to-end distance" ($L_c = N_{res}{}^* \, b_o$). The tabulated values of the $p(r)$ plot can be downloaded using the "Download P(r) Plot values" button as a tab separated file that can be opened as a spreadsheet for graphing using independent software.

3.3 Defining the L_p value for your linker (advanced users)

The "C_{eff} Calculator" app allows full control over the parameters of the WLC model including the "Persistence length (L_p)" and the "Size of one amino acid (b_o)". As default the app uses standard values for polypeptides of $L_p = 4$ Å and $b_o = 3.8$ Å (Zhou, 2004). The length of the monomer unit (b_o) is a fixed value, defined by the peptide backbone, and should be the same for all protein linkers. The b_o parameter should not be confused with the Kuhn length, a parameter related to persistent length and often termed b in the literature. The b_o parameter should thus rarely have to be adjusted, but is left variable to allow the app to be used for other polymers such as polyethyleneglycol ($L_p = 3.8$ Å, $b_o = 2.8$ Å) (Kienberger et al., 2000; Oesterhelt, Rief, & Gaub, 1999), single-stranded DNA ($L_p = 20$ Å, $b_o = 6.7$ Å) (Roth, Glick Azaria, Girshevitz, Bitler, & Garini, 2018) or double-stranded DNA ($L_p = 500$ Å, $b_o = 3.4$ Å) (Rubinstein & Colby, 2003; Seol, Li, Nelson, Perkins, & Betterton, 2007). The persistence length that best fits a specific set of linkers might vary depending on the sequence composition of the linker. While the WLC model deals with a homopolymer of fixed persistence length, real linkers are heteropolymers and therefore the persistence length might change along the chain depending on the conformational restrictions imposed by rigid amino acids such as proline or small structured regions within the linker. However, many linker systems can be satisfactorily modeled by using a unique L_p value. While there is currently no established method to calculate the L_p value from sequence composition alone, we provide some useful guidance for choosing the L_p value. Most measurements of L_p in linkers with low (<20%) proline content, yielded $L_p = 3$–6 Å, including a collection of protein loops and flexible linkers that had values ranging $L_p = 3$–4 Å (Bertagna et al., 2008; Borcherds et al., 2017;

van Rosmalen et al., 2017; Zhou, 2004). In glycine-serine (GS) linkers, which are widely used in biotechnological and molecular biology applications, the apparent L_p value ranges between 4 and 6 Å depending on the glycine content of the linker, with G4S2 linkers having $L_p = 3.7$ Å and all-serine (S6) linkers having $L_p = 6.2$ Å (Evers et al., 2006; van Rosmalen et al., 2017). Based on this, we suggest using a standard value of $L_p = 4$ Å as a guideline. This value can be modified to $L_p = 3$ Å if the linker under study is very flexible (for example if it has high, >30%, glycine content) and can be modified to $L_p = 6$ Å when glycine content is low (<10%). Larger L_p values ranging 9–12 Å might be suited when studying linkers where proline content is very high (>20%) (Walther et al., 2007). The true persistence length is usually unknown, so we recommend exploring a range of L_p values that might be suited to the system with $L_p = 4$ Å as the baseline value. In many cases, the main conclusions derived from the C_{eff} analysis will be independent of the exact L_p value used.

4. Modeling intramolecular and bivalent binding

In this section, we will provide examples of how WLC modeling helps understand the functional properties of systems tethered by flexible linkers. The first example will consider intra-molecular inhibition, when a protein is tethered to its own inhibitor. Many examples of this exist, for example the auto-inhibited form of tyrosine kinases such as cAbl where the myristoylated N-terminus latches onto the kinase domain preventing activation (Hantschel et al., 2003; Zhou, 2003b) and Src kinases, where the phosphorylated C-terminal tail binds to the SH2 domain establishing the auto-inhibited state (Harrison, 2003). The second example will consider binding of two interaction modules joined by a flexible linker to a bivalent partner protein. Bivalency enhances affinity through avidity and results in an interaction that is stronger than its constituent parts. The avidity enhancement depends both on the strength of the monovalent interactions and the linker architecture.

4.1 Auto-inhibition in MdmX

To illustrate the use of linker modeling, we will discuss the example of MdmX intramolecular inhibition, which decreases the ability of MdmX to bind the trans-activation domain (TAD) of p53 (Borcherds et al., 2017; Chen et al., 2015). p53TAD binds to the folded MdmX N-terminal domain (NTD) through a helical motif. However, the

MdmX NTD is tethered by a flexible linker to a "WW" motif that also binds at the p53TAD binding site, creating an auto-inhibited conformation, and this lowers MdmX affinity for p53 (Fig. 5A).

4.1.1 Defining the linker length and r_o for MdmX

To analyze the intramolecular inhibition of MdmX and how this impacts on p53TAD binding affinity, we need to determine parameters that characterize this system. The flexibility of the linker was established both from disorder prediction and from nuclear magnetic resonance experiments (Borcherds et al., 2017). The beginning of the linker was identified as the residue immediately following the last structured residue (109) in the MdmX NTD crystal structure (PDB:3DAB). The end of the linker was taken to be the residue preceding the start of the "WW" binding motif, determined from NMR experiments where residue 195 was the first peak affected by binding of the "WW" motif to the MdmX NTD (Borcherds et al., 2017). Therefore, the linker extends from residues 110 to 194 of MdmX, and has a length 85 residues (Fig. 5A) (Borcherds et al., 2017).

Since the complex of MdmX with the WW motif was not known, we used the structure of MdmX in complex with the p53TAD to estimate r_o by assuming both binding sites of the WW motif and p53TAD be overlapping and binding in the same orientation as the p53TAD. This assumption was based on the fact that both motifs could displace each other in competition experiments. In this case, the relevant distance was between the C-terminal end of the folded MdmX domain (site of linker attachment) to the N-terminal end of the p53TAD motif (PDB ID: 3DAB). This procedure yielded a distance between binding sites (r_o) of 38 Å (Fig. 3A).

4.1.2 Estimating the p(r) and C_{eff} for the MdmX WW motif

The WLC model was used to calculate the extension of the MdmX linker and the effective concentration of the "WW" motif at its binding site. The MdmX linker had a low proline content and could be classified as a moderately charged polyampholyte with intermediate hydropathy values, suggesting its dimensions might be moderately compact (Borcherds et al., 2017). For initial modeling, we used an L_p value of 3 Å and a linker length of 85 residues to obtain the probability distribution of end-to-end distances of the linker. Because the true L_p value of the linker was unknown we also tested the model using $L_p = 4$ Å and the main conclusions were unchanged, suggesting either parameter described the system with acceptable accuracy. The probability distribution of the natural linker peaked at 36 Å, closely

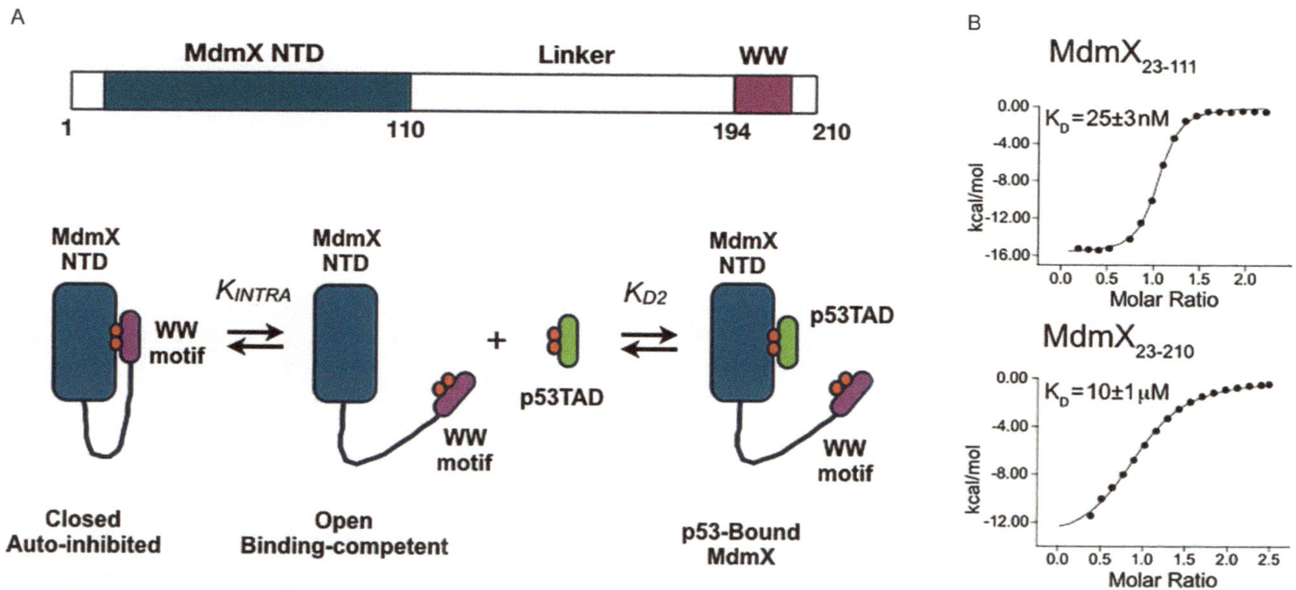

Fig. 5 *Modeling intramolecular inhibition in MdmX.* (A) MdmX binding to p53TAD can be represented by a coupled equilibrium between the "closed" and "open" conformations, followed by p53TAD binding to the binding-competent "open" conformation. The notation of the binding constants is described in the text. (B) Direct ITC binding titrations of p53TAD into the MdmX(23 – 110) construct containing only the MdmX domain, versus titration of p53TAD into the MdmX(23–210) construct containing the MdmX domain followed by the linker and the "WW" motif. *Adapted with permission from Borcherds, W., Becker, A., Chen, L., Chen, J., Chemes, L. B., & Daughdrill, G. W. (2017). Optimal affinity enhancement by a conserved flexible linker controls p53 mimicry in MdmX.* Biophysical Journal, 112*(10), 2038–2042. https://doi.org/10.1016/j.bpj.2017.04.017.*

matching the separation between binding sites (38 Å, Fig. 3A). As expected by this agreement, the effective concentration of the "WW" motif at the p53TAD binding site calculated from the WLC model was 2.12 mM, very close to the maximal C_{eff} that can be reached in the system (2.24 mM for a 69-residue linker) (Fig. 3B). This agreement suggests the system evolved to optimize the auto-inhibitory interaction, as supported by the strong conservation of the MdmX linker length (Borcherds et al., 2017).

4.1.3 Calculating populations of "open" and "closed" or auto-inhibited MdmX

In absence of ligands, MdmX forms an equilibrium between an "open" conformation where the "WW" motif diffuses freely within a volume defined by its tether, and an auto-inhibited or "closed" conformation where the "WW" motif is bound to the p53TAD binding groove of MdmX (Fig. 3A, Fig. 5A). By tethering the "WW" motif to the MdmX domain, the linker turns the binding reaction intramolecular and thus independent of protein concentration. The equilibrium constant K_{INTRA} that governs the equilibrium between the "open" and "closed" states when the "WW" motif is tethered can be calculated from the following relationship:

$$K_{INTRA} = \frac{C_{eff}}{K_{D1}} = \frac{MdmX_{CLOSED}}{MdmX_{OPEN}} \tag{4}$$

Where K_{D1} is the bimolecular dissociation constant for binding of the untethered "WW" motif to the MdmX domain and C_{eff} is the effective concentration enforced by the linker. As both C_{eff} and K_{D1} are in molar units, K_{INTRA} is unimolecular and unit-less, and thus the equilibrium between the "closed" and "open" conformations does not depend on protein concentration. The bimolecular affinity, K_{D1}, was measured to 8.3 ± 0.1 μM by determining the binding affinity of a short peptide containing the "WW" motif to the MdmX domain. Using $C_{eff}=2.12$ mM as predicted by the app yields $K_{INTRA}=255$. From Eqs. (5) and (6), the fractional populations of the "open" and "closed" MdmX conformations can be calculated. The very high K_{INTRA} value indicates that the system is strongly auto-inhibited, with only 0.4% MdmX molecules in the binding-competent "open" conformation and 99,6% of MdmX in the "closed" auto-inhibited state.

$$f_{MdmX(CLOSED)} = \frac{K_{INTRA}}{1 + K_{INTRA}} \tag{5}$$

$$f_{MdmX(OPEN)} = \frac{1}{1 + K_{INTRA}} \tag{6}$$

4.1.4 Modeling the competition between p53TAD and the MdMX "WW" motif

The effect of auto-inhibition on binding of p53TAD to MdmX can be modeled as two linked reactions involving the MdmX "closed"-"open" equilibrium followed by p53TAD binding to the "open" conformation (Fig. 5A). The magnitude of this inhibition can be observed as a 400-fold reduction in the binding affinity of p53TAD for MdmX (23 – 210), a fragment containing the MdmX domain tethered to the "WW" motif ($K_D = 10 \pm 1\,\mu M$), compared to p53 affinity of the isolated MdmX domain lacking the "WW" motif ($K_{D2} = 25 \pm 3\,nM$) (Fig. 5B). The complete binding equilibrium is shown in Fig. 5A, and the global dissociation constant can be described by:

$$K_{D,AUTO} = K_{INTRA} * K_{D2} \tag{7}$$

Where $K_{D,AUTO}$ is the global dissociation constant for binding of p53TAD to MdmX, K_{INTRA} is the unimolecular equilibrium constant describing the "closed"-"open" equilibrium, and K_{D2} is the dissociation constant for binding of p53TAD to the isolated MdmX domain ("open" conformation). The reduction of p53TAD binding affinity by auto-inhibition, can be compared to that expected based on the WLC model. Using the value of $K_{INTRA} = 255$ (Section 4.1.3), the K_{D_AUTO} value was $9.4 \pm 1.5\,\mu M$. The excellent agreement between this value and the experimentally measured affinity of p53TAD for the tethered MdmX construct ($10 \pm 1\,\mu M$) suggests that the MdmX linker acts as a flexible entropic tether that optimizes MdmX auto-inhibition. In this system, higher C_{eff} values lead to stronger MdmX intramolecular interactions and weaken p53TAD association to MdmX.

4.2 Calculating affinity enhancement in bivalent interactions

The WLC model can also be used to describe the enhancement caused by tethering of two binding motifs or domains in bivalent interactions. Tethered bivalent interactions can involve a domain-linker-domain, domain-linker-motif or motif-linker-motif architecture. Tethering of two binding motifs can enhance the global binding affinity by several orders of magnitude, and for each system there will be an optimal linker length that

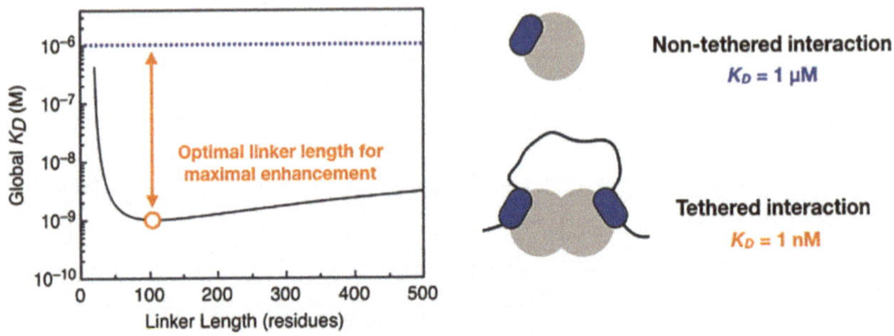

Fig. 6 Modeling tethering in bivalent interactions. The figure shows schematically the enhancement of global affinity in a bivalent interaction and its dependence on linker length. For simplicity, we showcase a system where two identical binding motifs (K_D=1 μM) shown in blue are tethered by a linker and bind identical sites in a receptor domain, shown in gray. The graph shows how tethering enhances the global binding affinity over a broad range of linker lengths, but there is an optimal linker length (103 residues) that maximizes the binding affinity. This system reaches a 1000-fold enhancement of global binding affinity (K_D=1 nM). In this theoretical example, r_o=50 Å and L_p=3 Å.

maximizes this enhancement (Fig. 6). In these systems, the global dissociation constant of the tethered bivalent construct is defined by:

$$K_{D,GLOBAL} = K_{D1} * K_{D2} * C_{eff}^{-1} \tag{8}$$

Where K_{D1} and K_{D2} are the dissociation constants of each untethered domain/motif and C_{eff} is the effective concentration enforced by the linker. The C_{eff} prediction using the WLC model allows calculation of the global binding affinity when it cannot be measured directly. Eq. (8) implies that affinity enhancement only occurs when C_{eff} is larger than the K_D of the weakest interaction. Under these conditions, the bivalent complex dominates over monovalent bound forms and the global affinity is higher than the monovalent interactions. If C_{eff} is less than the K_D of the weakest interaction, bivalent binding will not be favored and thus there will be no avidity enhancement (Sørensen et al., 2019).

When global affinity can be measured, WLC modeling allows comparison of the experimental (C_{eff}=K_{D1}*K_{D2}/$K_{D,GLOBAL}$) and predicted C_{eff} values. A close correspondence would indicate that the linker behaves as an entropic tether, while differences will point to additional factors governing the binding affinity in the system. For example, in the Oct1POU system (Klemm & Pabo, 1996; Zhou, 2001), tethering of the POU_H (K_{D1}=150 nM) and POU_S (K_{D2}=1.7 μM) DNA binding domains by a flexible linker yielded a picomolar affinity complex with DNA ($K_{D,GLOBAL}=71$ pM), yielding an experimental C_{eff} value $K_{D1}{*}K_{D2}/K_{D,GLOBAL}=3.6$ mM. C_{eff} can also be predicted from the distance between the last residue of POU_S and the first residue of POU_H in the Oct-1 POU structure (27.6 Å) and the linker length (26 residues) The predicted C_{eff} of 5.8 mM agrees well with experiment, which indicates that the linker behaves largely as an entropic tether. In this system there is cooperativity between the POU_H and POU_S sites. To account for this, the affinity of the POU_S domain was measured by titrating a POU_H/DNA complex (Klemm & Pabo, 1996). In the Notch intracellular receptor domain (NICD), the RAM motif and the ANK domain are tethered by a flexible linker to form a bivalent complex with the CSL transcription factor that regulates Notch-dependent genes (Bertagna et al., 2008; Sherry, Johnson, Hatem, Majumdar, & Barrick, 2015). The length of the linker was shown to be tuned to the optimal value for affinity enhancement, increasing the weak binding of the $NICD_{ANK}$ domain to CSL (Bertagna et al., 2008). In this case, the WLC model only partially accounts for the dependence of transcriptional activity on linker length. The deviations observed upon deletion of certain linker regions, revealed that some linker regions mediate sequence-specific effects on transcriptional activation (Sherry et al., 2015). A striking example of affinity enhancement is in the viral E1A protein, where entropic tethering of two linear motifs underlies the hijack mechanism used by E1A to bind tightly to Rb and displace the E2F transcription factor and induce cell proliferation. The evolutionary conservation of this mechanism suggests that linker dimensions might be tuned through evolution to support tethering functions.

5. Limitations of WLC models

The worm-like chain modeling assumes that the linkers behave as uniform and feature-less chains. This is a useful first approximation in many systems, but will not be appropriate in others. It is therefore important to critically assess whether the WLC is appropriate for your system. In the following, we will discuss reasons that some intrinsically disordered regions will be poorly modeled by WLC models.

The WLC and other simplified polymer modes of IDPs are homopolymers and thus assume uniform behavior for all linker segments. This is often a good approximation if the linker region has properties close to average. Yet, proteins are heteropolymers and sequence compositions can deviate from average, which will affect the compaction of an IDR. The sequence composition of the linkers can be analyzed with computational tools such as e.g. CIDER (http://pappulab.wustl.edu/CIDER/) (Holehouse, Das, Ahad, Richardson, & Pappu, 2017). The strongest determinant of the compaction of IDRs is their net charge as disordered proteins expand at high net charge due to charge-charge repulsion (Hofmann et al., 2012; Mao, Crick, Vitalis, Chicoine, & Pappu, 2010; Marsh & Forman-Kay, 2010; Müller-Späth et al., 2010). Proteins with highly charged linkers may still be described by homopolymer models (Sørensen & Kjaergaard, 2019), but in these cases the persistence length of the chain should be increased. The relationship between sequence composition and persistence length is still work in progress, so any modeling of linkers with highly divergent sequence properties will likely be less precise. Similarly, IDRs with a high proline content are more rigid and extended, and should thus be modeled with elevated persistence lengths, as described in Section 3 (Walther et al., 2007). Conversely, IDRs can contract if they contain a high density of aromatic residues (Martin et al., 2020), have a high proportion of glycine residues (van Rosmalen et al., 2017) or are strong polyampholytes (Sørensen & Kjaergaard, 2019). Many natural linkers contain helical segments (George & Heringa, 2002) and helical linkers such as $(EAAAK)_n$ are often used in fusion proteins to separate functional domains (Arai, Ueda, Kitayama, Kamiya, & Nagamune, 2001). Regions with high helical propensity can often be identified from sequence using, e.g., AGADIR (http://agadir.crg.es/) (Lacroix, Viguera, & Serrano, 1998) although the fractional helicity is highly context specific and thus difficult to predict. Therefore it is difficult identify WLC parameters that reliably model helical linkers although a higher L_p is expected.

A key function of simplified models is to identify cases where experiment deviates from the idealized predictions. Such deviations can highlight features of the linker evolved to regulate the attached proteins that may not be obvious from examination of the sequence, and may reveal additional regulatory elements and partial structuring of the linker (Sherry et al., 2015). Deviations from average sequence composition can be handled within the framework of an idealized chain, however the unequal distribution of chemical properties through-out the chain poses a more fundamental problem. A patchy distribution of charges can lead to strong interactions between patches of opposite charge (Das & Pappu, 2013), and patchy chains are thus more compact than well-mixed sequences of similar composition. As the effective concentration depends on the end-to-end distance, it may be particularly sensitive to end-to-end collapse driven by terminal patches of opposite charge (Sherry, Das, Pappu, & Barrick, 2017). Linkers with strongly patchy charge distributions can be identified, e.g., using the "Linear net charge per residue"-plot produced by CIDER. Another source of deviations from ideality is interactions with the tethered domains, which may affect the conformational properties of the linker (Mittal et al., 2018). Furthermore, it is not trivial to define where interaction motifs ends as regions flanking the core motif can contribute (Bugge et al., 2020), or if several linkers are involved they may interact with each other or with the folded partner domains (Sherry et al., 2015, 2017; Sørensen et al., 2019). This complicates both the definition of where the linker starts and what the affinity of the monovalent interaction is. Similar deviations can be seen when the binding sites of a bivalent interaction affect each other through allostery mediated by the folded partner. In such cases, predictions of avidity should use the affinity of one ligand in the presence of the other, as exemplified in Section 4.2.

6. Summary

Disordered linkers are common in proteins and need to be quantitatively accounted for to understand phenomena such as auto-inhibition, avidity and tethered enzymes. We present a user-friendly framework for modeling disordered linkers using worm-like chains using a freely available app that can be run without a need for skills in programming or math. Such modeling will be useful for generating baseline expectations for how a system should behave if the linker was indeed a passive tether. In turn, deviations from ideal behavior can help identify cryptic regulatory elements embedded in linkers. Despite the simplifying assumptions involved, polymer models remain a robust tool to account for the behavior of flexible protein linkers. We illustrate how linker modeling can be used to make quantitative predictions for how the length and compactness of linkers will affect protein interactions. Modeling is also key to the rational design of new multidomain proteins such as biosensors or multivalent drugs, and the C_{eff} Calculator app provides the optimal linker length to be used in these kinds of applications. The main weakness of the protocol presented here is that it describes heteropolymeric proteins as uniform homopolymers. The present protocol can thus be improved by accounting better for the sequence-specific effects on linker compaction, initially by automatically adjusting L_p based on sequence composition, and in the long term by abandoning the homopolymer assumption entirely. However, such improvements require a better understanding of sequence-to-ensemble relationships in IDPs.

Acknowledgments

We thank Cesar Leonetti for developing the web deployment and server infrastructure for the "C_{eff} Calculator" web app. JG is a postdoctoral fellow and LBC is an independent researcher from Consejo Nacional de Investigaciones Científicas y Técnicas (CONICET), Argentina. The work was supported by Agencia Nacional de Promoción Científica y Tecnológica (ANPCyT) grant #PICT 2017-1924 to LBC and grants from the "Young Investigator Program" of the Villum Foundation and PROMEMO – Center for Proteins in Memory, a Center of Excellence funded by the Danish National Research Foundation (grant number DNRF133) to MK.

References

Arai, R., Ueda, H., Kitayama, A., Kamiya, N., & Nagamune, T. (2001). Design of the linkers which effectively separate domains of a bifunctional fusion protein. *Protein Engineering Design and Selection*, *14*, 529–532. https://doi.org/10.1093/protein/14.8.529.

Bertagna, A., Toptygin, D., Brand, L., & Barrick, D. (2008). The effects of conformational heterogeneity on the binding of the notch intracellular domain to effector proteins: A case of biologically tuned disorder. *Biochemical Society Transactions*, *36*(2), 157–166. https://doi.org/10.1042/BST0360157.

Borcherds, W., Becker, A., Chen, L., Chen, J., Chemes, L. B., & Daughdrill, G. W. (2017). Optimal affinity enhancement by a conserved flexible linker controls p53 mimicry in MdmX. *Biophysical Journal*, *112*(10), 2038–2042. https://doi.org/10.1016/j.bpj.2017.04.017.

Brown, C. J., Johnson, A. K., Dunker, A. K., & Daughdrill, G. W. (2011). Evolution and disorder. *Current Opinion in Structural Biology*, *21*(3), 441–446. https://doi.org/10.1016/j.sbi.2011.02.005. Frontiers in Molecular Biosciences, 7, 110. https://doi.org/10.3389/fmolb.2020.00110.

Bugge, K., Brakti, I., Fernandes, C. B., Dreier, J. E., Lundsgaard, J. E., Olsen, J. G., et al. (2020). Interactions by disorder–A matter of context. *Frontiers in Molecular Biosciences*, 7. https://doi.org/10.3389/fmolb.2020.00110, 110.

Chang, W., Cheng, J., Allaire, J., Xie, Y., & McPherson, J. (2020). *Shiny: Web application framework for R. (1.5.0) [computer software]*.

Chen, L., Borcherds, W., Wu, S., Becker, A., Schonbrunn, E., Daughdrill, G. W., et al. (2015). Autoinhibition of MDMX by intramolecular p53 mimicry. *Proceedings of the National Academy of Sciences of the United States of America*, *112*(15), 4624–4629. https://doi.org/10.1073/pnas.1420833112.

Das, R. K., & Pappu, R. V. (2013). Conformations of intrinsically disordered proteins are influenced by linear sequence distributions of oppositely charged residues. *Proceedings of the National Academy of Sciences of the United States of America*, *110*(33), 13392–13397. https://doi.org/10.1073/pnas.1304749110.

Dyla, M., & Kjaergaard, M. (2020). Intrinsically disordered linkers control tethered kinases via effective concentration. *Proceedings of the National Academy of Sciences of the United States of America*, *117*(35), 21413–21419. https://doi.org/10.1073/pnas.2006382117.

El-Gebali, S., Mistry, J., Bateman, A., Eddy, S. R., Luciani, A., Potter, S. C., et al. (2019). The Pfam protein families database in 2019. *Nucleic Acids Research*, *47*(D1), D427–D432. https://doi.org/10.1093/nar/gky995.

Evers, T. H., van Dongen, E. M. W. M., Faesen, A. C., Meijer, E. W., & Merkx, M. (2006). Quantitative understanding of the energy transfer between fluorescent proteins connected via flexible peptide linkers. *Biochemistry*, *45*(44), 13183–13192. https://doi.org/10.1021/bi061288t.

George, R. A., & Heringa, J. (2002). An analysis of protein domain linkers: Their classification and role in protein folding. *Protein Engineering, Design and Selection*, *15*, 871–879. https://doi.org/10.1093/protein/15.11.871.

Hantschel, O., Nagar, B., Guettler, S., Kretzschmar, J., Dorey, K., Kuriyan, J., et al. (2003). A myristoyl/phosphotyrosine switch regulates c-Abl. *Cell*, *112*(6), 845–857. https://doi.org/10.1016/S0092-8674(03)00191-0.

Harrison, S. C. (2003). Variation on an Src-like theme. *Cell*, *112*(6), 737–740. https://doi.org/10.1016/S0092-8674(03)00196-X.

Hofmann, H., Soranno, A., Borgia, A., Gast, K., Nettels, D., & Schuler, B. (2012). Polymer scaling laws of unfolded and intrinsically disordered proteins quantified with single-molecule spectroscopy. *Proceedings of the National Academy of Sciences of the United States of America*, *109*(40), 16155–16160. https://doi.org/10.1073/pnas.1207719109.

Holehouse, A. S., Das, R. K., Ahad, J. N., Richardson, M. O. G., & Pappu, R. V. (2017). CIDER: Resources to analyze sequence-ensemble relationships of intrinsically disordered proteins. *Biophysical Journal*, *112*(1), 16–21. https://doi.org/10.1016/j.bpj.2016.11.3200.

Kienberger, F., Pastushenko, V. P., Kada, G., Gruber, H. J., Riener, C., Schindler, H., et al. (2000). Static and dynamical properties of single poly(ethylene glycol) molecules investigated by force spectroscopy. *Single Molecules*, *1*, 123–128.

Klemm, J. D., & Pabo, C. O. (1996). Oct-1 POU domain-DNA interactions: Cooperative binding of isolated subdomains and effects of covalent linkage. *Genes & Development*, *10*, 27–36.

Krishnamurthy, V. M., Semetey, V., Bracher, P. J., Shen, N., & Whitesides, G. M. (2007). Dependence of effective molarity on linker length for an Intramolecular protein – ligand system. *Journal of the American Chemical Society*, *129*(5), 1312–1320. https://doi.org/10.1021/ja066780e.

Lacroix, E., Viguera, A. R., & Serrano, L. (1998). Elucidating the folding problem of α-helices: Local motifs, long-range electrostatics, ionic-strength dependence and prediction of NMR parameters. *Journal of Molecular Biology*, *284*(1), 173–191. https://doi.org/10.1006/jmbi.1998.2145.

Mack, E. T., Snyder, P. W., Perez-Castillejos, R., Bilgiçer, B., Moustakas, D. T., Butte, M. J., et al. (2012). Dependence of avidity on linker length for a bivalent ligand–bivalent receptor model system. *Journal of the American Chemical Society*, *134*(1), 333–345. https://doi.org/10.1021/ja2073033.

Mao, A. H., Crick, S. L., Vitalis, A., Chicoine, C. L., & Pappu, R. V. (2010). Net charge per residue modulates conformational ensembles of intrinsically disordered proteins. *Proceedings of the National Academy of Sciences of the United States of America*, *107*(18), 8183–8188. https://doi.org/10.1073/pnas.0911107107.

Marchler-Bauer, A., Derbyshire, M. K., Gonzales, N. R., Lu, S., Chitsaz, F., Geer, L. Y., et al. (2015). CDD: NCBI's conserved domain database. *Nucleic Acids Research*, *43*(D1), D222–D226. https://doi.org/10.1093/nar/gku1221.

Marsh, J. A., & Forman-Kay, J. D. (2010). Sequence determinants of compaction in intrinsically disordered proteins. *Biophysical Journal*, *98*(10), 2383–2390. https://doi.org/10.1016/j.bpj.2010.02.006.

Martin, E. W., Holehouse, A. S., Peran, I., Farag, M., Incicco, J. J., Bremer, A., et al. (2020). Valence and patterning of aromatic residues determine the phase behavior of prion-like domains. *Science*, *367*(6478), 694–699. https://doi.org/10.1126/science.aaw8653.

Mittal, A., Holehouse, A. S., Cohan, M. C., & Pappu, R. V. (2018). Sequence-to-conformation relationships of disordered regions tethered to folded domains of proteins. *Journal of Molecular Biology*, *430*(16), 2403–2421. https://doi.org/10.1016/j.jmb.2018.05.012.

Mizianty, M. J., Peng, Z., & Kurgan, L. (2013). MFDp2: Accurate predictor of disorder in proteins by fusion of disorder probabilities, content and profiles. *Intrinsically Disordered Proteins*, *1*(1), e24428. https://doi.org/10.4161/idp.24428.

Müller-Späth, S., Soranno, A., Hirschfeld, V., Hofmann, H., Rüegger, S., Reymond, L., et al. (2010). Charge interactions can dominate the dimensions of intrinsically disordered proteins. *Proceedings of the National Academy of Sciences of the United States of America*, *107*(33), 14609–14614.

Nielsen, J. T., & Mulder, F. A. A. (2019). Quality and bias of protein disorder predictors. *Scientific Reports*, *9*(1), 5137. https://doi.org/10.1038/s41598-019-41644-w.

Oesterhelt, F., Rief, M., & Gaub, H. E. (1999). Single molecule force spectroscopy by AFM indicates helical structure of poly(ethylene-glycol) in water. *New Journal of Physics*, *1*, 6. https://doi.org/10.1088/1367-2630/1/1/006.

Roth, E., Glick Azaria, A., Girshevitz, O., Bitler, A., & Garini, Y. (2018). Measuring the conformation and persistence length of single-stranded DNA using a DNA origami structure. *Nano Letters*, *18*, 6703–6709. https://doi.org/10.1021/acs.nanolett.8b02093.

Rubinstein, M., & Colby, R. H. (2003). *Polymer physics* (1st ed.). Oxford University Press.

Schimmel, P. R., & Flory, P. J. (1967). Conformational energy and Configurational statistics of poly-L-proline. *Proceedings of the National Academy of Sciences of the United States of America*, *58*(1), 52–59.

Scott, J. D., & Pawson, T. (2009). Cell signaling in space and time: Where proteins come together and when They're apart. *Science*, *326*(5957), 1220–1224. https://doi.org/10.1126/science.1175668.

Seol, Y., Li, J., Nelson, P. C., Perkins, T. T., & Betterton, M. D. (2007). Elasticity of short DNA molecules: Theory and experiment for contour lengths of 0.6–7 μm. *Biophysical Journal*, *93*, 4360–4373. https://doi.org/10.1529/biophysj.107.112995.

Sherry, K. P., Das, R. K., Pappu, R. V., & Barrick, D. (2017). Control of transcriptional activity by design of charge patterning in the intrinsically disordered RAM region of the notch receptor. *Proceedings of the National Academy of Sciences of the United States of America*, *114*(44), E9243–E9252. https://doi.org/10.1073/pnas.1706083114.

Sherry, K. P., Johnson, S. E., Hatem, C. L., Majumdar, A., & Barrick, D. (2015). Effects of linker length and transient secondary structure elements in the intrinsically disordered notch RAM region on notch Signaling. *Journal of Molecular Biology*, *427*(22), 3587–3597. https://doi.org/10.1016/j.jmb.2015.09.001.

Sørensen, C. S., Jendroszek, A., & Kjaergaard, M. (2019). Linker dependence of avidity in multivalent interactions between disordered proteins. *Journal of Molecular Biology*, *431*(24), 4784–4795. https://doi.org/10.1016/j.jmb.2019.09.001.

Sørensen, C. S., & Kjaergaard, M. (2019). Effective concentrations enforced by intrinsically disordered linkers are governed by polymer physics. *Proceedings of the National Academy of Sciences of the United States of America*, *116*(46), 23124–23131. https://doi.org/10.1073/pnas.1904813116.

Speltz, E. B., & Zalatan, J. G. (2020). The relationship between effective molarity and affinity governs rate enhancements in tethered kinase–substrate reactions. *Biochemistry*, *59*(23), 2182–2193. https://doi.org/10.1021/acs.biochem.0c00205.

Szabo, B., Horvath, T., Schad, E., Murvai, N., Tantos, A., Kalmar, L., et al. (2019). Intrinsically disordered linkers impart processivity on enzymes by spatial confinement of binding domains. *International Journal of Molecular Sciences*, *20*(9), 2119. https://doi.org/10.3390/ijms20092119.

van Dongen, E. M. W. M., Evers, T. H., Dekkers, L. M., Meijer, E. W., Klomp, L. W. J., & Merkx, M. (2007). Variation of linker length in ratiometric fluorescent sensor proteins allows rational tuning of Zn(II) affinity in the picomolar to femtomolar range. *Journal of the American Chemical Society, 129*(12), 3494–3495. https://doi.org/10.1021/ja069105d.

van Rosmalen, M., Krom, M., & Merkx, M. (2017). Tuning the flexibility of Glycine-serine linkers to allow rational design of multidomain proteins. *Biochemistry, 56*(50), 6565–6574. https://doi.org/10.1021/acs.biochem.7b00902.

Walther, K. A., Grater, F., Dougan, L., Badilla, C. L., Berne, B. J., & Fernandez, J. M. (2007). Signatures of hydrophobic collapse in extended proteins captured with force spectroscopy. *Proceedings of the National Academy of Sciences of the United States of America, 104*(19), 7916–7921. https://doi.org/10.1073/pnas.0702179104.

Zhou, H.-X. (2001). The affinity-enhancing roles of flexible linkers in two-domain DNA-binding proteins. *Biochemistry, 40*(50), 15069–15073. https://doi.org/10.1021/bi015795g.

Zhou, H.-X. (2003a). Quantitative account of the enhanced affinity of two linked scFvs specific for different epitopes on the same antigen. *Journal of Molecular Biology, 329*(1), 1–8. https://doi.org/10.1016/S0022-2836(03)00372-3.

Zhou, H.-X. (2003b). How often does the myristoylated N-terminal latch of c-Abl come off? *FEBS Letters, 552*(2–3), 160–162. https://doi.org/10.1016/S0014-5793(03)00911-6.

Zhou, H.-X. (2004). Polymer models of protein stability, folding, and interactions. *Biochemistry, 43*(8), 2141–2154. https://doi.org/10.1021/bi036269n.

CHAPTER SEVEN

ER/K-link—Leveraging a native protein linker to probe dynamic cellular interactions

Tejas M. Gupte, Michael Ritt, and Sivaraj Sivaramakrishnan*

Department of Genetics, Cell Biology, and Development, College of Biological Sciences, University of Minnesota, Minneapolis, MN, United States

*Corresponding author: e-mail address: sivaraj@umn.edu

Contents

Methods in Enzymology, Volume 647
ISSN 0076-6879
https://doi.org/10.1016/bs.mie.2020.10.002

Abstract

ER/K α-helices are a subset of single alpha helical domains, which exhibit unusual stability as isolated protein secondary structures. They adopt an elongated structural conformation, while regulating the frequency of interactions between proteins or polypeptides fused to their ends. Here we review recent advances on the structure, stability and function of ER/K α-helices as linkers (ER/K linkers) in native proteins. We describe methodological considerations in the molecular cloning, protein expression and measurement of interaction strengths, using sensors incorporating ER/K linkers. We highlight biological insights obtained over the last decade by leveraging distinct biophysical features of ER/K-linked sensors. We conclude with the outlook for the use of ER/K linkers in the selective modulation of dynamic cellular interactions.

1. Structure, stability and functional characteristics of ER/K linkers

Seminal characterization of ER/K linker structure and function have been previously reviewed in Swanson et al. (Swanson & Sivaramakrishnan, 2014). This section highlights the significance of protein α-helices in folded proteins, followed by recent advances in our structural understanding of Single Alpha Helical (SAH) domains, resources that have emerged to identify SAH domains in native proteins, and concludes with three current studies that feature unique functional characteristics of ER/K linkers in cellular proteins.

1.1 α-helices are a dominant structural element in proteins

α-helices, β-sheets and random coils are the most common elements of secondary structure in proteins. α-helices are formed and maintained by backbone interactions parallel to the primary axis of the helix. These interactions are hydrogen bonds between the carbonyl oxygen and amino nitrogen of the *i*th and $i+4$th amino acids. The side chains of all residues in the α-helix are directed outwards and away from the helical axis, and the occurrence of polar or charged side chains in the helix can facilitate additional interactions with other side chains in the helix or with other elements outside of the helical structure, imparting further stability (Pauling, Corey, & Branson, 1951). Consequently, α-helices are the most commonly occurring secondary structure, representing 30% of the structure of the average globular protein (Pace & Scholtz, 1998). In globular domains, α-helices can also pack with β-sheets in different arrangements. Among these, orientation of the α-helix along the strands of β-sheets is energetically most favored, followed in stability by a perpendicular orientation of the helical axis to the β-strands (Chou, Némethy, Rumsey, Tuttle, & Scheraga, 1985). α-helices are also the core component of coiled-coil domains and of transmembrane bundles. Coiled-coil domains consist of two to seven α-helices, where the separate helices stabilize each other through hydrophobic patches of side chain interactions following a "knobs-into-holes" motif (Crick, 1953). Multiple α-helices composed primarily of hydrophobic amino acids can also give rise to transmembrane bundles, through protein-protein and protein-lipid interactions (Lee, 2003).

1.2 Stability of a protein α-helix

The stability of a protein α-helix is affected by the number of amino acids involved in backbone interactions and the helical propensity of each amino acid. The helix propensity (Pα) is denoted on a scale of 0 to 1 and expresses the tendency of an amino acid to occur in an α-helix. Not considering the ends of a helix, Alanine residues have the highest propensity of occurrence in the central, solvent exposed residues, and Glycine residues have the lowest helix propensity, while Proline residues tend to disrupt helices (Pace & Scholtz, 1998). Intra-helical interactions between side-chains also contribute to the helical state and their complete absence destabilizes the helix (Zangi, 2014). One extreme of the stabilization of an α-helix is a scenario in which the strength and number of intra-helical interactions are sufficient to maintain the helical state of a peptide, even when isolated from the tertiary folded context of a protein i.e., interactions with other secondary structural elements are not required to stabilize the α-helix. The C-peptide of ribonuclease A was observed to form such an α-helix when isolated in aqueous solution (Bierzynski, Kim, & Baldwin, 1982). Intra-helical stabilization through side-chain interactions was tested by engineering Alanine-based peptides of 16 or 17 residues, containing three pairs of Glu-Lys residues (Marqusee & Baldwin, 1987). Based on the spacing between the Glu and Lys residues, the peptides were designated as either $i+4$ or $i+3$. $i+4$ denotes that the Glu and Lys are separated by three residues (EAAAK), with the subsequent Glu following immediately (i.e. EAAAKEAAAK...); while in the $i+3$ peptides, the Glu and Lys were separated by two residues with one Ala residue separating the subsequent Glu $(EAAKA)_3$. This study revealed that the $(i, i+4)$ spacing (i.e. $(EAAAK)_3$) yielded significantly higher helicity as compared to $(i, i+3)$ spacing ($(EAAKA)_3$) between the Glu and Lys residues. It was hypothesized that the $i+4$ spacing provided better stability to the helix due to the Glu (E) and Lys (K) side chain rotamers (Ponder & Richards, 1987). Extending this observation, efforts were made to understand α-helix formation through charge interactions, independent of the greater helical propensity of Alanines (Lyu, Gans, & Kallenbach, 1992). 18-mer peptides were synthesized that contained repeating units of either E_4K_4 or E_2K_2. Though both peptides had the same chemical composition, only the E_4K_4 peptide formed a stable α-helix (Lyu et al., 1992). The periodicity of a protein α-helix (3.5 residues per turn) would permit stabilizing electrostatic interactions between E and K in the E_4K_4 but not in the E_2K_2 peptide, consistent with the observations from (i, $i+4$) helices.

Collectively, these studies on the intrinsic stability of α-helices led to the identification of other peptides that exist as isolated stable single α-helices or SAH domains. The high content of charged and polar residues in these peptides provides additional stability to the helix, and (i, $i+4$) spacing between side chain pairs provides optimal helicity. Such stable peptides are enriched in Glu (E), Lys (K) and Arg (R) arranged in a manner that they exhibit salt bridges between side-chains at the i, $i+3$ and i, $i+4$ position, in addition to the backbone hydrogen bonds. These intra-helix salt bridges provide additional stability allowing the SAH peptides to retain their secondary structure without relying on interactions with other regions of a larger protein structure (Scholtz & Baldwin, 1992). Even peptides as short as five to eight residues with i, $i+3$ and i, $i+4$ salt bridges were shown to exhibit high helicity (Sommese, Sivaramakrishnan, Baldwin, & Spudich, 2010). Investigations into the differences in contributions of Glu-Lys and Glu-Arg pairs to α-helix stabilization were performed with short (Sommese et al., 2010) and long peptides (Wolny et al., 2017). Among the short peptides, E_4R_4 peptides were found to have higher helix content than E_4K_4 peptides (Sommese et al., 2010). More recently, a 98 residue polypeptide was created that was composed of repeating units of the 7-residue sequence AEEEXXX, where X is K or R. Only the Lys-rich sequences EK3 (AEEEKKK) or EK2R1 (AEEEKRK) were found to form a SAH in solution. Increasing the Arg content in the peptide (EK1R2—AEEERKR, or ER3—AEEERRR) resulted in aggregation of the synthesized domain. Molecular modeling and PDB analysis indicated that Glu and Arg were more dynamic than Glu-Lys pairs (Wolny et al., 2017). As a result, a greater fraction of the charged side-chains were involved in intra-molecular salt bridges, which dramatically reduced polar interactions between the peptide and water molecules. These factors combined to limit the solubility of the Glu-Arg peptides.

Similar SAH domains occur naturally in a variety of proteins, often as linkers of disparate domains (Peckham, 2011; Swanson &

Sivaramakrishnan, 2014). To distinguish SAH domains in general from those that contain primarily E, R and K residues, the term ER/K α-helix was introduced (Sivaramakrishnan, Spink, Sim, Doniach, & Spudich, 2008) and it is referred to as an ER/K linker when it is fused to polypeptides at both of its ends (Sivaramakrishnan et al., 2008). The relevance of Lys and Arg content is reflected in the sequences of native ER/K linkers, which contain either Lys or a mixture of Arg and Lys (Sivaramakrishnan & Spudich, 2011). Consistent with the increase in stability arising from Glu-Arg pairs, K→R substitutions in the naturally occurring myosin VI SAH, which constitutes an ER/K α-helix, led to an increase in thermostability (Wolny et al., 2017). Overall, increasing the R:K ratio of ER/K linkers enhances helix stability while potentially compromising protein function through aggregation.

Further experimental insights into the contribution of Arg and Lys residues to the stability of native SAH domains emerged from a recent NMR study (Batchelor et al., 2019). The authors validated that residues 858–935 of murine Myosin 7a (M7a) indeed formed a SAH in solution, as was originally predicted by Yang et al. (2009). NMR experiments were used to understand the nature of bonds that contributed to conformational stability, and very few "through" hydrogen bond couplings were detected, indicating that the M7a SAH domain is stabilized either by a dynamic network of ion pairs, or by a preference for weaker, solvent-bridged ion pairs on the outside of the helix. Molecular dynamics (MD simulations) were used to test these mutually exclusive possibilities. The simulations revealed that each Arg residue had the ability to interact with a larger network of Glu side chains than a corresponding Lys residue. For example, in a representative sequence ExxRxxEE starting at the *i*th position, the central Arg residue ($i+3$) can form simultaneous bonds with Glu residues at both the $i+6$ and $i+7$ positions. The same Arg residue can dynamically alter its rotamer conformation to form simultaneous bonds with Glu at i and $i+7$ positions. This dynamic flexibility of the ion pair network arising from Glu-Arg interactions significantly enhances the stability of the SAH domain. Lys residues do not display this dynamic stability in their ion pair network, contributing to the stability of ER/K α-helices to a lesser extent.

1.3 Identifying SAH domains

SAH domains have very distinct properties compared to other α-helices such as coiled-coil repeats. The functional differences between SAH domains and other α-helices can significantly affect interpretation of protein function. Hence, for biologists attempting functional characterization of a protein of interest, a quick check for the presence of SAH domains would be useful. However, many structural analysis tools that provide protein domain analysis misidentify SAH domains as either coiled-coil sequences or as being related to other protein families. For example, the MOTIF tool (https://www.genome.jp/tools-bin/search_motif_lib) indicates that the SAH domain of Myosin VI is related to the Macoilin family of transmembrane proteins. Two in silico tools—CSAH—for charged single alpha helices (Süveges, Gáspári, Tóth, & Nyitray, 2009) and Waggawagga-CLI (Simm & Kollmar, 2018)—have been created with the aim of predicting the propensity of an amino acid sequence to form SAH domains.

CSAH is applied to amino acid sequences to predict the length and probability that a contiguous stretch of amino acids could form a single α-helical segment. There are two methods available for this tool, SCAN4CSAH and FT_CHARGE. SCAN4CSAH detects ionic interactions between the *i*th and i + 3rd residues, and the *i*th and i + 4th residues to score the sequence for α-helix stabilization. The second method used by CSAH is FT_CHARGE, in which Fourier transformation is applied to the ionic state of individual side chains in the sequence to identify repeated charge patterns, separated by seven residues. Quality control for these methods involves checking the Chou-Fasman Helical propensity parameters (Pα) to eliminate hits that are unlikely to form α-helices. This tool has been provided as a web server (http://csahserver.itk.ppke.hu/cgi-bin/csahserver.cgi) that predicts and identifies SAH-containing proteins and analyzes full UniProt submissions to update the repertoire of proteins containing SAH domains on a monthly basis (Gáspári, Süveges, Perczel, Nyitray, & Tóth, 2012).

A second tool, Waggawagga-CLI provides a genome-wide prediction of SAH-containing proteins (Simm & Kollmar, 2018). This method first uses the *helical net* diagram which represents the peptide sequence and residue level interactions in a two-dimensional format, along with the helical axis (Dunnill, 1968). Next, the interactions between residues are drawn and their effects are calculated. Three types of interactions are considered (1) binary interactions between residues at the $(i, i+3)$ and $(i, i+4)$ positions, (2) occurrence of pairs of hydrophobic amino acids in the $(i+3, i+6)$, $(i+3, i+7)$, $(i+4, i+7)$ and $(i+4, i+8)$ positions and (3) occurrence of pairs of oppositely charged amino acids at the positions listed above. The binary interactions are classified as having stabilizing, α-helix supporting or destabilizing effects. The stabilizing interactions are further differentiated as being strong, medium, or weak. The interactions are scored for defined window sizes, summed, and normalized to an ideal SAH sequence (e.g. EEEKKK). A default score is set as the cut-off to determine the likelihood of any amino acid being part of the SAH domain. For the entire SAH domain composed of multiple such windows, three criteria are evaluated to determine the possibility of SAH formation—(1) a sequence longer than 14 amino acids, (2) at least 80% amino acids have SAH score more than the cut-off and (3) amino acids at the beginning and end must have an SAH score greater than the cut-off value. This approach differs from the SCAN4CSAH algorithm in that it does not consider destabilizing interactions between identically charged residues in $i+3$, $i+4$ positions and between oppositely charged amino acids at i, $i+1$, $i+2$ positions. The Waggawagga-CLI approach has increased the frequency of detection of SAH domains from 0.2% by CSAH to between 0.5% and 3.5% of the total proteins in the genomes of different organisms. However, experimental validation of these computationally detected SAH domains will be necessary to compare the accuracy of the two methods.

1.4 Functional significance of SAH domains in native proteins

Characterization of the SAH domains of muscle caldesmon, apoptotic protein PDCD5, ribosomal protein L9 and myosin VI have been discussed previously (Swanson & Sivaramakrishnan, 2014). Subsequently, SAH domains have been identified in other proteins including Rpn12 subunit of the proteasome lid (Tomko et al., 2015), MFAP1 and Snu23 proteins that interact with Prp38 (Ulrich, Seeger, Schütze, Bartlick, & Wahl, 2016), Myosin X (Wolny et al., 2014), inner centromere protein INCENP (Samejima et al., 2015), and the Gag protein of Murine Leukemia Virus (MLV) (Doležal et al., 2016). Here we summarize the findings of mechanosensitivity of the SAH domain in Myosin X and INCENP, and the rotational flexibility enabled by the SAH domain in the Gag protein.

Myosin X was shown to contain an ER/K-rich SAH domain, linking the motor domain at the N-terminus of the protein to a coiled-coil, and a FERM domain at the C-terminus (Knight et al., 2005). A recent study by Wolny et al., provides experimental evidence that the SAH domain of myosin X behaves like a constant force spring, undergoing mechanical extension. A 97-residue peptide from myosin X was verified to be an SAH based on helicity, non-cooperative thermal unfolding and high efficiency of refolding (Wolny et al., 2014). Atomic force microscopy (AFM) was used to record the changes in the peptide upon application of a constant force. Using this technique, the SAH domain unfolded at forces

<30 pN in a non-cooperative manner. Simulations suggested that the application of a deforming force leads to mechanical extension of the SAH domain at constant force. Further, simulations also indicated that the domain would refold when the force was removed and the structure allowed to shorten. In the cell, myosin X functions as dimer and plays a key role in the biogenesis of filopodia through actin convergence, contributing to the formation of characteristic parallel actin bundles observed in filopodia (He, Sakai, Tsukasaki, Watanabe, & Ikebe, 2017), and through cargo transport into the filopodia (Kerber & Cheney, 2011). Unfolding of the SAH domains would allow both myosin heads in the dimer to stay attached to separate actin filaments as they are being pulled apart during actin remodeling. In the absence of this feature, the force experienced during this actin reorganization could cause either the motor domain to unfold and detach, or the coiled-coil dimerization domain to unwind, dissociating the functional dimer. Physical extension of the SAH domain would also allow myosin X to overcome physical obstructions introduced by actin cross-linking proteins during cargo transport along parallel actin bundles in the filopodia. The constant force spring characteristic identified for the myosin X SAH domain is likely to be relevant for other ER/K-linkers in mechanosensitive cytoskeletal proteins.

The SAH domain of INCENP regulates Aurora-B kinase activation during mitosis. INCENP and Aurora-B function together in the Chromosomal Passenger Complex (CPC), which regulates mitosis by removing improper attachments at the kinetochore, and also by activating the spindle assembly checkpoint (Carmena, Wheelock, Funabiki, & Earnshaw, 2012). A central domain of INCENP (503–715) was previously suggested to form a coiled-coil (Mackay, Eckley, Chue, & Earnshaw, 1993). However, coiled-coil formation in INCENP would enforce Aurora-B dimerization, leading to unregulated kinase activation through transphosphorylation (Sessa et al., 2005). Further, given the elongated structure of INCENP, the rigidity of a coiled-coil would restrict the ability of Aurora-B kinase to access its substrates in the outer kinetochore. These disagreements were resolved by a recent study demonstrating that the putative coiled-coil in INCENP is in fact a SAH domain (Samejima et al., 2015). The monomeric nature of the SAH domain would not support spontaneous formation of Aurora-B dimers. Further, the N-terminus of the INCENP SAH domain binds to microtubules, and the authors hypothesize that forces arising from microtubule—kinetochore interaction could cause mechanical unfolding of the SAH domain. The extension from this unfolding would allow

Aurora-B to access substrates in the outer kinetochore, further away from its CPC tethering site. Thus, the presence of a SAH domain in the central region of INCENP prevents the formation of obligate Aurora-B dimers, and indicates a mechanism for force/tension-dependent phosphorylation of substrates in the outer kinetochore. However, the factors that limit trans-phosphorylation of Aurora-B following the proximity induced recruitment to microtubule-bound INCENP-SAH remain unclear.

In select retroviruses, a SAH domain is critical for virion assembly, where its rotational flexibility facilitates the conformational changes associated with capsid maturation. The retroviral genome of murine leukemia virus (MLV) is enclosed by nucleocapsid domains (NC), which are connected to the capsid domain (CA) by a stretch of amino acids termed the charged assembly helix (CAH) (Cheslock et al., 2003). The structure-function relationships between these domains are critical for successful viral assembly, since deletion of CAH drastically reduces the number of assembled virions. A previous model assumed that the CAH would assemble as a coiled-coil and facilitate virus assembly. However, this model does not explain critical features in the maturation of the MLV capsid. In MLV, the interfaces of CA domains are largely constant between the immature and mature capsids (Qu et al., 2018). Hence, the assumption that the CAH segment constitutes a coiled-coil was evaluated through biophysical characterization by Doležal et al. This region of 41 amino acids exhibits helicity, thermal unfolding and refolding, and conformation consistent with a SAH, while remaining monomeric in solution (Doležal et al., 2016). A coiled-coil conformation of the CAH would have greater rigidity whereas a SAH linking CA and NC domains is theorized to facilitate rotation of the protein domains enabling capsid maturation.

2. ER/K-linked sensors—Mechanism, molecular cloning, protein expression, and measurement of interaction strengths

A subset of SAH domains contain a characteristic motif, which comprises approximately four glutamic acid residues (E_4) followed by approximately four basic residues that are either arginine or lysine $(R/K)_4$. A repeating sequence of this $E_4(R/K)_4$ motif creates a SAH termed an ER/K-helix. Naturally occurring ER/K helices of known length have been repurposed to tether two protein/polypeptide domains together (Sivaramakrishnan & Spudich, 2011). In this context, the ER/K helix is termed the ER/K linker. In this section, we describe the use of ER/K linkers to detect and modulate existing protein-protein interactions through the use of engineered sensors.

2.1 Sensor mechanism

The ER/K α-helix functions as a worm-like chain (WLC) with rare, stochastic breaks proportional to helix length (Sivaramakrishnan et al., 2008; Sivaramakrishnan & Spudich, 2011). Initial biophysical characterization of the ER/K α-helix using circular dichroism and small-angle x-ray scattering (SAXS) revealed an elongated structure consistent with an isolated, stable, extended α-helix (Sivaramakrishnan et al., 2008). Subsequent single molecule optical tweezers measurements, combined with biophysical modeling suggested that the ER/K α-helix behaves like a worm-like chain (WLC) with persistence length ($L_P \sim 15$ nm) (Sivaramakrishnan et al., 2009). For an ideal WLC with lengths comparable to L_P the probability of proteins at the ends interacting with each other is negligible (probability $\sim 10^{-8}$) (Mehraeen, Sudhanshu, Koslover, & Spakowitz, 2008). Consequently, if ER/K linkers of length 10–30 nm behaved as an ideal WLC, it would function simply as a spacer for proteins/peptides at its ends. However, tethering a well-characterized interacting pair (CaM-peptide) at the ends of ER/K α-helix demonstrated a robust calcium-dependent interaction as measured by a fluorophore FRET pair fused to the interacting proteins (eCFP-CaM; Citrine-C15W CaM interacting peptide) (Fig. 1A and B). Control constructs that lacked the peptide displayed insignificant FRET regardless of Ca^{2+} concentration. The dependence of FRET upon an interaction between the moieties at the ends of the linker is consistent with linker lengths substantially greater than the Förster's radius of the FRET pair (R_o for CFP/YFP ~ 5 nm). Hence, while the ER/K linker does function as a spacer, it can also facilitate protein-protein interactions with proteins/peptides tethered to its ends. Given this divergence from an ideal WLC, it was proposed that rare stochastic breaks in the elongated α-helix could enable interactions. Regardless of underlying mechanism, subsequent studies have repeatedly demonstrated the ability of the ER/K linker to facilitate interactions.

To understand the relationship between FRET ratio and the interaction state of proteins at the ends of the ER/K linker, the interacting proteins and their associated fluorophores (eCFP-CaM; Citrine-C15W) were separated by TEV protease treatment targeting a TEV-site engineered at the

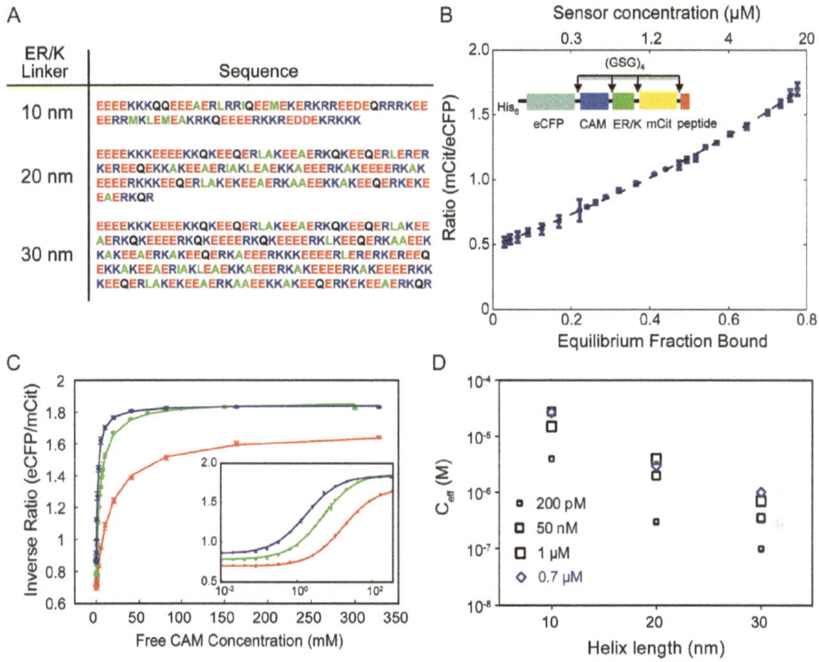

Fig. 1 ER/K linkers modulate the effective concentration of protein-protein interactions. (A) Sequence of 10, 20 and 30 nm linkers used in ER/K-linked sensors. (B) FRET ratio (mCit acceptor/eCFP donor) of ER/K-linked CaM-peptide sensor correlates linearly with fraction of sensors with proteins in the bound state. (C, D) Effective concentrations (C_{eff}) engineered by 10, 20 and 30 nm ER/K linkers determined from (C) concentration-dependent quenching of FRET by unlabeled, free CaM protein. (D) C_{eff} for CaM interaction with peptides of indicated binding affinity (K_D listed) and 10, 20 and 30 nm ER/K linker length. For all experiments: $N=4$, and data are shown as mean ± SD. *Panels (A–D) adapted from Sivaramakrishnan, S., & Spudich, J. A. (2011). Systematic control of protein interaction using a modular ER/K α-helix linker.* Proceedings of the National Academy of Sciences of the United States of America, 108*(51), 20467–20472.*

N-terminus of the ER/K linker (Sivaramakrishnan & Spudich, 2011). A dilution series of the TEV-cleaved sensor was used to measure the concentration dependence of FRET ratio of the equivalent bi-molecular interaction (Fig. 1B). Fitting this concentration-dependent measurement to the classic quadratic equation of bi-molecular association yields a linear relationship between sensor FRET ratio and the fraction of protein in the bound state (Fig. 1B). Given that the elongated ER/K linker maintains minimal FRET in the absence of an interaction, these measurements demonstrate that the sensor FRET ratio is a surrogate for the fraction of sensors with the flanking protein/peptides in the interacting state.

To assess the effective concentration engineered by the ER/K linkers, increasing concentrations of unlabeled CaM were added to quench sensor FRET in a dose-dependent manner (Fig. 1C). Given the linear relationship between sensor FRET ratio and fraction of protein in the interacting state, these dose-response curves were sufficient to measure the effective concentration (*Ceff*) of sensors engineered with different length ER/K linkers and CaM interacting peptides with varying affinity (Fig. 1D). Regardless of the CaM-peptide interaction affinity, increasing ER/K linker length systematically decreased *Ceff*, from ~10 μM for the 10 nm linker, to ~1 μM for the 20 nm linker, and ~100 nM for the 30 nm linker. To dissect the role of the ER/K linker on the kinetics of the interaction (on versus off rates), the rate of dissociation of a high affinity peptide (Trp3) from CaM was measured by quenching the interaction with a vast excess of unlabeled CaM (300 μM >> *Ceff*). Sensors displayed very similar dissociation rates, regardless of ER/K linker length, demonstrating that the linker controls association (k_{on}) rather than dissociation (k_{off}) rates (Sivaramakrishnan & Spudich, 2011). While the interacting proteins have distinct shapes and sizes, resulting in distinct configurations of interactions, 12 residue-long GSG linkers were included between all protein domains (including between the fluorophores and ER/K linker) to allow rotational flexibility. Consequently, the limiting factor for the protein-protein interaction are the structural properties of the ER/K linker. Extensive biophysical characterization has established that the ensemble conformation of the ER/K α-helix is consistent with a worm-like chain with persistence length (L_p) ~ 15 nm (Sivaramakrishnan et al., 2009). For ER/K linkers with lengths comparable to L_p, the ER/K linker makes an effective spacer between proteins tethered at its ends. However, infrequent stochastic breaks in the ER/K α-helix can also facilitate an interaction between proteins tethered to its ends. Given the relative homogeneity of ER/K linker sequence (E_4—R_4/K_4), the frequency of breaks is proportional to linker length, whereas the probability that the ends will be within 2 nm distance of each other, and thus allow an interaction, is inversely proportional to the square of the linker length, as the breaks are unlikely to be spatially coordinated. These factors together yield the observed inverse relationship between k_{on} and C_{eff} on one hand, and linker length on the other (Fig. 1D). Further, given that the ER/K linker length is significantly larger than most protein domains/polypeptides (>10 nm compared to ~3 nm for a 66 kDa globular protein such as Bovine Serum Albumin), the linker length dependence of association rates and consequently effective concentrations is likely to persist across protein-protein interactions.

2.2 Sensor plasmid design

ER/K-linked sensors are created by cloning distinct protein domains/polypeptide components into sequential sites of the multiple-cloning-site (MCS) of different vectors. The diversity of plasmid constructs available enables rapid generation of sensor plasmids for new protein-protein interactions using different expression systems. A range of sensor plasmids generated by the Protein Acrobatics Lab, along with detailed plasmid restriction maps are available through AddGene (https://www.addgene.org/Sivaraj_Sivaramakrishnan/). In all layouts of ER/K-linked sensors, the individual components are all separated from each other by 2–4 (Gly-Ser-Gly) linker segments. These short linkers impart rotational flexibility, distinct from the distance constraints programmed by the ER/K linker. This rotational flexibility is designed to allow the proteins tethered at either end of the linker to interact without being sterically limited by the other components.

Within the broad framework of sensor designs with the ER/K linker in the middle, flanked by a peptide-fluorophore combination on either side, multiple variations have been reported. In the earliest work that reported development of an ER/K-linked sensor, the ER/K linker was cloned between *Eco*RI and *Asc*I, with an N-terminal TEV protease site (ENFLYQG/S) (Sivaramakrishnan & Spudich, 2011). Treatment with bacterially expressed TEV protease generates the equivalent bi-molecular interaction through selective proteolytic cleavage at the protease site. The TEV protease site can be used to compare FRET between the unimolecular context of the ER/K-linked sensor and the bi-molecular protein-protein interaction following TEV treatment. It can also be used as a purification tool, especially in context of catalytically active enzymes, whose activity and consequently expression are actively suppressed. For instance, the isolated kinase domain of constitutively active kinases such as PKCs express poorly by themselves, and are therefore expressed as part of an ER/K-linked sensor linked to a substrate peptide or pseudosubstrate domain (Sommese & Sivaramakrishnan, 2016). Following expression as a tandem protein, TEV treatment is used to cleave, separate, and purify the kinase domain for biochemical applications.

The repeating nature of the ER/K sequence makes plasmid DNA constructs susceptible to recombination during plasmid amplification in *E. coli*. Recombination deficient bacterial strains (*Stbl2* and *JM109*) have been used selectively for DNA amplification, especially when plasmid recombination is detected in vector digests or DNA sequencing. To minimize recombination effects, most sensor plasmid clones have maintained the original

*Eco*RI-*Asc*I fragment of the ER/K linker with the N-terminal TEV site. Rather than PCR amplify the ER/K linker, restriction digests have been used to laterally transfer this fragment to other plasmid vectors. Nonetheless, plasmid designs containing *Bam*HI/*Asc*I sites as well as *Not*I/*Avr*II sites have subsequently been generated to avoid the limitations posed by the presence of *Eco*RI in either the plasmid vector backbone or the polypeptides being studied. *Eco*RI is a 6-cutter and a widely used restriction enzyme, whereas *Asc*I is an 8-cutter and is relatively infrequent.

N- and C-terminal protein fusions to an ER/K linker involve a diversity of multiple cloning sites (MCS) to accommodate both the interacting polypeptide and fluorophores. The restriction site layout in the original ER/K-linked sensors was subsequently used to create constructs linking a variety of protein domains as depicted in Fig. 2 and outlined next. Sensors for monitoring the conformational state of Focal Adhesion Kinase (FAK) were created by including from N-to-C an mCitrine (mCit), the FERM domain of FAK, the ER/K linker, an mCerulean (mCer) and the catalytic domain of FAK, in both pBiex1 (insect cell expression) and pcDNA/FRT (mammalian cell) expression vectors (Ritt, Guan, & Sivaramakrishnan, 2013). An alternate design incorporated (from N-to-C) a GPCR between *Hin*dIII-*Xba*I, followed by mCit between *Xba*I and *Eco*RI. As in the previous case, the ER/K linker was cloned between *Eco*RI/*Asc*I, followed by mCer between *Asc*I/*Pac*I, and finally the Gα subunit or a C-terminal peptide encompassing the α5 helix of the Gα subunit (α5-peptide) between *Pac*I/*Not*I (Malik et al., 2013). In the GPCR sensors, both fluorophores flanked the ER/K linker. An alternate 'fluorophores-outside' design was adopted in sensors linking the PKC catalytic domain to substrate peptides (Sommese & Sivaramakrishnan, 2016). Yet another variation incorporates two copies of the FRET acceptor, one on each side of the peptide moiety at the C-terminus of the ER/K linker (Ritt & Sivaramakrishnan, 2016).

2.3 Sensor expression and purification from bacterial and insect cells

ER/K-linked sensors have been engineered in commonly used vectors for protein expression and purification including pET47b and pBiex1 that allow for expression and purification from bacteria and Sf9 cells. For purification purposes, the original ER/K linker based Ca^{2+} sensors were cloned with an N-terminal hexaHistidine-tag and purified using Ni-NTA chromatography (Sivaramakrishnan & Spudich, 2011). Ni-NTA purification requires multiple chromatography steps, given the non-specific binding of proteins

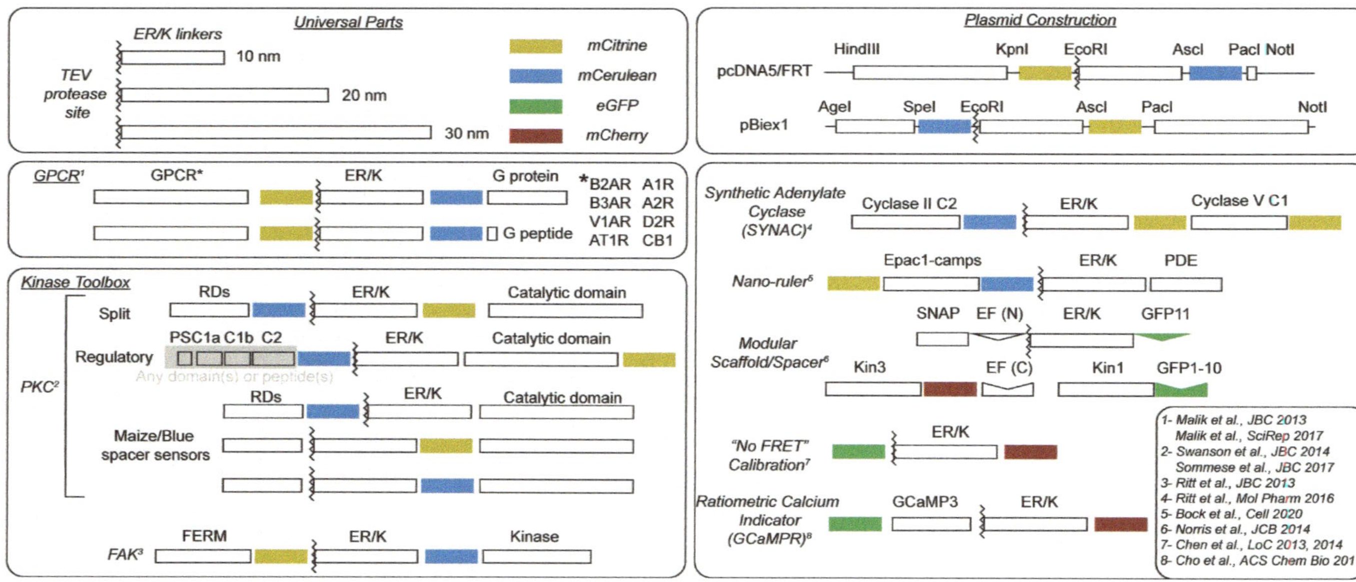

Fig. 2 Schematics of sensors utilizing ER/K linkers. Generalized schematics of published sensor proteins that incorporate the ER/K linker. ER/K linkers have successfully been used to generate sensors employed to investigate cellular interactions involving GPCRs, kinases, motors, and second messenger signaling. ER/K-linked sensors are constructed so that the domains can be interchanged with relative ease during molecular cloning using restriction sites between protein elements (top right). Select citations for each sensor (superscript numbering) are also listed (bottom right).

containing more than one histidine to the Ni-NTA column (Bornhorst & Falke, 2000). The consequent trade-off between yield and protein quality necessitates large expression titers afforded by bacterial expression. Nonetheless, limitations in protein folding and post-translational modifications require sensor expression in insect and mammalian culture system. In contrast, the FLAG purification strategy yields relatively homogenous sensor protein (>75% assessed by Coomasie staining) with modest expression titres (30mL Sf9 cultures). A 30mL culture of Sf9 cells, containing $3–5 \times 10^7$ cells at harvest can provide on average ~100μL of 1μM protein (yields are limited by the interacting polypeptides rather than the ER/K linker or fluorophores). Further, the high specificity of the anti-FLAG antibody compared to Ni-NTA provides relatively pure and intact protein (>80% as assessed by fluorescence gel scans) in a single step, without the use of chromatography equipment. A fused FLAG-hexaHis sequence allows purification with FLAG and antibody labeling or detection with hexaHis (Hariadi, Cale, & Sivaramakrishnan, 2014). Both N-terminal (Ritt et al., 2013) and C-terminal (Ritt & Sivaramakrishnan, 2016) FLAG-tags have been used, with equal efficacy for purifying ER/K-linked sensors.

Sensor plasmids are transiently expressed in Sf9 cells using Escort IV transfection reagent (Sigma). Individual FRET measurements in a fluorimeter (Fluoromax 4; Horiba) require 50μL of 5–50nM sensor, when used in conjunction with quartz microcuvettes. Thus, one 30mL culture is sufficient for ~40–400 individual readings. Alternately, fluorescence spectra can be measured in a plate reader format, which allows more conditions to be sampled in a shorter time, and with greater consistency (Lee et al., 2017). However, this method typically requires 50μL sample per well at a concentration of 100nM due to the relatively lower sensitivity of most plate reader instrumentation compared to individual cuvette measurements.

2.4 Sensor expression in mammalian cells

For experiments involving the use of ER/K-linked sensors in mammalian cells, pcDNA5/FRT plasmids containing the mammalian pCMV promoter were used (Malik et al., 2013; Ritt et al., 2013). Fugene HD (Promega) and XtremeGene HP (Roche) were used in early studies to promote high levels of expression that can be tuned by a combination of expression time and plasmid/transfection reagent titers (Malik et al., 2013). Appropriate cellular localization of the sensors is sensitive to many variables including cell passage number, length of expression and level of sensor expression. Assessing sensor localization is especially important when dealing with GPCR signaling with extracellular water-soluble ligand application (Malik et al., 2013) and also in monitoring changes in sensor relocalization in response to stimuli (Sommese, Ritt, Swanson, & Sivaramakrishnan, 2017). GPCR sensors with $\alpha5$ peptides derived from distinct Gα subunits displayed a wide variation in optimal expression time required to obtain significant expression, while maintaining plasma membrane localization (assessed by fluorescence microscopy of live cells). The variations in optimal expression times dictated that experiments be set-up at multiple overlapping time points, with subsequent selection of matched expression levels prior to experiments. As a cost-effective alternative to expensive transfection reagents, polyethyleneimine (PEI) mediated transfection was successfully used with ER/K-linked sensors (Gupte, Ritt, Dysthe, Malik, & Sivaramakrishnan, 2019). While this requires removal of PEI at 4h post-transfection, the use of PEI routinely provides >80% transfection efficiency (fraction of cells expressing sensor assessed by fluorescence microscopy) with appropriate membrane localization of ER/K-linked GPCR sensors.

2.5 Sensor measurements

Fluorescence emission spectra are routinely used to calculate a FRET ratio. For commonly used pairs of FRET probes, eCFP-mCitrine and mCerulean-mCitrine, the donor excitation wavelength is set to 430nm and an emission scan spanning 450nm to 600nm is obtained (Sivaramakrishnan & Spudich, 2011). The fluorescence intensity is noted at the emission maxima of the donor (475nm, designated emDonorMax) and at the emission maxima of the acceptor (525nm, emAcceptorMax). FRET ratio is the ratio of emAcceptorMax to emDonorMax. The FRET ratio has been shown to linearly correlate with the fraction of sensors in which the polypeptides flanking the linker are in the bound (interacting) state (Sivaramakrishnan & Spudich, 2011). Addition of one of the partners as a non-fluorescent variant leads to a loss of FRET and has been used to estimate the dissociation constant of the interaction (Sivaramakrishnan & Spudich, 2011). An alternate method relies on the use of time-resolved FRET or TCSPC (time correlated single photon counting), rather than steady-state FRET measurements. Based on light sources and spectral overlap, ER/K-linked sensors incorporating eGFP and mCherry, as FRET donor and acceptor, were employed with a pulsed diode laser and photomultiplier module to calculate the amplitude weighted lifetime $<\tau_{FRET}>$ and the efficiency of energy transfer as $1-<\tau_{FRET}>/\tau_{eGFP}$ (Sommese & Sivaramakrishnan, 2016).

3. Biological insights using ER/K-linked sensors

The application of ER/K-linked sensors has provided numerous insights into protein interactions underlying signal transduction. Here, we highlight distinct features of ER/K-linked sensors that have been used to circumvent the limitations of traditional biophysical and cell biological techniques.

3.1 Pairwise comparison of the relative strengths of weak interactions in live cells

The application of ER/K-linked sensors in detecting G protein selective GPCR conformations and engineering differential levels of FAK activation have been reviewed previously (Swanson & Sivaramakrishnan, 2014). While the prototypical GPCR, β2-AR, is conventionally thought of as a Gs-coupled receptor, various results indicate the possibility of Gi-coupling as well (Daaka, Luttrell, & Lefkowitz, 1997; Pierce, Premont, & Lefkowitz, 2002). It was therefore suggested that β2-AR could adopt multiple conformations, including a Gi-selective conformation (Kobilka, 2011). The α5 helix of the Gα subunit captures the majority of structural contacts (~75%) between receptor and G protein, and its engagement is essential for selective G protein activation (Chung et al., 2011). However, the receptor-Gα peptide interaction is inherently weak ($K_D > 10\,\mu M$) and transient (Gupte et al., 2019), and receptor conformation is influenced by the lipid microenvironment in live cells (Bruzzese, Gil, Dalton, & Giraldo, 2018; Strohman et al., 2019). Hence, control over effective concentration provided by the ER/K linker, enabled pair-wise comparison of the strength of the interaction of β2-AR with native peptides derived from α5 helices of distinct Gα subunits. As expected, the beta-blocker metoprolol diminished the β2-AR-Gαs peptide interaction (Fig. 3A and B), consistent with its previously reported role as an inverse agonist (Galandrin & Bouvier, 2006). In contrast, metoprolol, but not the full agonist isoproterenol, stimulated the β2-AR-Gαi peptide interaction. Hence, the use of an ER/K-linked sensor provided support for an inverse agonist at β2-AR inducing/stabilizing a Gi-selective conformation (Malik et al., 2013).

3.2 Identifying structural interaction hot-spots through site-directed-mutagenesis

While most GPCRs signal primarily through a single cognate G protein, the mechanistic basis for selectively activating the cognate versus structurally similar non-cognate G proteins remains an outstanding challenge. A seminal crystal structure of β2-AR in complex with cognate Gs provided high-resolution information about the interface for this interaction (Rasmussen et al., 2011). More recently, high-resolution structures for other cognate GPCR-G protein interactions have emerged from a combination of X-ray crystallography and cryo-electron microscopy (García-Nafría & Tate, 2019). However, given that interactions with non-cognate G proteins are significantly weaker (Gupte et al., 2019), there are currently no high-resolution structures of non-cognate GPCR-G protein interactions. In the absence of such structural information, molecular dynamics (MD) simulations were used to gain comparative insights into a cognate Gs receptor, β2-AR and a cognate Gq receptor, V1aR. MD simulations identified distinct orientations of the α5 helix between cognate and non-cognate interactions. To gain insights into potential structural hot-spots that drive G protein selectivity, ER/K-linked sensors were used to examine the effects of switching hot-spots from cognate to non-cognate using site-directed mutagenesis (Sandhu et al., 2019; Semack, Sandhu, Malik, Vaidehi, & Sivaramakrishnan, 2016). The pairwise control over effective concentration provided by the ER/K linker, revealed three hot-spot residues that are necessary and sufficient to tune selectivity between cognate and non-cognate G proteins (Fig. 3C–E). While the α5 helix has been extensively reported as a determinant of G protein selectivity, the combination of MD simulations and ER/K-linked sensors was able to precisely identify a structural basis of this phenomenon.

3.3 Determine the cause-effect relationships in GPCR priming

The cellular environment allows interactions of GPCRs with cognate G proteins, which are canonically activated by the ligand-bound GPCR. The GPCR can also interact with non-cognate G proteins, though the outcome of this weak, transient interaction is not known. Using an ER/K linker based sensor, we showed that facilitating non-cognate GPCR-G protein interactions paradoxically stimulated signaling downstream of the canonical pathway, a phenomenon that was termed GPCR priming (Gupte, Malik, Sommese, Ritt, & Sivaramakrishnan, 2017). However, GPCR networks are notorious for cross-talk, as diverse upstream activation events can converge on similar downstream factors. For example, activation of Gq and Gi pathways can result in an increase in cytosolic calcium. While Gαq directly stimulates PLC-β, to influence inositol triphosphate (IP_3) and IP_3-receptor (IP_3R), the Gi effect is mediated by the Gβγ dimer that

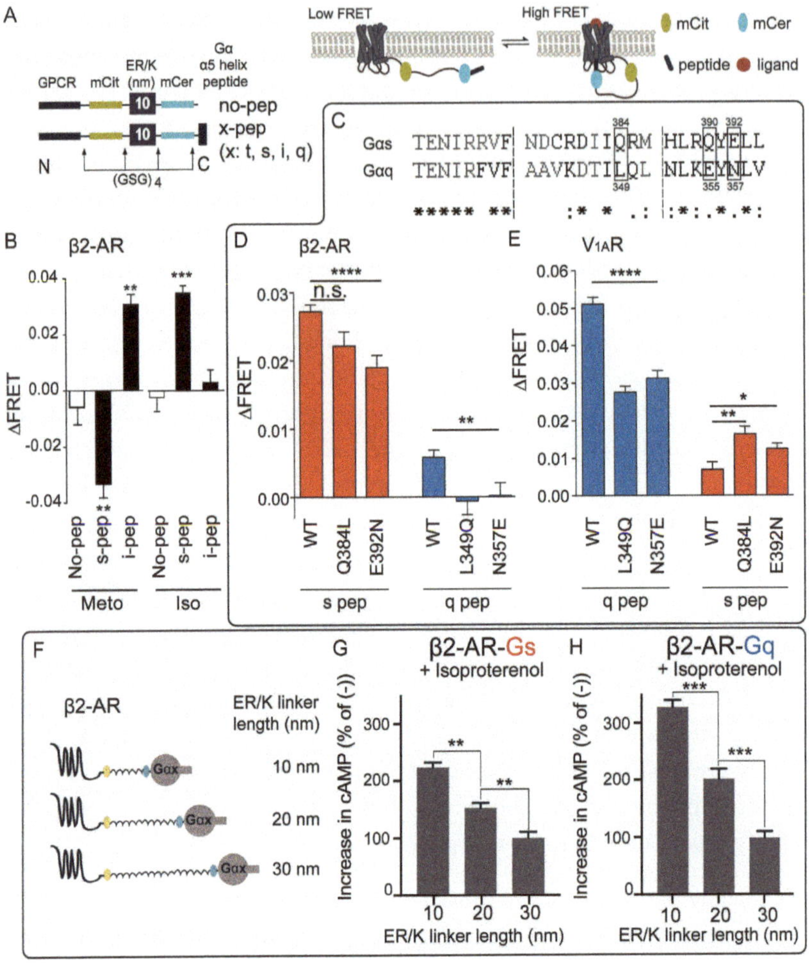

Fig. 3 GPCR biology investigated through the use of ER/K linkers. (A) Detailed schematic design of GPCR-Gα α5-helix peptide (Gα peptide) sensors (left). Cartoon rendering of GPCR-Gα peptide sensor interactions (right). (B) Metoprolol stimulation of the indicated GPCR-Gα peptide sensors. Metoprolol (Meto), an inverse agonist of β1-AR signaling in the heart, suppresses the β2-AR-Gαs peptide interaction while stimulating interaction between β2-AR and Gαi peptide. In contrast, isoproterenol (Iso), a full agonist is shown to stimulate only the canonical β2-AR-Gαs peptide interaction. (C) Sequence alignment of Gαs and Gαq α5-helix, from which the Gα peptides are derived. (D, E) The indicated residues in panel C have been mutated to their counterpart residue from the other peptide. β2AR, is a Gαs coupled receptor (D) and shows diminished interaction with S-peptide mutated to look like Q-peptide. V1AR, a Gαq coupled receptor (E) shows enhanced interaction to S-peptide mutated to look like Q-peptide. (F) Cartoon rendering of GPCR-Gα protein sensors with different length linkers. These protein

functionally dissociates from its active Gαi counterpart, and interacts with IP_3R (Zeng et al., 2003). Similarly, while the activation of Gα12/13 subunits exhibits cytoskeletal reorganization through Rho proteins, an increase in cAMP through Gαs activation can also lead to changes in cytoskeletal architecture (Ganguly, Saxena, & Chattopadhyay, 2011). Hence, there was a possibility that the increase in cAMP observed from tethering non-cognate Gαq to β2-AR (i.e. GPCR priming), arose from the convolution of downstream events. To address potential cross-talk between signaling pathways, the ER/K linker-length dependence of effective concentration of the intra-molecular interaction was leveraged. Specifically, three different length linkers (10, 20 and 30 nm) were used to engineer apparent concentrations of ~10 μM, 1 μM, and 100 nM respectively, while maintaining the stoichiometry of expression of GPCR and G protein (Sivaramakrishnan & Spudich, 2011). While high apparent concentrations (~10 μM) resulted in enhanced cAMP upon tethering Gαq, compared to the Gαs subunit, low concentrations (~100 nM) resulted in comparable responses that are indistinguishable from β2-AR sensors without any Gα subunit (Fig. 3F–H). Such experiments indicate that the modulation of downstream signaling results from an intra-molecular interaction between β2-AR and the tethered Gα subunit, and is not a consequence of the convolution of downstream signaling (Gupte et al., 2017).

sensors have the full-length G protein instead of the Gα α5-helix alone. (G, H) β2-AR sensors tethered to either full length Gαs (G) or full length Gαq (H) with varying length ER/K linkers. Both sensors show decreases in cAMP production with increasing length of ER/K linkers. For all experiments: *N* ≥ 3, *, *P* ≤ *0.05*; **, *P* ≤ *0.01*; ***, *P* ≤ *0.001*; ****, *P* ≤ *0.0001*; *n.s.* not significant. Significance determined using Student's *t*-test or ANOVA with Tukey's post-hoc test where appropriate. *Panels (A, left) and (B) adapted from Malik, R. U., Ritt, M., DeVree, B. T., Neubig, R. R., Sunahara, R. K., Sivaramakrishnan, S. (2013). Detection of G protein-selective G protein-coupled receptor (GPCR) conformations in live cells.* Journal of Biological Chemistry, 288*(24), 17167–17178. © the American Society for Biochemistry and Molecular Biology. Panels (A, right), and (C–E) adapted from Semack, A., Sandhu, M., Malik, R. U., Vaidehi, N., Sivaramakrishnan, S. (2016). Structural elements in the Gαs and Gαq C-termini that mediate selective GPCR signaling.* Journal of Biological Chemistry, 291*(34), 17929–17940. © the American Society for Biochemistry and Molecular Biology. Panels (F–H) adapted from Gupte, T. M., Malik, R. U., Sommese, R. F., Ritt, M., Sivaramakrishnan, S. (2017). Priming GPCR signaling through the synergistic effect of two G proteins.* Proceedings of the National Academy of Sciences of the United States of America, 114*(14), 3756–3761.*

3.4 Quantifying intra-molecular interactions in multi-domain proteins

Classical PKCs (cPKC) are lipid and Ca^{2+}-dependent AGC kinases that play critical roles in modulating a number of diverse physiological and pathophysiological processes, including cardiovascular performance and tumor generation (Newton, 2018). Despite their therapeutic significance, there are currently no isoform-specific inhibitors that can be used to selectively target disease states. The lack of isoform selectivity stems from inhibitor binding to a structurally conserved ATP-binding site within the kinase (Huse & Kuriyan, 2002). In contrast, biological modulation of kinase activity is often achieved through regulatory interactions. Multi-domain kinases, such as PKC have evolved intra-molecular regulatory domain interactions that are isoform-specific, with sequence-divergent binding interfaces (Steinberg, 2008). However, therapeutic targeting of these binding interfaces is limited by the knowledge of inter-domain interactions under basal and activated conditions. The kinase and regulatory domains are serially connected by flexible linkers that contribute to the conformational heterogeneity of the full-length protein, and have stymied efforts to obtain high-resolution structural information on inter-domain interactions. In the absence of structures of full-length PKCs, there has been significant confusion on the relative significance of intra-molecular regulatory interactions necessary to maintain basal auto-inhibition (Antal, Callender, Kornev, Taylor, & Newton, 2015; Leonard, Róycki, Saidi, Hummer, & Hurley, 2011; Sommese et al., 2017). Further, regulatory interactions can be interdependent and multiplexed, clouding interpretation of biochemical measurements of binary interactions between individual domains. Given these limitations, ER/K-linked sensors were used to probe the relative strengths of interactions between combinations of domains of PKCα under basal conditions (Sommese et al., 2017) (Fig. 4A). Sensor measurements revealed that the C1a regulatory domain functioned cooperatively with the pseudosubstrate domain to suppress basal kinase activity. The use of ER/K-linked sensors provided the added advantage of enabling recombinant kinase expression for measurement of protein interactions. The kinase domain alone expresses poorly in insect cells, whereas tethering it to regulatory domains, in particular the pseudosubstrate, using the ER/K linker substantially enhanced protein expression (~100 nM vs >2 μM from a 30 mL Sf9 culture). These ER/K-linked proteins were digested with TEV-protease to cleave an engineered protease site at the N-terminus of the ER/K linker, yielding a bi-molecular pair of kinase and regulatory domains fused respectively to a FRET acceptor and donor. Given the equal stoichiometry derived from the co-expression of the bi-molecular combination, measuring the FRET ratio in a serial dilution of the TEV digested fragments was sufficient to quantify an equilibrium dissociation constant for the kinase-regulatory domain combination (Sivaramakrishnan & Spudich, 2011; Sommese et al., 2017) (Fig. 4B).

3.5 Mapping inter-molecular interactions in PKC oligomers

Lipid/Ca^{2+} activation of PKCs results in their activation through membrane translocation. Studies using cross-linking, enzyme activity, and bi-molecular FRET sensors both *in vitro* and in live cells have shown that classical PKCs self-associate with high apparent affinity following activation (~5 nM) (Bonny et al., 2016; Slater et al., 2002; Swanson et al., 2014). Further, dynamic light scattering and fluorescence microscopy measurements revealed that they form distinct self-assembly states depending on the activating stimuli (Ca^{2+}, DAG, PIP_2) (Swanson et al., 2016). Macromolecular self-assembly has been observed for other signaling proteins including Ras (nano-clusters) (Tian et al., 2007) and Lck (Rossy, Owen, Williamson, Yang, & Gaus, 2013). In each instance, the amorphous nature of self-assembly precludes the use of traditional high-resolution structural biology techniques to characterize inter-molecule interactions. Further, potential synergies between weak inter-domain interactions complicate interpretation of domain-deletion experiments. In this context, a toolbox of ER/K-linked sensors provided a coarse map of key inter-domain interactions that drive distinct assembly states of activated PKCα. PKCα contains five conserved domains (PS, C1a, C1b, C2, and kinase) separated by four variable linker domains (V1–V4) (Steinberg, 2008). An ER/K linker flanked by mCerulean and mCitrine fluorophores was inserted individually, after each of the first four conserved domains (Swanson et al., 2014). FRET ratio measurements of recombinant sensor protein, expressed and purified from Sf9 insect cells, reported on the strength of the interaction between combinations of domains flanking the ER/K linker. These combinations capture potential synergies between interacting domains, such as PS and C1a/C1b interacting with the kinase domain (Swanson et al., 2014) (Fig. 4C). The use of fluorophores at both ends of the ER/K linker does not distinguish between intra- and inter-molecular interactions. Hence, bi-molecular pairs of sensors containing only mCerulean or mCitrine, either at the same or opposite ends of the ER/K-linker can be used to assess intra- vs inter-domain interactions in a protein oligomer. For instance, such a strategy revealed a folded-back interaction within an activated PKC oligomer (Swanson et al., 2014) (Fig. 4D–F). Last but not least, varying ER/K linker length tunes intra-molecular but not inter-molecular interaction and distinguish their relative roles in the context of an oligomer. Varying ER/K linker length from 10 to 30 nm, systematically reduces the concentration of an intra-molecular C1a-C2 interaction and diminishes oligomerization assessed by the number of fluorescent C1a-C2 punctae visualized by fluorescence microscopy (Swanson et al., 2016) (Fig. 4E–G). Taken together, these examples highlight features of ER/K-linked sensors that uniquely enable structural characterization of amorphous oligomeric assemblies.

3.6 Allosteric effects of small-molecule inhibitors on weak protein–protein interactions

Small-molecule inhibitors of enzymes such as protein kinases often target the nucleotide-binding site. While high affinity competitive displacement of nucleotide effectively shuts down kinase activity, this targeting mode suffers from a lack of inhibitor-kinase selectivity due to the structural conservation of the kinase domain for efficient catalysis (Taylor & Kornev, 2011). Further, while each individual kinase phosphorylates a range of protein substrates, only those substrates responsible for disease-associated signaling should be targeted to avoid side effects from disrupting normal physiological functions. Kinases bind substrates though a kinase-specific recognition motif surrounding the phosphorylated residue (Miller & Turk, 2018). Hence, small-molecule inhibitors that disrupt the kinase-substrate interaction have the potential to enhance both kinase and pathway specificity. Nonetheless, the kinase-substrate interaction is inherently weak and transient and involves an intrinsically disordered peptide sequence on the substrate. Consequently, the kinase-substrate interaction is not easily accessible to biophysical approaches. In this context, the ER/K-linked sensors have provided a robust assay to profile the relative interaction strengths of distinct substrate peptides for PKCα. Measurements using ER/K-linked sensors demonstrated an inverse correlation between kinase-substrate interaction strength and specific activity (Sommese & Sivaramakrishnan, 2016) (Fig. 5A and B). The sensors were also leveraged using site-directed mutagenesis to identify key interaction hot-spots on the substrate peptide that were predicted by molecular dynamics (MD) simulations (Lee et al., 2017). By integrating MD

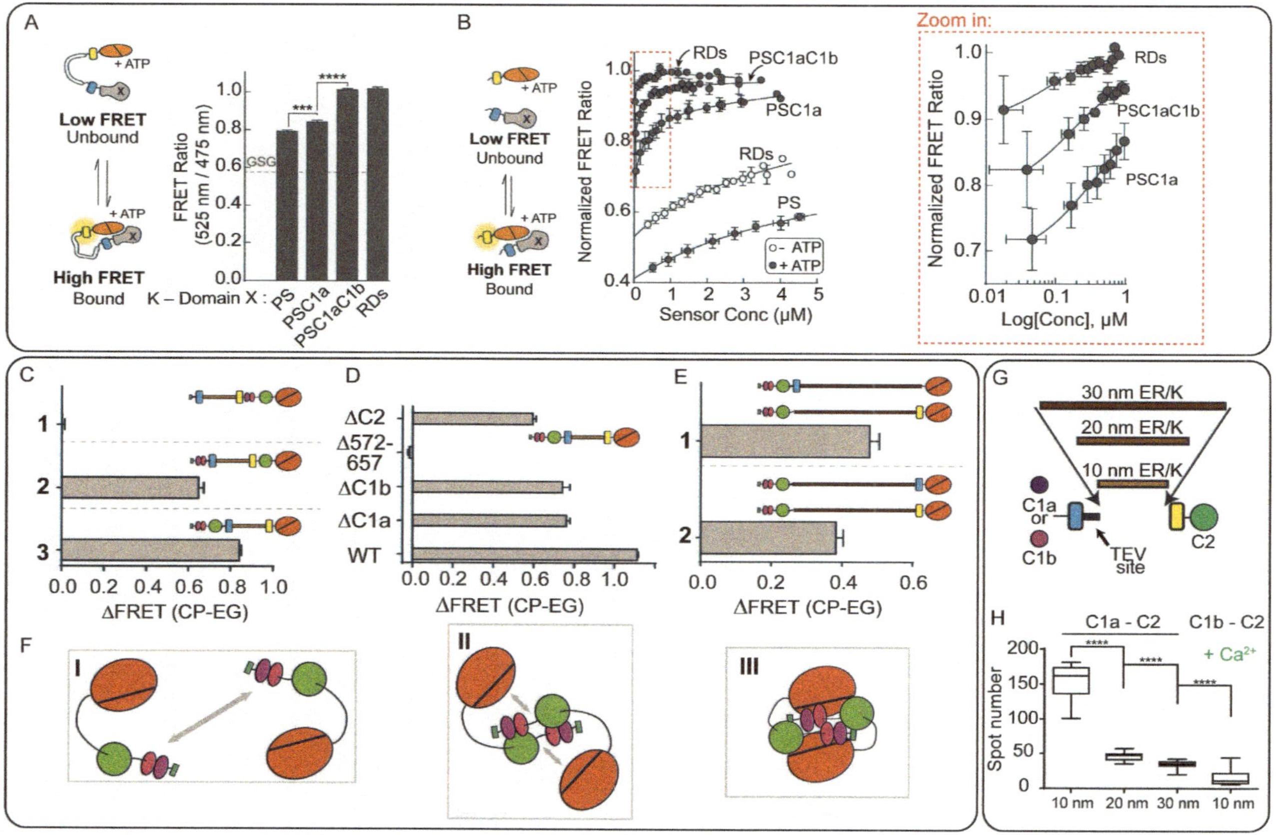

Fig. 4 See figure legend on next page.

simulations and FRET measurements, we were able to delineate distinct mechanisms that dictate substrate binding strength and kinase activity.

ER/K-linked FRET sensors that probe the kinase-substrate interaction have been used to profile the allosteric effects of chemically distinct

Fig. 4 Investigating PKC structure and function using ER/K linkers. (A) Cartoon diagram depicting interactions between regulatory domains and kinase domains when tethered with the ER/K linker (left). The observed FRET ratio for sensors with the kinase domain (K) and the indicated domains (right). PS—pseudosubtrate, C1a, C1b—diacylglycerol-sensitive regulatory domains, RDs—includes all regulatory domains (PS-C1a-C1b-C2). (B) Cartoon diagram depicting bi-molecular interactions between the regulatory domains and kinase domain derived by proteolytic cleavage of a TEV-protease site at the N-terminus of the ER/K linker (left). FRET ratio of the indicated pairings between regulatory domains and kinase domain (center). The region highlighted by the red box is depicted in the zoom-in on the right. (C–E) Change in FRET ratio in the indicated PKC sensors (ΔFRET) following activation. CP (calcium and PMA stimulation); EG (buffer containing EGTA). (C) By inserting the ER/K linker at various points in the protein, the domains that participate in the interaction can be discerned. 1—Sensor with pseudo-substrate separated from the regulatory domains. 2—pseudo-substrate and C1a/b separated from the C2 and kinase domains. 3—regulatory domains separated from the kinase domain. (D) PKC sensors with various truncations. The changes in ΔFRET following PKC activation indicated the relative importance of each excluded domain to the interaction. For instance, residues 572–657 that constitute the C-terminus of the kinase domain are essential to stimulate interactions upon PKC activation. (E) The change in FRET ratio for a pair of PKC sensors, each containing 30 nm linkers, and either an mCitrine or mCerulean fluorophore. By locating the fluorophores at opposite ends of the ER/K linker (1), a greater change in FRET is observed upon calcium and PMA stimulation than when they are located on the same side of the linker (2), indicating that the interaction between molecules likely involves close interaction between the regulatory and kinase domains on at least two separate proteins. (F) A proposed model for the interactions probed by the experiments in panels C–E. (G) Diagram of the sensors used in panel H. These sensors contain either the C1a or C1b domains linked to C2 with varying length ER/K linkers. (H) Decrease in observed spots in a microscopy assay using the indicated sensors depicted in (G). C1a, but not C1b aggregates in response to calcium stimulation in a manner proportional to the length of the ER/K linker. For all experiments: $N \geq 3$, **, $P \leq 0.01$; ****, $P \leq 0.0001$. Significance determined using ANOVA with Tukey's post-hoc test. *Panels (A) and (B) adapted from Sommese, R. F., Ritt, M., Swanson, C. J., Sivaramakrishnan, S. (2017). The role of regulatory domains in maintaining auto-inhibition in the multi-domain kinase PKCα.* Journal of Biological Chemistry, *292(7), 2873–2880.* © *the American Society for Biochemistry and Molecular Biology. Panels (C–F) adapted from Swanson, C. J., Ritt, M., Wang, W., Lang, M. J., Narayan, A., Tesmer, J. J., et al. (2014). Conserved modular domains team up to latch-open active protein kinase Cα.* The Journal of Biological Chemistry, *289(25), 17812–17829.* © *the American Society for Biochemistry and Molecular Biology. Panels (G) and (H) adapted from Swanson, C. J., Sommese, R. F., Petersen, K. J., Ritt, M., Karslake, J., Thomas, D. D., et al. (2016). Calcium stimulates self-assembly of protein kinase C α in vitro.* PLoS One, 11, *e0162331. https://doi.org/10.1371/journal.pone.0162331.*

small-molecule kinase inhibitors (SMKIs). Specifically, butyl-derivatives of staurosporine, such as Bisindolylmaleimde I (BimI), were found to allosterically disrupt the kinase-substrate interaction (Ma et al., 2018). In contrast to ATP analog inhibitors such as sangivamycin, BimI and sotrastaurin function as bitopic inhibitors that competitively displace ATP and substrate peptide (Fig. 5C). The ability of ER/K-linked sensors to profile the comparative effects of SMKIs on the weak kinase-substrate peptide interaction, combined with their utility in plate reader format (Lee et al., 2017) make them suitable for screening efforts to identify allosteric modulators of enzymes such as protein kinases.

3.7 ER/K linker as a FRET-free spacer in Ca^{2+} biosensors

Monitoring changes in the ubiquitous intracellular second messenger Ca^{2+} is of broad interest in cell biology. Ratiometric Ca^{2+} sensing dyes can be limited by the need to selectively deliver them to cells of interest, cellular toxicity through the chelation of endogenous Ca^{2+} by high affinity dyes, and the use of phototoxic UV excitation wavelengths. These limitations drove the development of genetically encoded Ca^{2+} sensors, which rely on a Ca^{2+}-sensing module from calmodulin (CaM) to undergo conformational changes in the presence of calcium, resulting in an increase in fluorescence from a circularly permuted GFP (cpGFP) (Nakai, Ohkura, & Imoto, 2001; Ohkura et al., 2012). The change in fluorescence intensity from such sensors, termed GCaMP, is quantitatively related to the change in cytosolic Ca^{2+}. However, the lack of basal fluorescence complicates distinction between changes in Ca^{2+} and sensor expression levels. Hence, a red-shifted fluorophore, mCherry was incorporated in the sensors to quantify expression (Al-Osta et al., 2018). Nonetheless, FRET between mCherry and cpGFP partially quenches GFP fluorescence, complicating quantitative interpretation of Ca^{2+} changes from the intensity data. Thus, there was a need to visualize the fluorescence of the Ca^{2+} sensor while preventing FRET between cpGFP and mCherry. In the absence of a significant interaction between proteins fused to the ends of an ER/K linker, its ensemble extended conformation minimizes FRET (Sivaramakrishnan et al., 2008). This property was used to design GCaMP-R, a Ca^{2+} biosensor in which the GCaMP module is linked to mCherry through a 30 nm ER/K linker (Cho et al., 2017). While minimizing FRET, tethering GCaMP to mCherry with the ER/K linker ensures 1:1 stoichiometry of expression enabling ratiometric imaging (GFP/mCherry) of changes in Ca^{2+}.

Avoiding FRET also significantly increased sensor brightness (notably at low calcium concentrations), the signal-to-noise ratio, and sensor dynamic range. This example illustrates the utility of the ER/K linker as an extended spacer to minimize FRET in the absence of a significant interaction, while maintaining stoichiometry of expression.

3.8 ER/K-linked sensors measure endogenous cellular activity

In general, ER/K-linked sensors have been used to report on the interaction of proteins tethered to the ends of the ER/K linker. A recent report from Maziarz et al. describes an ER/K-linked sensor designed to report the activity of endogenous G proteins (Maziarz et al., 2020). The sensor uses the ER/K linker as a spacer to separate a bioluminescent donor (NLuc) and a resonance energy acceptor (YFP). This BRET module with ER/K linker and YFP (BERKY) was inserted between a lipid anchor and KB-1753, a synthetic peptide that selectively binds activated (GTP loaded) Gαi subunits. BERKY sensors transfected into cells report a significantly faster activation kinetics of endogenous G proteins ($t_{1/2}$ ~250 ms) compared to exogenously expressed counterparts ($t_{1/2}$ ~1300 ms). The faster activation potentially stems from the rapid access of the GPCR to a localized endogenous G protein pool, whereas exogenous G protein activation is limited by the progressive disengagement of their endogenous counterparts. The BERKY sensors illustrate the utility and significance of ER/K linkers in measuring endogenous cellular activity, with broader application to other cellular interactions.

3.9 Visualizing cyclic AMP (cAMP) compartmentalization using ER/K linkers as "nanorulers"

cAMP is a universal second messenger that activates numerous signaling pathways. To address signaling specificity, cAMP signaling is thought to be compartmentalized through its hydrolysis by phosphodiesterases (PDEs) (Buxton & Brunton, 1983; Nikolaev, Gambaryan, Engelhardt, Walter, & Lohse, 2005; Omori & Kotera, 2007). However, compartmentalized signaling is inconsistent with the rapid diffusion of small molecules such as cAMP combined with the relatively low activity of PDEs. To address the potential for nanometer-scale confinement of cAMP, Bock et al. fused the catalytic domain of PDEs to an EPAC biosensor using 10 and 30 nm ER/K linkers (Bock et al., 2020). The ER/K linker serves as a nanoruler that reports on the cAMP concentration at a defined distance from the

PDE anchors. As a control, a direct fusion of PDE and EPAC was expressed in cells and did not report any measurable signal from the EPAC sensor following activation of cellular cAMP. In contrast, tethering EPAC to PDE4A1 with a 10 nm ER/K linker resulted in a sensor response comparable with that elicited by co-expression of EPAC and PDE4A1. This result suggests that PDE4A1 is capable of sustaining a nanometer scale compartment devoid of cAMP. Interestingly, the degree of confinement is isoform-specific, as PDE2 is capable of suppressing sensor response even when tethered to it with a 30 nm ER/K linker. These observations using the ER/K linker as a nanoruler, combined with both computational and experimental measurements of cAMP diffusion in cells suggest a "buffered diffusion" model, wherein under basal conditions cAMP is buffered by cytosolic components resulting in slow diffusion. In turn, PDEs acts on cAMP in their immediate vicinity and in conjunction with buffering by cytosolic cAMP binding sites create nanometer-sized cAMP compartments.

4. Future perspectives

Cell signaling processes contain transient, selective protein-protein interactions that together transduce diverse extracellular stimuli to distinct physiological responses. Disrupting pathophysiological signaling pathways, without perturbing normal cell function, requires the development of small-molecule or biologic-based therapies that can selectively target pathological interactions. However, the similarity in sequence and structure of the human members of signaling protein superfamilies such as GPCRs (~800), G proteins (~20), and kinases (~500) has confounded our understanding of the structural determinants of signaling specificity and challenges ongoing efforts to selectively target them. Hence, drug discovery efforts are increasingly focused on identifying allosteric modulators that target distal, non-conserved sites to gain specificity. In parallel, reducing the harmful side-effects of target-selective therapies requires tuning responses rather than switching on/off signaling pathways. Addressing these challenges requires technologies that can robustly monitor and modulate weak, transient interactions in cells. In this context, the natively occurring ER/K α-helical linker provides an accessible, genetically encoded approach to control the stoichiometry and effective concentration of protein-protein interactions both in isolated recombinant preparations and in live cells. Over the last decade, ER/K-linked sensors that monitor and modulate protein-protein interactions have repeatedly demonstrated their niche in profiling

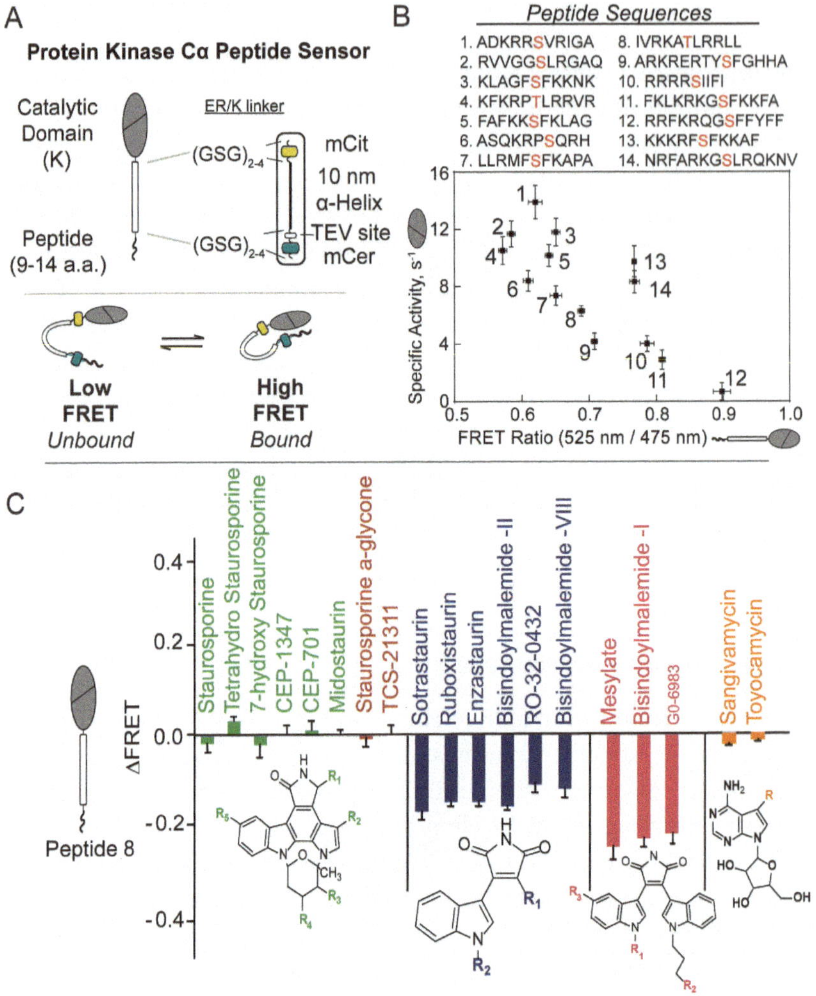

Fig. 5 Substrate peptide-kinase interactions and their allosteric modulation by small molecules probed using ER/K linkers. (A) Cartoon schematic of PKC substrate peptide biosensor (top) and mechanism of action (bottom). (B) Correlation between kinase activity and FRET ratio observed for 14 different peptides derived from phosphorylated substrates of PKC (phosphorylated Ser/Thr residue is highlighted in red). Lower FRET ratios (affinities) for peptides were observed to correlate with higher activity. (C) Families of structurally similar small molecule kinase inhibitors led to similar observed changes in FRET ratio in PKC biosensors. Small molecules with a purported bitopic mode of binding (navy, magenta) were more effective in disrupting the kinase/substrate interaction. For all experiments: $N \geq 3$, and data are shown as mean $\pm$ SE. *Panels (A) and (B) adapted from Sommese, R. F., Sivaramakrishnan, S., et al. (2016). Substrate affinity differentially influences protein kinase C regulation and inhibitor potency.* Journal of Biological Chemistry, 291*(42), 21963–21970. © the American Society for Biochemistry and Molecular Biology. Panel (C) adapted with permission from Ma, N., Lippert, L. G., Devamani, T., Levy, B., Lee, S., Sandhu, M., et al. (2018). Bitopic inhibition of ATP and substrate binding in Ser/Thr kinases through a conserved allosteric mechanism.* Biochemistry, *57, 6387–90. https:/ /doi.org/10.1021/acs.biochem.8b00729. Copyright (2018) American Chemical Society.*

modest changes in protein interactions in response to perturbations induced by mutagenesis, post-translational modifications, and exogenous small-molecules or biologics. Technical advances in the dynamic range of sensor readout, cell-free assays that capture the complexity of the cellular interaction milieu, and integration with protein evolution strategies will further enhance the utility of the ER/K-linker technology in drug discovery efforts to identify selective modulators that alleviate pathologic cellular processes.

Acknowledgments

We acknowledge funding from the NIH (R35-GM126940 and R01-GM117923 to S.S.)

References

Al-Osta, I., Mucha, M., Pereda, D., Piqué-Gili, M., Okorocha, A. E., Thomas, R., et al. (2018). Imaging calcium in hippocampal presynaptic terminals with a ratiometric calcium sensor in a novel transgenic mouse. *Frontiers in Cellular Neuroscience*, *12*, 209. https://doi.org/10.3389/fncel.2018.00209.

Antal, C. E., Callender, J. A., Kornev, A. P., Taylor, S. S., & Newton, A. C. (2015). Intramolecular C2 domain-mediated autoinhibition of protein kinase C βII. *Cell Reports*, *12*, 1252–1260. https://doi.org/10.1016/j.celrep.2015.07.039.

Batchelor, M., Wolny, M., Baker, E. G., Paci, E., Kalverda, A. P., & Peckham, M. (2019). Dynamic ion pair behavior stabilizes single -helices in proteins. *The Journal of Biological Chemistry*, *294*, 3219–3234. https://doi.org/10.1074/jbc.RA118.006752.

Bierzynski, A., Kim, P. S., & Baldwin, R. L. (1982). A salt bridge stabilizes the helix formed by isolated C-peptide of RNase A. *Proceedings of the National Academy of Sciences of the United States of America*, *79*, 2470–2474. https://doi.org/10.1073/pnas.79.8.2470.

Bock, A., Annibale, P., Konrad, C., Hannawacker, A., Anton, S. E., Maiellaro, I., et al. (2020). Optical mapping of cAMP signaling at the nanometer scale. *Cell*, *182*, 1–12. https://doi.org/10.1016/j.cell.2020.07.035.

Bonny, M., Hui, X., Schweizer, J., Kaestner, L., Zeug, A., Kruse, K., et al. (2016). C2-domain mediated nano-cluster formation increases calcium signaling efficiency. *Scientific Reports*, *6*, 1–11. https://doi.org/10.1038/srep36028.

Bornhorst, J. A., & Falke, J. J. (2000). Purification of proteins using polyhistidine affinity tags. *Methods in Enzymology*, *326*, 245–254. https://doi.org/10.1016/s0076-6879(00)26058-8.

Bruzzese, A., Gil, C., Dalton, J. A. R., & Giraldo, J. (2018). Structural insights into positive and negative allosteric regulation of a G protein-coupled receptor through protein-lipid interactions. *Scientific Reports*, *8*, 1–14. https://doi.org/10.1038/s41598-018-22735-6.

Buxton, I., & Brunton, L. (1983). Compartments of cyclic AMP and protein kinase in mammalian cardiomyocytes. *The Journal of Biological Chemistry*, *258*(17), 10233–10239. https://pubmed.ncbi.nlm.nih.gov/6309796/. (Accessed 3 September 2020).

Carmena, M., Wheelock, M., Funabiki, H., & Earnshaw, W. C. (2012). The chromosomal passenger complex (CPC): From easy rider to the godfather of mitosis. *Nature Reviews. Molecular Cell Biology*, *13*, 789–803. https://doi.org/10.1038/nrm3474.

Cheslock, S. R., Poon, D. T. K., Fu, W., Rhodes, T. D., Henderson, L. E., Nagashima, K., et al. (2003). Charged assembly helix motif in murine leukemia virus capsid: An important region for virus assembly and particle size determination. *Journal of Virology*, *77*, 7058–7066. https://doi.org/10.1128/jvi.77.12.7058-7066.2003.

Cho, J. H., Swanson, C. J., Chen, J., Li, A., Lippert, L. G., Boye, S. E., et al. (2017). The GCaMP-R family of genetically encoded ratiometric calcium indicators. *ACS Chemical Biology*, *12*, 1066–1074. https://doi.org/10.1021/acschembio.6b00883.

Chou, K. C., Némethy, G., Rumsey, S., Tuttle, R. W., & Scheraga, H. A. (1985). Interactions between an α-helix and a β-sheet. Energetics of α β packing in proteins. *Journal of Molecular Biology*, *186*, 591–609. https://doi.org/10.1016/0022-2836(85)90133-0.

Chung, K. Y., Rasmussen, S. G. F., Liu, T., Li, S., Devree, B. T., Chae, P. S., et al. (2011). Conformational changes in the G protein Gs induced by the β 2 adrenergic receptor. *Nature*, *477*, 611–617. https://doi.org/10.1038/nature10488.

Crick, F. H. C. (1953). The packing of α-helices: Simple coiled-coils. *Acta Crystallographica*, *6*, 689–697. https://doi.org/10.1107/s0365110x53001964.

Daaka, Y., Luttrell, L. M., & Lefkowitz, R. J. (1997). Switching of the coupling of the β2-adrenergic receptor to different g proteins by protein kinase A. *Nature*, *390*, 88–91. https://doi.org/10.1038/36362.

Doležal, M., Hadravová, R., Kožíšek, M., Bednárová, L., Langerová, H., Ruml, T., et al. (2016). Functional and structural characterization of novel type of linker connecting capsid and nucleocapsid protein domains in murine leukemia virus. *The Journal of Biological Chemistry*, *291*, 20630–20642. https://doi.org/10.1074/jbc.M116.746461.

Dunnill, P. (1968). The use of helical net-diagrams to represent protein structures. *Biophysical Journal*, *8*, 865–875. https://doi.org/10.1016/S0006-3495(68)86525-7.

Galandrin, S., & Bouvier, M. (2006). Distinct signaling profiles of β1 and β2 adrenergic receptor ligands toward adenylyl cyclase and mitogen-activated protein kinase reveals the pluridimensionality of efficacy. *Molecular Pharmacology*, *70*, 1575–1584. https://doi.org/10.1124/mol.106.026716.

Ganguly, S., Saxena, R., & Chattopadhyay, A. (2011). Reorganization of the actin cytoskeleton upon G-protein coupled receptor signaling. *Biochimica et Biophysica Acta—Biomembranes*, *1808*, 1921–1929. https://doi.org/10.1016/j.bbamem.2011.04.001.

García-Nafría, J., & Tate, C. G. (2019). Cryo-EM structures of GPCRs coupled to G s, G i and G o. *Molecular and Cellular Endocrinology*, *488*, 1–13. https://doi.org/10.1016/j.mce.2019.02.006.

Gáspári, Z., Süveges, D., Perczel, A., Nyitray, L., & Tóth, G. (2012). Charged single alpha-helices in proteomes revealed by a consensus prediction approach. *Biochimica et Biophysica Acta—Proteins & Proteomics*, *1824*, 637–646. https://doi.org/10.1016/j.bbapap.2012.01.012.

Gupte, T. M., Malik, R. U., Sommese, R. F., Ritt, M., & Sivaramakrishnan, S. (2017). Priming GPCR signaling through the synergistic effect of two G proteins. *Proceedings of the National Academy of Sciences of the United States of America*, *114*, 3756–3761. https://doi.org/10.1073/pnas.1617232114.

Gupte, T. M., Ritt, M., Dysthe, M., Malik, R. U., & Sivaramakrishnan, S. (2019). Minute-scale persistence of a GPCR conformation state triggered by non-cognate G protein interactions primes signaling. *Nature Communications*, *10*, 1–15. https://doi.org/10.1038/s41467-019-12755-9.

Hariadi, R. F., Cale, M., & Sivaramakrishnan, S. (2014). Myosin lever arm directs collective motion on cellular actin network. *Proceedings of the National Academy of Sciences of the United States of America*, *111*, 4091–4096. https://doi.org/10.1073/pnas.1315923111.

He, K., Sakai, T., Tsukasaki, Y., Watanabe, T. M., & Ikebe, M. (2017). Myosin X is recruited to nascent focal adhesions at the leading edge and induces multi-cycle filopodial elongation. *Scientific Reports*, 7, 1–12. https://doi.org/10.1038/s41598-017-06147-6.

Huse, M., & Kuriyan, J. (2002). The conformational plasticity of protein kinases. *Cell*, *109*, 275–282. https://doi.org/10.1016/S0092-8674(02)00741-9.

Kerber, M. L., & Cheney, R. E. (2011). Myosin-X: A MyTH-FERM myosin at the tips of filopodia. *Journal of Cell Science, 124*, 3733–3741. https://doi.org/10.1242/jcs.023549.

Knight, P. J., Thirumurugan, K., Xu, Y., Wang, F., Kalverda, A. P., Stafford, W. F., et al. (2005). The predicted coiled-coil domain of myosin 10 forms a novel elongated domain that lengthens the head. *The Journal of Biological Chemistry, 280*, 34702–34708. https://doi.org/10.1074/jbc.M504887200.

Kobilka, B. K. (2011). Structural insights into adrenergic receptor function and pharmacology. *Trends in Pharmacological Sciences, 32*, 213–218. https://doi.org/10.1016/j.tips.2011.02.005.

Lee, A. G. (2003). Lipid-protein interactions in biological membranes: A structural perspective. *Biochimica et Biophysica Acta—Biomembranes, 1612*, 1–40. https://doi.org/10.1016/S0005-2736(03)00056-7.

Lee, S., Devamani, T., Song, H. D., Sandhu, M., Larsen, A., Sommese, R., et al. (2017). Distinct structural mechanisms determine substrate affinity and kinase activity of protein kinase Cα. *The Journal of Biological Chemistry, 292*, 16300–16309. https://doi.org/10.1074/jbc.M117.804781.

Leonard, T. A., Róycki, B., Saidi, L. F., Hummer, G., & Hurley, J. H. (2011). Crystal structure and allosteric activation of protein kinase C βiI. *Cell, 144*, 55–66. https://doi.org/10.1016/j.cell.2010.12.013.

Lyu, P. C., Gans, P. J., & Kallenbach, N. R. (1992). Energetic contribution of solvent-exposed ion pairs to alpha-helix structure. *Journal of Molecular Biology, 223*, 343–350. https://doi.org/10.1016/0022-2836(92)90735-3.

Ma, N., Lippert, L. G., Devamani, T., Levy, B., Lee, S., Sandhu, M., et al. (2018). Bitopic inhibition of ATP and substrate binding in Ser/Thr kinases through a conserved allosteric mechanism. *Biochemistry, 57*, 6387–6390. https://doi.org/10.1021/acs.biochem.8b00729.

Mackay, A. M., Eckley, D. M., Chue, C., & Earnshaw, W. C. (1993). Molecular analysis of the INCENPs (inner centromere proteins): Separate domains are required for association with microtubules during interphase and with the central spindle during anaphase. *The Journal of Cell Biology, 123*, 373–385. https://doi.org/10.1083/jcb.123.2.373.

Malik, R. U., Ritt, M., DeVree, B. T., Neubig, R. R., Sunahara, R. K., & Sivaramakrishnan, S. (2013). Detection of G protein-selective G protein-coupled receptor (GPCR) conformations in live cells. *The Journal of Biological Chemistry, 288*, 17167–17178. https://doi.org/10.1074/jbc.M113.464065.

Marqusee, S., & Baldwin, R. L. (1987). Helix stabilization by Glu- … Lys+ salt bridges in short peptides of de novo design. *Proceedings of the National Academy of Sciences of the United States of America, 84*, 8898–8902. https://doi.org/10.1073/pnas.84.24.8898.

Maziarz, M., Park, J. C., Leyme, A., Marivin, A., Garcia-Lopez, A., Patel, P. P., et al. (2020). Revealing the activity of trimeric G-proteins in live cells with a versatile biosensor design. *Cell, 182*, 770–785.e16. https://doi.org/10.1016/j.cell.2020.06.020.

Mehraeen, S., Sudhanshu, B., Koslover, E. F., & Spakowitz, A. J. (2008). End-to-end distribution for a wormlike chain in arbitrary dimensions. *Physical Review. E, Statistical, Nonlinear, and Soft Matter Physics*, 77, 061803. https://doi.org/10.1103/PhysRevE.77.061803.

Miller, C. J., & Turk, B. E. (2018). Homing in: Mechanisms of substrate targeting by protein kinases. *Trends in Biochemical Sciences, 43*, 380–394. https://doi.org/10.1016/j.tibs.2018.02.009.

Nakai, J., Ohkura, M., & Imoto, K. (2001). A high signal-to-noise Ca^{2+} probe composed of a single green fluorescent protein. *Nature Biotechnology, 19*, 137–141. https://doi.org/10.1038/84397.

Newton, A. C. (2018). Protein kinase C: Perfectly balanced. *Critical Reviews in Biochemistry and Molecular Biology, 53*, 208–230. https://doi.org/10.1080/10409238.2018.1442408.

Nikolaev, V. O., Gambaryan, S., Engelhardt, S., Walter, U., & Lohse, M. J. (2005). Real-time monitoring of the PDE2 activity of live cells: Hormone-stimulated cAMP hydrolysis is faster than hormone-stimulated cAMP synthesis. *The Journal of Biological Chemistry*, *280*, 1716–1719. https://doi.org/10.1074/jbc.C400505200.

Ohkura, M., Sasaki, T., Sadakari, J., Gengyo-Ando, K., Kagawa-Nagamura, Y., Kobayashi, C., et al. (2012). Genetically encoded green fluorescent Ca^{2+} indicators with improved detectability for neuronal Ca^{2+} signals. *PLoS One*, *7*, e51286. https://doi.org/10.1371/journal.pone.0051286.

Omori, K., & Kotera, J. (2007). Overview of PDEs and their regulation. *Circulation Research*, *100*, 309–327. https://doi.org/10.1161/01.RES.0000256354.95791.f1.

Pace, C. N., & Scholtz, J. M. (1998). A helix propensity scale based on experimental studies of peptides and proteins. *Biophysical Journal*, *75*, 422–427. https://doi.org/10.1016/s0006-3495(98)77529-0.

Pauling, L., Corey, R. B., & Branson, H. R. (1951). The structure of proteins; two hydrogen-bonded helical configurations of the polypeptide chain. *Proceedings of the National Academy of Sciences of the United States of America*, *37*, 205–211. https://doi.org/10.1073/pnas.37.4.205.

Peckham, M. (2011). Coiled coils and SAH domains in cytoskeletal molecular motors. *Biochemical Society Transactions*, *39*, 1142–1148. https://doi.org/10.1042/BST0391142.

Pierce, K. L., Premont, R. T., & Lefkowitz, R. J. (2002). Seven-transmembrane receptors. *Nature Reviews. Molecular Cell Biology*, *3*, 639–650. https://doi.org/10.1038/nrm908.

Ponder, J. W., & Richards, F. M. (1987). Tertiary templates for proteins. Use of packing criteria in the enumeration of allowed sequences for different structural classes. *Journal of Molecular Biology*, *193*, 775–791. https://doi.org/10.1016/0022-2836(87)90358-5.

Qu, K., Glass, B., Dolezal, M., Schur, F. K. M., Murciano, B., Rein, A., et al. (2018). Structure and architecture of immature and mature murine leukemia virus capsids. *Proceedings of the National Academy of Sciences of the United States of America*, *115*, E11751–E11760. https://doi.org/10.1073/pnas.1811580115.

Rasmussen, S. G. F., Devree, B. T., Zou, Y., Kruse, A. C., Chung, K. Y., Kobilka, T. S., et al. (2011). Crystal structure of the β 2 adrenergic receptor-Gs protein complex. *Nature*, *477*, 549–557. https://doi.org/10.1038/nature10361.

Ritt, M., Guan, J. L., & Sivaramakrishnan, S. (2013). Visualizing and manipulating focal adhesion kinase regulation in live cells. *The Journal of Biological Chemistry*, *288*, 8875–8886. https://doi.org/10.1074/jbc.M112.421164.

Ritt, M., & Sivaramakrishnan, S. (2016). Correlation between activity and domain complementation in adenylyl cyclase demonstrated with a novel fluorescence resonance energy transfer sensor. *Molecular Pharmacology*, *89*, 407–412. https://doi.org/10.1124/mol.115.101626.

Rossy, J., Owen, D. M., Williamson, D. J., Yang, Z., & Gaus, K. (2013). Conformational states of the kinase Lck regulate clustering in early T cell signaling. *Nature Immunology*, *14*, 82–89. https://doi.org/10.1038/ni.2488.

Samejima, K., Platani, M., Wolny, M., Ogawa, H., Vargiu, G., Knight, P. J., et al. (2015). The inner centromere protein (INCENP) coil is a single α-helix (SAH) domain that binds directly to microtubules and is important for chromosome passenger complex (CPC) localization and function in mitosis. *The Journal of Biological Chemistry*, *290*, 21460–21472. https://doi.org/10.1074/jbc.M115.645317.

Sandhu, M., Touma, A. M., Dysthe, M., Sadler, F., Sivaramakrishnan, S., & Vaidehi, N. (2019). Conformational plasticity of the intracellular cavity of GPCR−G-protein complexes leads to G-protein promiscuity and selectivity. *Proceedings of the National Academy of Sciences of the United States of America*, *116*, 11956–11965. https://doi.org/10.1073/pnas.1820944116.

Scholtz, J. M., & Baldwin, R. L. (1992). The mechanism of alpha-Helix formation by peptides. *Annual Review of Biophysics and Biomolecular Structure*, *21*, 95–118. https://doi.org/10.1146/annurev.bb.21.060192.000523.

Semack, A., Sandhu, M., Malik, R. U., Vaidehi, N., & Sivaramakrishnan, S. (2016). Structural elements in the Gαs and Gβq C termini that mediate selective G protein-coupled receptor (GPCR) signaling. *The Journal of Biological Chemistry*, *291*, 17929–17940. https://doi.org/10.1074/jbc.M116.735720.

Sessa, F., Mapelli, M., Ciferri, C., Tarricone, C., Areces, L. B., Schneider, T. R., et al. (2005). The allosteric activators of Aurora B activation by INCENP and inhibition by hesperadin. *Molecular Cell*, *18*, 379–391. https://doi.org/10.1016/j.molcel.2005.03.031.

Simm, D., & Kollmar, M. (2018). Waggawagga-CLI: A command-line tool for predicting stable single α-helices (SAH-domains), and the SAH-domain distribution across eukaryotes. *PLoS One*, *13*, e0191924. https://doi.org/10.1371/journal.pone.0191924.

Sivaramakrishnan, S., Spink, B. J., Sim, A. Y. L., Doniach, S., & Spudich, J. A. (2008). Dynamic charge interactions create surprising rigidity in the ER/K α-helical protein motif. *Proceedings of the National Academy of Sciences of the United States of America*, *105*, 13356–13361. https://doi.org/10.1073/pnas.0806256105.

Sivaramakrishnan, S., & Spudich, J. A. (2011). Systematic control of protein interaction using a modular ER/K α-helix linker. *Proceedings of the National Academy of Sciences of the United States of America*, *108*, 20467–20472. https://doi.org/10.1073/pnas.1116066108.

Sivaramakrishnan, S., Sung, J., Ali, M., Doniach, S., Flyvbjerg, H., & Spudich, J. A. (2009). Combining single-molecule optical trapping and small-angle X-ray scattering measurements to compute the persistence length of a protein ER/K α-Helix. *Biophysical Journal*, *97*, 2993. https://doi.org/10.1016/J.BPJ.2009.09.009.

Slater, S. J., Seiz, J. L., Cook, A. C., Buzas, C. J., Malinowski, S. A., Kershner, J. L., et al. (2002). Regulation of PKCα activity by C1-C2 domain interactions. *The Journal of Biological Chemistry*, *277*, 15277–15285. https://doi.org/10.1074/jbc.M112207200.

Sommese, R. F., Ritt, M., Swanson, C. J., & Sivaramakrishnan, S. (2017). The role of regulatory domains in maintaining autoinhibition in the multidomain kinase PKCα. *The Journal of Biological Chemistry*, *292*, 2873–2880. https://doi.org/10.1074/jbc.M116.768457.

Sommese, R. F., & Sivaramakrishnan, S. (2016). Substrate affinity differentially influences protein kinase C regulation and inhibitor potency. *The Journal of Biological Chemistry*, *291*, 21963–21970. https://doi.org/10.1074/jbc.M116.737601.

Sommese, R. F., Sivaramakrishnan, S., Baldwin, R. L., & Spudich, J. A. (2010). Helicity of short E-R/K peptides. *Protein Science*, *19*, 2001–2005. https://doi.org/10.1002/pro.469.

Steinberg, S. F. (2008). Structural basis of protein kinase C isoform function. *Physiological Reviews*, *88*, 1341–1378. https://doi.org/10.1152/physrev.00034.2007.

Strohman, M. J., Maeda, S., Hilger, D., Masureel, M., Du, Y., & Kobilka, B. K. (2019). Local membrane charge regulates β2 adrenergic receptor coupling to Gi3. *Nature Communications*, *10*, 1–10. https://doi.org/10.1038/s41467-019-10108-0.

Süveges, D., Gáspári, Z., Tóth, G., & Nyitray, L. (2009). Charged single α-helix: A versatile protein structural motif. *Proteins: Structure, Function, and Bioinformatics*, *74*, 905–916. https://doi.org/10.1002/prot.22183.

Swanson, C. J., Ritt, M., Wang, W., Lang, M. J., Narayan, A., Tesmer, J. J., et al. (2014). Conserved modular domains team up to latch-open active protein kinase Cα. *The Journal of Biological Chemistry*, *289*, 17812–17829. https://doi.org/10.1074/jbc.M113.534750.

Swanson, C. J., & Sivaramakrishnan, S. (2014). Harnessing the unique structural properties of isolated α-helices. *The Journal of Biological Chemistry*, *289*, 25460–25467. https://doi.org/10.1074/jbc.R114.583906.

Swanson, C. J., Sommese, R. F., Petersen, K. J., Ritt, M., Karslake, J., Thomas, D. D., et al. (2016). Calcium stimulates self-assembly of protein kinase C α in vitro. *PLoS One, 11*, e0162331. https://doi.org/10.1371/journal.pone.0162331.

Taylor, S. S., & Kornev, A. P. (2011). Protein kinases: Evolution of dynamic regulatory proteins. *Trends in Biochemical Sciences, 36*, 65–77. https://doi.org/10.1016/j.tibs.2010.09.006.

Tian, T., Harding, A., Inder, K., Plowman, S., Parton, R. G., & Hancock, J. F. (2007). Plasma membrane nanoswitches generate high-fidelity Ras signal transduction. *Nature Cell Biology, 9*, 905–914. https://doi.org/10.1038/ncb1615.

Tomko, R. J., Taylor, D. W., Chen, Z. A., Wang, H. W., Rappsilber, J., & Hochstrasser, M. (2015). A single α helix drives extensive remodeling of the proteasome lid and completion of regulatory particle assembly. *Cell, 163*, 432–444. https://doi.org/10.1016/j.cell.2015.09.022.

Ulrich, A. K. C., Seeger, M., Schütze, T., Bartlick, N., & Wahl, M. C. (2016). Scaffolding in the spliceosome via single α helices. *Structure, 24*, 1972–1983. https://doi.org/10.1016/j.str.2016.09.007.

Wolny, M., Batchelor, M., Bartlett, G. J., Baker, E. G., Kurzawa, M., Knight, P. J., et al. (2017). Characterization of long and stable de novo single alpha-helix domains provides novel insight into their stability. *Scientific Reports*, 7, 44341. https://doi.org/10.1038/srep44341.

Wolny, M., Batchelor, M., Knight, P. J., Paci, E., Dougan, L., & Peckham, M. (2014). Stable single α-helices are constant force springs in proteins. *The Journal of Biological Chemistry, 289*, 27825–27835. https://doi.org/10.1074/jbc.M114.585679.

Yang, Y., Baboolal, T. G., Siththanandan, V., Chen, M., Walker, M. L., Knight, P. J., et al. (2009). A FERM domain autoregulates Drosophila myosin 7a activity. *Proceedings of the National Academy of Sciences of the United States of America, 106*, 4189–4194. https://doi.org/10.1073/pnas.0808682106.

Zangi, R. (2014). Side-chain-side-chain interactions and stability of the helical state. *Physical Review. E, Statistical, Nonlinear, and Soft Matter Physics, 89*, 012723. https://doi.org/10.1103/PhysRevE.89.012723.

Zeng, W., Mak, D. O. D., Li, Q., Shin, D. M., Foskett, J. K., & Muallem, S. (2003). A new mode of Ca^{2+} signaling by G protein-coupled receptors: Gating of IP3 receptor Ca^{2+} release channels by Gβγ. *Current Biology, 13*, 872–876. https://doi.org/10.1016/S0960-9822(03)00330-0.

CHAPTER EIGHT

Design of helical linkers for fusion proteins and protein-based nanostructures

Ryoichi Arai[a,b,*]
[a]Department of Biomolecular Innovation, Institute for Biomedical Sciences, Shinshu University, Ueda, Nagano, Japan
[b]Department of Applied Biology, Faculty of Textile Science and Technology, Shinshu University, Ueda, Nagano, Japan
[*]Corresponding author: e-mail address: rarai@shinshu-u.ac.jp

Contents

Abstract

The construction of recombinant fusion/chimeric proteins has been widely used for expression of soluble proteins and protein purification in a variety of fields of protein engineering and biotechnology. Fusion proteins are constructed by the linking of two protein domains with a peptide linker. The selection of a linker sequence is important for the construction of stable and bioactive fusion proteins. Empirically designed linkers are generally classified into two categories according to their structural features: flexible linkers and rigid linkers. Rigid linkers with the α-helix-forming sequences $A(EAAAK)_nA$ ($n = 2$–5) were first designed about two decades ago to control the distance between two protein domains and to reduce their interference. Thereafter, the helical linkers have been applied to the construction of many fusion proteins to improve

Methods in Enzymology, Volume 647
ISSN 0076-6879
https://doi.org/10.1016/bs.mie.2020.10.003

expression and bioactivity. In addition, the design of fusion proteins that self-assemble into supramolecular complexes is useful for nanobiotechnology and synthetic biology. A protein that forms a self-assembling oligomer was fused by a rigid helical linker to another protein that forms another self-assembling oligomer, and the fusion protein symmetrically self-assembled into a designed protein nanoparticle or nanomaterial. Moreover, to construct chain-like polymeric nanostructures, extender protein nanobuilding blocks were designed by tandemly fusing two dimeric *de novo* proteins with helical or flexible linkers. The linker design of fusion proteins can affect conformation and dynamics of self-assembling nanostructures. The present review and methods focus on useful helical linkers to construct bioactive fusion proteins and protein-based nanostructures.

1. Introduction

Protein fusion has become an indispensable tool in a variety of fields of protein engineering and biotechnology. The construction of recombinant fusion/chimeric proteins has been widely used to increase the expression of soluble proteins and to facilitate protein purification (Terpe, 2003). Also, the fusion of two or more protein domains can enhance bioactivities and generate novel functional combinations with a wide range of biotechnological and biopharmaceutical applications (Yu, Liu, Kim, & Lee, 2015). The construction of a fusion protein involves the linking of two or more proteins or domains of proteins by peptide linkers. As an essential component of recombinant fusion proteins, the selection of the linker sequence is particularly important for the construction of stable and bioactive fusion proteins (Chen, Zaro, & Shen, 2013). The general properties of linkers derived from natural multi-domain proteins can be considered as the foundation for linker design (Argos, 1990; George & Heringa, 2002). Empirically designed linkers are generally classified into two categories according to their structural features: flexible linkers and rigid linkers (Fig. 1). Besides the basic role in linking the functional domains together, linkers can offer many other advantages for the production of fusion proteins, such as improving biological activity, increasing expression yield, and achieving desirable pharmacokinetic profiles (Chen et al., 2013).

As suggested by George & Heringa (2002), many natural linkers exhibit α-helical structures. An α-helical structure is rigid and stable, with intra-segment hydrogen bonds and a closely packed backbone. Therefore, stiff α-helical linkers can act as rigid spacers between protein domains.

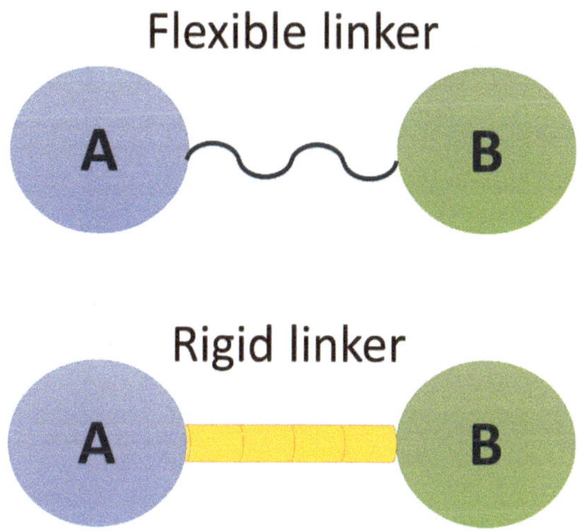

Fig. 1 Schematics of fusion proteins constructed from a protein A and a protein B with peptide linkers. Empirically designed linkers are generally classified into two categories according to their structural features: flexible linkers and rigid linkers.

An empirical rigid linker with the α-helix-forming sequence of $A(EAAAK)_nA$ ($n = 2$–5) was first designed and introduced to a fusion protein by Arai et al. and "linker engineering" was proposed to rationally design linkers for multifunctional fusion proteins (Arai, Ueda, Kitayama, Kamiya, & Nagamune, 2001). Thereafter, helical linkers have been applied to the construction of many recombinant fusion proteins (Amet, Lee, & Shen, 2009; Bai & Shen, 2006; Chen et al., 2013; Kobayashi et al., 2018). In addition, a hybrid linker of helical and flexible segments, $(GSG)_6A(EAAAK)_6A(GSG)_6A(EAAAK)_6A(GSG)_6$, was successfully applied to construction of Förster resonance energy transfer (FRET)-based fluorescent sensor proteins (Boersma, Zuhorn, & Poolman, 2015; Golynskiy, Rurup, & Merkx, 2010). Another type of rigid linkers has a Pro-rich sequence, $(XP)_n$, with X designating any amino acid, preferably Ala, Lys, or Glu. The presence of Pro in non-helical linkers can increase the stiffness and allow for separation of the protein domains (Bhandari, Levine, Trayer, & Yeadon, 1986; McCormick, Thomas, & Heath, 2001; Turner, Russell, Williamson, & Guest, 1993).

In the present review and methods, I focus on design and applications of helical linkers for fusion proteins and protein-based nanostructures.

2. Design of helical linkers for fusion proteins

2.1 Development of helical linkers for fusion proteins

A few decades ago, several studies on linkers for multi-domain proteins have been reported (Alfthan, Takkinen, Sizmann, Soderlund, & Teeri, 1995; Argos, 1990; Crasto & Feng, 2000; Robinson & Sauer, 1998). These early studies mainly suggested that flexibility and hydrophilicity of linkers were important not to disturb functions of domains. However, there were several examples in the literature where the use of flexible linkers resulted in loss of biological activity or poor expression yields. For instance, the immunoglobulin binding ability of a streptococcal protein G C1 domain in a protein G–*Vargula* luciferase fusion protein was not recovered by inserting a flexible linker (GGGGS) (Maeda et al., 1997). In addition, a transferrin (Tf)–granulocyte colony stimulating factor (G-CSF) fusion protein failed to be expressed with a flexible linker $(GGGGS)_3$ (Amet et al., 2009). The ineffectiveness of flexible linkers in these instances was attributed to the inefficient separation of the protein domains or their interference with each other. Under these situations, rigid linkers should be applied to keep a spatial distance between the domains and to maintain their independent functions.

Therefore, Arai et al. designed the helical linkers, $A(EAAAK)_nA$ ($n = 2–5$), to spatially separate bifunctional domains of a fusion protein (Arai et al., 2001). The linkers forming a monomeric hydrophilic α-helix were designed according to the earlier study on a short peptide forming a monomeric α-helix. The helix-forming peptide, AEAAAKEAAAKEAAAKA [$(i+4)$E,K], which was stabilized by Glu^-–Lys^+ salt bridges, showed ~80% helicity (Marqusee & Baldwin, 1987). To evaluate separation of the protein domains, the helical linkers were inserted between enhanced blue fluorescent protein (EBFP) and enhanced green fluorescent protein (EGFP), and the fluorescent resonance energy transfer (FRET) efficiency between EBFP and EGFP was measured. The FRET efficiency decreased as the length of helical linkers increased, indicating that the helical linkers can control the distance between the protein domains by changing repetitions of the EAAAK motif (Table 1). Compared to flexible linkers with the same length, the helical linkers resulted in a lower FRET efficiency when inserted into EBFP–EGFP fusion proteins, indicating that the helical linkers can spatially separate the two domains effectively. Moreover, the conformations of the EBFP–EGFP fusion proteins with the helical and flexible linkers were evaluated using small-angle X-ray scattering (SAXS) (Arai, Wriggers, Nishikawa, Nagamune, & Fujisawa, 2004).

Table 1 Evaluation of distances between two domains (EBFP and EGFP) of EBFP–linker–EGFP fusion proteins by fluorescence energy transfer (FRET) (Arai et al., 2001).

Linkers	F_{DA}/F_D	$E_T = 1 - F_{DA}/F_D$	$r = R_0(1/E_T - 1)^{1/6}$ (Å)
SL	0.739	0.261	50.0
FL3	0.735	0.265	49.8
FL4	0.752	0.248	50.5
HL2	0.766	0.234	51.2
HL3	0.811	0.189	53.5
HL4	0.864	0.136	57.1
HL5	0.910	0.090	61.8
PA	0.868	0.132	57.5

The fluorescence intensity of EBFP (donor) at 444 nm of its emission peak was measured with 380 nm excitation at 25 °C. For the measurement, 2 μM of each fusion protein in PBS (pH 7.4) was used. F_{DA}, fluorescence intensity of donor in the presence of acceptor (The fusion proteins were not treated with trypsin.); F_D, fluorescence intensity of donor in the absence of acceptor (The fusion proteins were treated with trypsin.); E_T, efficiency of FRET; r, estimated distance between EBFP and EGFP assuming that the critical Förster distance R_0 of the pair EBFP–EGFP is 42 Å. PA: B domain of staphylococcal protein A. Reprinted with permission from (Arai et al., 2001).

The SAXS experiment and analysis indicated that the fusion proteins with the helical linkers formed more elongated conformations as compared to those with the flexible linkers and the helical linkers probably connected the two domains diagonally (Fig. 2). The results demonstrated that the longer helical linkers can separate two domains of fusion proteins more effectively and the linker design affects the global conformations of the fusion proteins.

2.2 Applications of helical linkers for fusion proteins

Expression and bioactivity of fusion proteins can be improved by linker engineering (Chen et al., 2013). The effect of linker insertion on expression level of fusion proteins was observed in transferrin (Tf) fusion proteins designed for Tf receptor-mediated protein drug oral delivery (Amet et al., 2009). A long helical linker $(H4)_2$, $A(EAAAK)_4ALEA(EAAAK)_4A$, was inserted into the fusion proteins between Tf and G-CSF or Tf and human growth hormone (hGH), greatly improving their expression level. In addition, the G-CSF–Tf fusion proteins with the helical linker $(H4)_2$ showed high activity in an NFS-60 cell proliferation assay, with an EC_{50} about 10-fold lower than that of the non-linker fusion protein (Bai & Shen, 2006). Also, the insertion of the helical linker $(H4)_2$ improved

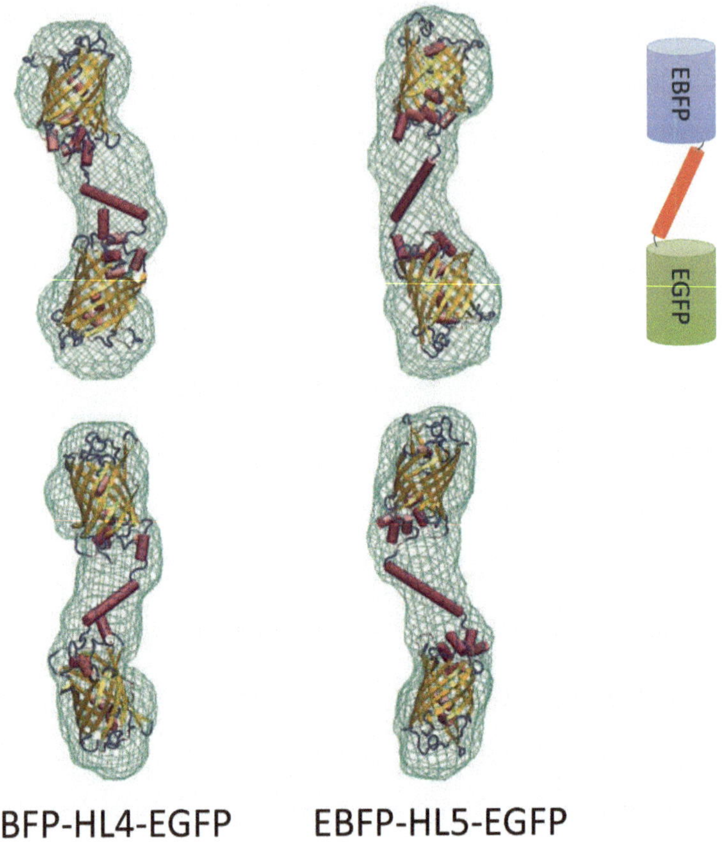

Fig. 2 Low-resolution models (wire frames) and high-resolution models (cartoon representation) of EBFP–HL4–EGFP and EBFP–HL5–EGFP based on SAXS analyses (Arai et al., 2004). Two different views are shown. Schematics of the fusion protein is shown on the right. *Reprinted with permission from (Arai et al., 2004). Copyright © 2004, Wiley-Liss.*

bioactivity of human growth hormone (hGH) and TF fusion proteins designated as hGH–$(H4)_2$–Tf and Tf–$(H4)_2$–hGH (Amet, Wang, & Shen, 2010).

Moreover, the linker optimization facilitated the development of bioactive and stable fusion proteins consisting of HSA and IFN-α2b. Three types of linkers including a flexible linker (GGGGS), a rigid Pro-rich linker (PAPAP) and a rigid helical linker (AEAAAKEAAAKA) were investigated as spacers to reduce interdomain interference (Zhao et al., 2008). With the linker insertion, the anti-viral activities of three fusion proteins were increased by 39% (flexible linker), 68% (Pro-rich linker) and 115% (helical linker) compared to the fusion protein without any linker. The enhanced bioactivity was likely attributed to the correct folding of the fusion protein and the proper separation of the protein domains after linker insertion. These studies indicate that the insertion of the helical linker is a feasible approach to improve expression and bioactivity of fusion proteins.

Furthermore, a hybrid linker of helical and flexible segments, $(GSG)_6A(EAAAK)_6A(GSG)_6A(EAAAK)_6A(GSG)_6$, was designed and applied to FRET-based fluorescent sensor proteins (fusion proteins between a pair of fluorescent proteins, Cerulean and Citrine) for antibody detection (Golynskiy et al., 2010) and quantification of macromolecular crowding in living cells (Boersma et al., 2015).

3. Design of linkers for protein-based nanostructures

3.1 Design of linkers for symmetrically self-assembling fusion proteins

The design of novel proteins that self-assemble into supramolecular complexes is useful for development in nanobiotechnology and synthetic biology (Arai, 2018; Kobayashi & Arai, 2017). In natural protein complexes (Ahnert, Marsh, Hernandez, Robinson, & Teichmann, 2015; Pieters, van Eldijk, Nolte, & Mecinovic, 2016), protein subunits self-assemble into symmetric polyhedral shapes, such as a tetrahedron, hexahedron, octahedron, dodecahedron, and icosahedron. Since symmetry provides a powerful tool for building protein-based nanostrucures, a general strategy using symmetry to construct protein nanomaterials as "nanohedra" was described about two decades ago (Padilla, Colovos, & Yeates, 2001) (Fig. 3A). In this strategy, a protein that naturally forms a self-assembling oligomer (e.g., dimer) is fused by a rigid helical linker to another protein that forms another self-assembling oligomer (e.g., trimer), and the fusion protein self-assembles with identical copies of itself into a designed nanohedral particle or material. Success in obtaining a well-ordered tetrahedral cage with the additional design relied on optimization of the helical linker region (Lai, Cascio, & Yeates, 2012). The mutations were designed based on inspection of the interfacial regions. One residue (Lys118) was mutated to a smaller alanine residue to avoid interference with the helical linker, and the other residues (Glu24 and Leu279) was mutated to add a favorable contact with residues on the helical linker (Lai, Tsai, Sawaya, Asturias, & Yeates, 2013). The nanohedra strategy allows for symmetrical construction of a wide variety of potentially useful protein-based nanomaterials (Yeates, Liu, & Laniado, 2016). Moreover, a

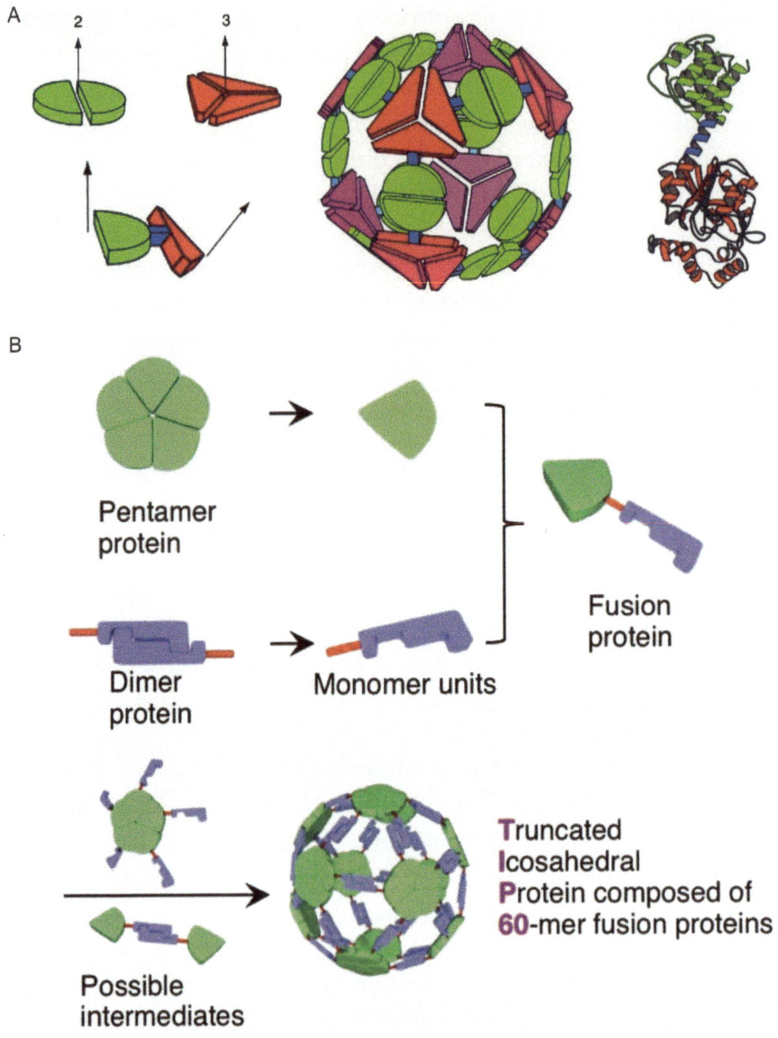

Fig. 3 Symmetrical self-assemblies of protein-based nanostructures. (A) A protein cage based on nanohedra strategy: a general strategy for designing fusion proteins that assemble into symmetric nanostructures (Padilla et al., 2001). (Left panel) The green semicircle represents a natural dimeric protein, whereas the red shape represents a trimeric protein. The symmetry axes of the natural oligomers are shown. The two natural proteins are combined by genetic methods into a single fusion protein. Each of the original natural proteins serves as an oligomerization domain in the designed fusion protein. A fusion protein is shown to illustrate that the oligomerization domains can be joined rigidly at a geometric angle. (Center panel) The fusion proteins self-assemble into

(Continued)

60-subunit truncated icosahedral protein nanoparticle (TIP60) was recently constructed from a designed fusion protein composed of a pentameric protein and a dimeric protein domain (Kawakami et al., 2018) (Fig. 3B). A selected short linker (RVL) of the TIP60 fusion protein may contribute to constructing a uniform structure of a truncated icosahedron (soccer-ball-like shape) as a rigid linker.

In addition, another simple, generalizable approach to assemble proteins into cage-like structures was described using short *de novo* designed coiled-coil domains to mediate assembly (Sciore et al., 2016). Eight copies of a C_3-symmetric trimeric esterase were assembled into a well-defined octahedral protein cage by appending a C_4-symmetric coiled-coil domain to the protein through a short flexible linker.

3.2 Design of linkers for protein nanobuilding blocks

Recently, Kobayashi et al. designed and constructed a protein nanobuilding block (PN-Block), WA20–foldon (Kobayashi et al., 2015), by fusing an intermolecularly folded dimeric *de novo* protein WA20 (Arai et al., 2012) and a trimeric foldon domain of T4 phage fibritin (Guthe et al., 2004) via a short linker (KLAAA). The WA20–foldon formed several types of self-assembling nanoarchitectures in multiples of 6-mers, including a barrel-like hexamer and a tetrahedron-like dodecamer (Fig. 4).

Moreover, to construct chain-like polymeric nanostructures, Kobayashi et al. designed *de novo* extender protein nanobuilding blocks (ePN-Blocks) by tandemly fusing two *de novo* binary-patterned proteins WA20 with

Fig. 3—Cont'd a cubic cage of a nanohedral structure. (Right panel) A ribbon diagram of a fusion protein showing one method for joining two oligomerization domains (red and green) in a relatively rigid fashion. One of the oligomerization domains must end in an α-helical conformation, and the other must begin in an α-helical conformation. The two protein domains are then linked by a short helical linker (blue) of amino acids that have a strong tendency to adopt an α-helical conformation. Thus, the two oligomerization domains are joined physically in a predictable orientation. (B) Schematics of design approach to produce a uniform protein nanoparticle, truncated icosahedral protein composed of 60-mer fusion proteins (TIP60) (Kawakami et al., 2018). Green-colored pentamer and blue-colored dimer proteins are corresponding to LSm and MyoX-coil, respectively. *Panel A: Reprinted with permission from (Padilla et al., 2001). Copyright © 2001, National Academy of Sciences, U.S.A. Panel B: Reprinted with permission from (Kawakami et al., 2018). Copyright © 2018, Wiley-VCH.*

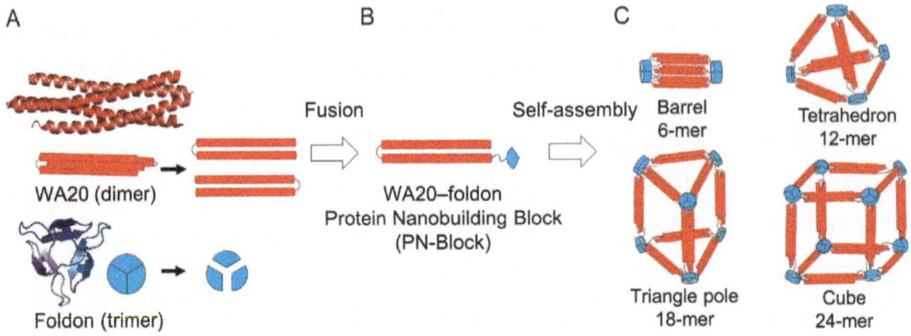

Fig. 4 Schematic diagrams of construction and assemblies of a WA20–foldon fusion protein as a protein nanobuilding block (PN-Block) (Kobayashi et al., 2015). (A) Ribbon representation and schematics of an intermolecularly folded dimeric *de novo* protein WA20 (PDB code: 3VJF) shown in red, and a trimeric foldon domain of T4 phage fibritin (PDB code: 1RFO) shown in blue. (B) Construction of the WA20-foldon fusion protein with a short linker. (C) Nanoarchitectures designed by expected self-assemblies of the WA20-foldon. In stable self-assembling complexes, the WA20–foldon is expected to form highly symmetric oligomers in multiples of 6-mer because of the combination of the WA20 dimer and the foldon trimer. *Reprinted with permission from (Kobayashi et al., 2015). Copyright © 2015, American Chemical Society.*

helical or flexible linkers (Kobayashi et al., 2018) (Fig. 5A – C). The ePN-Blocks with long helical or flexible linkers were well expressed in soluble fractions of *Escherichia coli*, and formed various structural homooligomers (Fig. 5D). Subsequently, heterooligomeric complexes were reconstructed from ePN-Blocks and stopper PN-Blocks (sPN-Blocks, i.e., original WA20) by denaturation and refolding (Fig. 5D). The small-angle X-ray scattering (SAXS) analyses showed that extender and stopper PN-Block (esPN-Block) heterocomplexes formed different types of extended chain-like conformations depending on their linker types: an esPN-Block heterocomplex with a helical linker $A(EAAAK)_4A$ (HL4) formed rigid conformation of a *Z*-like shape (Fig. 5E) and an esPN-Block heterocomplex with a flexible linker $(GGGGS)_4$ (FL4) formed dynamic conformations including flexible V-like shapes (Fig. 5F). These results indicate that the linker design significantly affects conformation and dynamics of fusion proteins and their complexes: helical linkers contribute to rigid conformation of protein-based nanostructures, whereas flexible linkers lead to dynamic and flexible conformations of nanostructures.

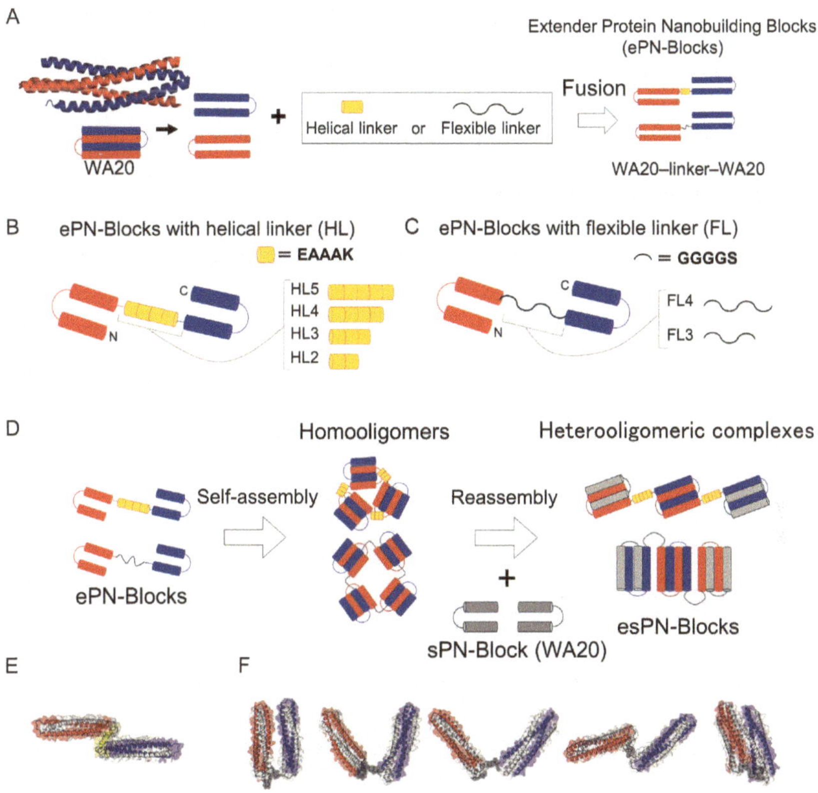

Fig. 5 Schematic diagrams of extender protein nanobuilding blocks (ePN-Blocks) (Kobayashi et al., 2018). (A) Construction of ePN-Blocks. Ribbon representation and schematics of an intermolecularly folded dimeric *de novo* protein WA20 (PDB code: 3VJF) are shown in red and blue. The ePN-Blocks were constructed by tandemly fusing two *de novo* WA20 proteins using various linkers (Arai et al., 2001, 2004). Helical and flexible linkers are shown as yellow rods and black lines, respectively. (B) Design of ePN-Blocks with helical linkers (HL). (C) Design of ePN-Blocks with flexible linkers (FL). (D) Schematics of self-assembly and reassembly of ePN-Blocks. The ePN-Blocks (red and blue) self-assemble into homooligomers. Furthermore, with the addition of stopper PN-Blocks (sPN-Blocks, *i.e.*, original WA20), heterooligomeric complexes (esPN-Blocks) of ePN-Blocks (red and blue) and sPN-Blocks (light gray) are reassembled by denaturation and refolding. (E) A rigid-body model structure of the esPN-Block (HL4) heterocomplex [e1s2(HL4)] composed of one ePN-Block and two sPN-Blocks. The model is shown as ribbon representation in translucent surface representation. The first WA20 domain (red), the helical linker (yellow), and the second WA20 domain (blue) of ePN-Block; sPN-Block (light gray). (F) Multi-state rigid-body model structures of the esPN-Block (FL4) heterocomplex [e1s2(FL4)] composed of one ePN-Block and two sPN-Blocks, based on SAXS analysis using MultiFoXS server (Schneidman-Duhovny, Hammel, Tainer, & Sali, 2016). The models are shown as ribbon representation in translucent surface representation. The first WA20 domain (red), the flexible linker FL4 (black), and the second WA20 domain (blue) of ePN-Block; sPN-Block (light gray). *Reprinted with permission from (Kobayashi et al., 2018). Copyright © 2018, American Chemical Society.*

4. Methods

4.1 Construction of expression plasmids of fusion proteins with linkers

4.1.1 Materials and reagents

In the cases of the fusion proteins of EBFP–linker–EGFP (Arai et al., 2001) and WA20–linker–WA20 (Kobayashi et al., 2018):

(1) Plasmid vectors: *e.g.*, pET-32b(+) (Novagen, Merck), pEGFP (Clontech, Takara Bio), pEBFP-N1 (Clontech, Takara Bio), pET3-WA20 (Arai et al., 2012).

(2) DNA fragments of linkers with appropriate restriction enzyme sites*:

Short linker (SL) (amino acid sequence: KLAAA).
aagcttgcggccgcg.
Helical linker 2 (HL2) (KLAEAAAKEAAAKAAA).
aagcttGCTGAGGCCGCTGCGAAAGAAGCTGCAGCG AAGgcggccgcg.
Helical linker 3 (HL3) (KLAEAAAKEAAAKEAAAKAAA).
aagcttGCTGAGGCCGCTGCGAAAGAAGCTGCAGCG AAGGAAGCTGCAGCGAAGgcggccgcg.
Helical linker 4 (HL4) (KLAEAAAKEAAAKEAAAKEAAAKAAA).
aagcttGCAGAGGCCGCGGCTAAGGAGGCCGCTGCG AAAGAAGCTGCAGCGAAGGAAGCTGCAGCG AAGgcggccgcg.
Helical linker 5 (HL5) (KLAEAAAKEAAAKEAAAKEAAAKEAAAKAAA).
aagcttGCCGAAGCAGCCGCAAAAGAGGCGGCGGCT AAGGAGGCCGCTGCGAAAGAAGCTGCAGCG AAGGAAGCTGCAGCGAAGgcggccgcg.
Flexible linker 3 (FL3) (KLGGGGSGGGGSGGGGSAAA).
aagcttGGTGGCGGTGGATCCGGTGGCGGTGGCTCC GGCGGTGGCGGTAGCgcggccgcg.
Flexible linker 4 (FL4) (KLSGGGGSGGGGSGGGGSGGGGSAAA).
aagcttTCCGGCGGGGGTGGATCCGGCGGGGGTGGA TCCGGTGGCGGTGGCTCCGGCGGTGGCGGT AGCgcggccgcg.

*The first amino-acid sequence KL is derived from the DNA sequence of *Hind* III site (aagctt). The last amino-acid sequence AAA is derived from the DNA sequence of the *Not* I site (gcggccgc).

(3) Regents for polymerase chain reaction (PCR): thermostable DNA polymerase with high fidelity, e.g., Pfu DNA polymerase (Agilent Technologies), KOD-Plus-Neo DNA polymerase (Toyobo); primers with appropriate restriction enzyme sites.

(4) Regents for DNA cloning: Restriction enzymes, DNA ligation reagents, *E. coli* competent cells.

4.1.2 Protocols

In the case of the fusion proteins of EBFP–linker–EGFP (Arai et al., 2001):

(1) A fusion expression system pET TRX Fusion System 32 (Novagen, Merck) with *E. coli* thioredoxin (Trx) was employed to enhance the solubility of the highly expressed proteins in *E. coli* cytoplasm. The DNA fragments encoding EBFP and EGFP were prepared from plasmids pEGFP and pEBFP-N1 by PCR using primers with appropriate restriction enzyme sites. The amplified fragments were digested by restriction enzymes and cloned into pET-32b(+) by DNA ligation to give the plasmid pET32-EBFP–SL–EGFP.

(2) The DNA fragments encoding the designed linkers were prepared by annealing and extension using the appropriate oligo-nucleotide primers with the cleavage sites for *Hind* III and *Not* I. These fragments were digested and cloned into pET32-EBFP–SL–EGFP between *Hind* III and *Not* I to give the plasmids pET32-EBFP–linker–EGFP (Fig. 6).

(3) Alternatively, restriction free cloning methods (Unger, Jacobovitch, Dantes, Bernheim, & Peleg, 2010; van den Ent & Lowe, 2006) can be used for plasmid construction instead of the conventional method by restriction enzymes and DNA ligation. Also, gene synthesis services can be utilized for not only preparing DNA fragments but also constructing a whole plasmid.

In the case of WA20–linker–WA20 (Kobayashi et al., 2018):

(1) A DNA fragment encoding the *de novo* protein WA20 was prepared from the plasmid pET3-WA20 (Arai et al., 2012) by PCR using T7 promoter primer and a primer with an appropriate restriction enzyme site (WA20RV_HindIII). The amplified fragment was digested using *Nde* I and *Hind* III and was cloned into pET32-EBFP–HL5–EGFP

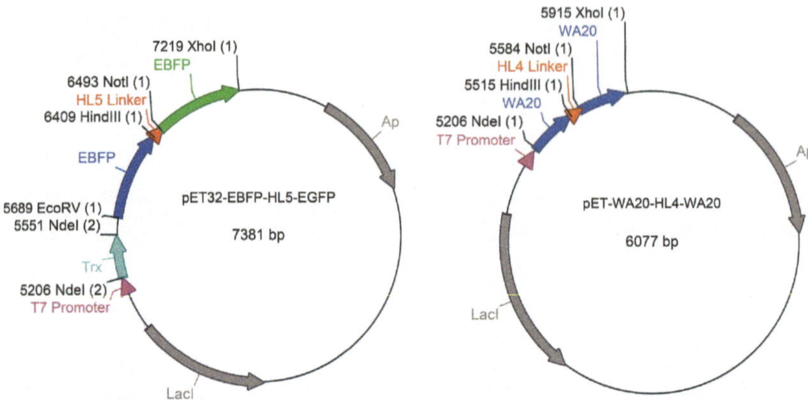

Fig. 6 Plasmid maps of pET32-EBFP–HL5–EGFP (Arai et al., 2001) and pET-WA20–HL4–WA20 (Kobayashi et al., 2018) for expression of the fusion proteins. The graphics were created using A plasmid Editor (ApE) (https://jorgensen.biology.utah.edu/wayned/ape/).

(Arai et al., 2001) between the *Nde* I and *Hind* III sites to construct the plasmid pET-WA20–HL5–GFP, in which the Trx tag and the EBFP gene were replaced with the WA20 gene. Another DNA fragment encoding WA20 was prepared from the plasmid pET3a-WA20 by PCR using primers with appropriate restriction enzyme sites (WA20FW_NotI and WA20RV_XhoI). The amplified fragment was then digested using *Not* I and *Xho* I and was cloned into pET-WA20–HL5–EGFP between the *Not* I and *Xho* I sites to give the expression plasmid pET-WA20–HL5–WA20 for an extender PN-Block (ePN-Block).

(2) DNA fragments encoding the other linker genes (HL2, HL3, HL4, FL3, and FL4) were prepared by digestion of the plasmid pET32-EBFP–linker–EGFP (Arai et al., 2001) with *Hind* III and *Not* I. The DNA fragments were inserted into pET-WA20–HL5–WA20 between the restriction sites to give the plasmids, pET-WA20–HL2–WA20, pET-WA20–HL3–WA20, pET-WA20–HL4–WA20 (Fig. 6), pET-WA20–FL3–WA20, and pET-WA20–FL4–WA20.

4.2 Expression and purification of fusion proteins

4.2.1 Materials and reagents

In the cases of the fusion proteins of EBFP–linker–EGFP (Arai et al., 2001) and WA20–linker–WA20 (Kobayashi et al., 2018):

(1) Expression plasmid vectors: pET32-EBFP–linker–EGFP or pET-WA20–linker–WA20

(2) *E. coli* for pET system protein expression: e.g., BL21 Star (DE3) (Invitrogen, Thermo Fisher Scientific), AD494(DE3)pLysS (Novagen, Merck)

(3) LB medium, appropriate antibiotics, isopropyl-β-D-thiogalactoside (IPTG)

(4) Immobilized metal affinity chromatography (IMAC) resin: *e.g.*, TALON metal affinity resin (Clontech, Takara Bio)

4.2.2 Protocols

In the case of the fusion proteins of EBFP–linker–EGFP (Arai et al., 2001):

(1) *E. coli* AD494(DE3)pLysS (Novagen) was transformed with the expression plasmids, and selected on LB agar plates containing 50 μg/mL ampicillin, 34 μg/mL chloramphenicol and 15 μg/mL kanamycin.

(2) For all cultivations, LB medium containing the same concentration of the three antibiotics was used. LB medium (1.5 L) was inoculated with 5 mL overnight culture at 30°C of each strain harboring pET32-EBFP–linker–EGFP containing the designed linkers, and cultured at 30°C for 7 h. At an OD_{600} of ≈0.5, IPTG was added to a final concentration of 1 mM to induce the expression of the fusion proteins, and cells were further cultured for 12 h at 16°C.

(3) Harvested cells were resuspended in sonication buffer (50 mM sodium phosphate, 300 mM NaCl, pH 7.0) and then disrupted by freeze–thaws and sonication. The supernatant was obtained after centrifugation twice at 20,000 × *g* for 20 min at 4°C. The fusion proteins which have the $(His)_6$-tag were purified using TALON metal affinity resin (Clontech) according to the manufacturer's protocol. The fractions with sufficient fluorescence activity were collected. The proteins were specifically digested with thrombin (Novagen) between Trx and EBFP, and Trx fragments were removed by size exclusion chromatography with Superdex 75 (Cytiva).

In the case of WA20–linker–WA20 (Kobayashi et al., 2018):

(1) WA20–linker–WA20 fusion proteins (ePN-Blocks) were expressed in *E. coli* BL21 Star (DE3) (Invitrogen) harboring pET-WA20–linker–WA20 in 2 L of LB broth (Lennox) (Nacalai tesque) containing 50 μg/mL ampicillin sodium salt at 37°C. Protein expression was induced using 0.2 mM IPTG at an optical density (OD_{600}) of about 0.8 at 600 nm, and cells were further cultured for 3–4 h at 37°C.

(2) Proteins were extracted from harvested cells by sonication in lysis buffer containing 50mM sodium phosphate buffer (pH7.0), 300mM NaCl, and 10% glycerol. Proteins were then purified using IMAC with TALON metal affinity resin (Clontech) according to the manufacturer's protocols. The equilibration/wash buffer contained 50mM sodium phosphate buffer (pH7.0) and 300mM NaCl, and the elution buffer contained 50mM sodium phosphate buffer (pH7.0), 300mM NaCl, 10% glycerol, and 250mM imidazole. Because many histidine residues are exposed on the surface of the WA20 structure (Arai et al., 2012), WA20 and ePN-Block proteins can bind TALON metal affinity resin even without a His-tag.

(3) Sodium dodecyl sulfate polyacrylamide gel electrophoresis (SDS-PAGE) analysis was performed according to the standard Laemmli procedure. The stacking gel contained 4.5% polyacrylamide, 125mM Tris–HCl (pH 6.8), and 0.1% SDS and the separating gel comprised 17.5% polyacrylamide, 125mM Tris–HCl (pH 8.8), and proteins were electrophoresed in running buffer containing 25mM Tris, 192mM glycine, and 0.1% SDS. For native PAGE analysis, stacking gels, separating gels (5% or 7.5% polyacrylamide), and running buffer were prepared as for SDS-PAGE, but with no SDS. Proteins in the gels were stained with Coomassie brilliant blue (CBB) R-250.

4.3 Measurement of fluorescence spectra and circular dichroism (CD) spectra

In the case of the fusion proteins of EBFP–linker–EGFP (Arai et al., 2001):

4.3.1 Materials and equipment

(1) Purified fusion protein samples: EBFP–linker–EGFP

(2) Fluorescence spectrophotometer: e.g., F-2000 (Hitachi)

(3) CD spectropolarimeter: e.g., J-725 (Jasco, Tokyo, Japan)

4.3.2 Protocols

(1) Fluorescence spectra and the intensity at 444nm (EBFP emission peak) of the fusion proteins were measured for evaluation of fluorescent resonance energy transfer (FRET) using a fluorescence spectrophotometer with 380nm excitation at 25°C

(2) The CD spectra of the fusion proteins were measured using a CD spectropolarimeter at 25°C

4.4 Small-angle X-ray scattering (SAXS) analyses of protein nanobuilding blocks

In the case of WA20–linker–WA20 (Kobayashi et al., 2018):

4.4.1 Materials and equipment

(1) Purified fusion protein samples: WA20–linker–WA20

(2) SAXS beamline at synchrotron facility: *e.g.*, KEK Photon Factory BL-10C (Tsukuba, Japan)

4.4.2 Protocols

(1) SAXS measurements were performed on several fraction samples of ePN-Block homooligomers and esPN-Block heterocomplexes after separation by size exclusion chromatography (SEC), sPN-Block (WA20), and chicken egg white lysozyme (FUJIFILM Wako Pure Chemical) in 20 mM HEPES buffer (pH 7.5) containing 100 mM NaCl, 200 mM ArgHCl, and 10% glycerol at 20°C using synchrotron radiation ($\lambda = 0.1488\,\text{nm}$) with a PILATUS3 2M detector (Dectris, Baden, Switzerland) at the Photon Factory BL-10C beamline (Igarashi et al., 2011). Two-dimensional scattering images were integrated into one-dimensional scattering intensities ($I(q)$) as a function of the magnitude of the scattering vector $q = (4\pi/\lambda)\sin(\theta/2)$ using FIT2D (Hammersley, Svensson, Hanfland, Fitch, & Häusermann, 1996) or SAngler (Shimizu et al., 2016), where θ is the total scattering angle.

(2) The scattering intensity for a colloidal dispersion is generally given by the product of the form factor $P(q)$ and structure factor $S(q)$. Hence, $I(q) = \text{n}\ P(q)\ S(q)$, where n is the number density of the particle. Because interparticle interactions such as the excluded volume effect and electrostatic interactions can be neglected at low protein and high salt concentrations, the structure factor was almost at unity, $I(q) \approx \text{n}\ P(q)$. Thus, the form factor is given by the Fourier transformation of the pair-distance distribution function $p(r)$, which expresses the size and shape of the particle as follows:

$$P(q) = 4\pi \int_0^{D_{\max}} p(r) \frac{\sin qr}{qr}\, dr,$$

where $D_{\max}$ is the maximum intraparticle distance. The indirect Fourier transformation (IFT) technique was used to calculate $p(r)$ for particles with a virtually model-free routine (Brunner-Popela & Glatter,

1997; Glatter, 1980; Glatter & Kratky, 1982). Forward scattering intensity $I(q \rightarrow 0)$ was extrapolated from SAXS data, and the radius of gyration R_g, was estimated using the Guinier approximation (Glatter & Kratky, 1982).

(3) Rigid-body models of esPN-Block heterocomplexes were constructed using the programs Coot (Emsley, Lohkamp, Scott, & Cowtan, 2010) and Foldit Standalone (Kleffner et al., 2017) based on the crystal structure of the *de novo* protein WA20 dimer (PDB code, 3VJF) (Arai et al., 2012) with consideration of N- and C-terminal directions. Rigid-body models were manually and iteratively refined to minimize differences between data calculated from models and those determined in SAXS experiments. The program CRYSOL (Svergun, Barberato, & Koch, 1995) in the ATSAS program suite (Franke et al., 2017) was used for evaluating the solution scattering from the models and fitting it to experimental scattering curves. Multi-state rigid-body model structures were constructed based on SAXS profiles using MultiFoXS server (Schneidman-Duhovny et al., 2016).

(4) The low-resolution dummy atom models were constructed from the SAXS data using *ab initio* shape modeling programs as following. Calculations of rapid *ab-initio* shape determination were performed 10 times by DAMMIF (Franke & Svergun, 2009) without a symmetry constraint, and the generated models were aligned and averaged by DAMAVER (Volkov & Svergun, 2003) in the ATSAS program suite (Franke et al., 2017) for small-angle scattering data analysis from biological macromolecules. The DAMAVER model was modified with fixed core by DAMSTART and further refinement of the model was performed by DAMMIN (Svergun, 1999). Superimposing model structures were performed by SUPCOMB (Kozin & Svergun, 2000).

5. Concluding remarks

Recombinant fusion proteins have been broadly used in a variety of fields of protein engineering and biotechnology. About two decades ago, "linker engineering" was proposed to design and optimize linkers for multifunctional fusion proteins, and the empirical rigid linker with the α-helix-forming sequence was designed to separate two protein domains and to control the distance between the domains (Arai et al., 2001). Thereafter, the helical linkers have been widely used to construct fusion proteins and to improve expression and bioactivity of fusion proteins. In addition, the design of fusion proteins that self-assemble into supramolecular complexes is useful for nanobiotechnology and synthetic biology. The linker design for self-assembling fusion proteins is important for not only construction but also conformation and dynamics of protein-based nanostructures. Appropriately designed fusion proteins of symmetric oligomers joined by a rigid α-helical linker can self-assemble into well-ordered nanostructures such as cages. In addition, a strategy to connect α-helices in protein domains with a spanning α-helical linker has been developed (Lai, Jiang, Chen, & Yeates, 2015). Furthermore, since rational and computational approaches for *de novo* protein design have greatly advanced in recent years (Huang, Boyken, & Baker, 2016), combination of computational and experimental approaches for linker engineering will open a way for the rational *de novo* design of multi-domain artificial proteins.

Acknowledgments

I thank all my lab members, our colleagues and collaborators, especially, Dr. Naoya Kobayashi at Institute for Molecular Science, Prof. Michael H. Hecht at Princeton University, Dr. Norifumi Kawakami at Keio University, Prof. Takaaki Sato at Shinshu University, and Prof. Teruyuki Nagamune at University of Tokyo for fruitful collaborations. The synchrotron X-ray experiments were performed at Photon Factory (PF), KEK, under the approval of PF program advisory committee (Proposal No. 2014G111, 2016G153, 2016G606, and 2018G634). We thank the beamline scientists and staff at PF, KEK. This work was supported by JSPS KAKENHI Grant Numbers JP16K05841, JP16H00761, JP17KK0104, and JP19H02522.

References

Ahnert, S. E., Marsh, J. A., Hernandez, H., Robinson, C. V., & Teichmann, S. A. (2015). Principles of assembly reveal a periodic table of protein complexes. *Science*, *350*, aaa2245.

Alfthan, K., Takkinen, K., Sizmann, D., Soderlund, H., & Teeri, T. T. (1995). Properties of a single-chain antibody containing different linker peptides. *Protein Engineering*, *8*, 725–731.

Amet, N., Lee, H. F., & Shen, W. C. (2009). Insertion of the designed helical linker led to increased expression of tf-based fusion proteins. *Pharmaceutical Research*, *26*, 523–528.

Amet, N., Wang, W., & Shen, W. C. (2010). Human growth hormone-transferrin fusion protein for oral delivery in hypophysectomized rats. *Journal of Controlled Release*, *141*, 177–182.

Arai, R. (2018). Hierarchical design of artificial proteins and complexes toward synthetic structural biology. *Biophysical Reviews*, *10*, 391–410.

Arai, R., Kobayashi, N., Kimura, A., Sato, T., Matsuo, K., Wang, A. F., et al. (2012). Domain-swapped dimeric structure of a stable and functional *de novo* four-helix bundle protein, WA20. *Journal of Physical Chemistry B*, *116*, 6789–6797.

Arai, R., Ueda, H., Kitayama, A., Kamiya, N., & Nagamune, T. (2001). Design of the linkers which effectively separate domains of a bifunctional fusion protein. *Protein Engineering*, *14*, 529–532.

Arai, R., Wriggers, W., Nishikawa, Y., Nagamune, T., & Fujisawa, T. (2004). Conformations of variably linked chimeric proteins evaluated by synchrotron X-ray small-angle scattering. *Proteins, 57*, 829–838.

Argos, P. (1990). An investigation of oligopeptides linking domains in protein tertiary structures and possible candidates for general gene fusion. *Journal of Molecular Biology, 211*, 943–958.

Bai, Y., & Shen, W. C. (2006). Improving the oral efficacy of recombinant granulocyte colony-stimulating factor and transferrin fusion protein by spacer optimization. *Pharmaceutical Research, 23*, 2116–2121.

Bhandari, D. G., Levine, B. A., Trayer, I. P., & Yeadon, M. E. (1986). 1H-NMR study of mobility and conformational constraints within the proline-rich N-terminal of the LC1 alkali light chain of skeletal myosin. Correlation with similar segments in other protein systems. *European Journal of Biochemistry, 160*, 349–356.

Boersma, A. J., Zuhorn, I. S., & Poolman, B. (2015). A sensor for quantification of macromolecular crowding in living cells. *Nature Methods, 12*, 227–229.

Brunner-Popela, J., & Glatter, O. (1997). Small-angle scattering of interacting particles. 1. Basic principles of a global evaluation technique. *Journal of Applied Crystallography, 30*, 431–442.

Chen, X., Zaro, J. L., & Shen, W. C. (2013). Fusion protein linkers: Property, design and functionality. *Advanced Drug Delivery Reviews, 65*, 1357–1369.

Crasto, C. J., & Feng, J. A. (2000). LINKER: A program to generate linker sequences for fusion proteins. *Protein Engineering, 13*, 309–312.

Emsley, P., Lohkamp, B., Scott, W. G., & Cowtan, K. (2010). Features and development of coot. *Acta Crystallographica Section D, 66*, 486–501.

Franke, D., Petoukhov, M. V., Konarev, P. V., Panjkovich, A., Tuukkanen, A., Mertens, H. D. T., et al. (2017). ATSAS 2.8: A comprehensive data analysis suite for small-angle scattering from macromolecular solutions. *Journal of Applied Crystallography, 50*, 1212–1225.

Franke, D., & Svergun, D. I. (2009). DAMMIF, a program for rapid ab-initio shape determination in small-angle scattering. *Journal of Applied Crystallography, 42*, 342–346.

George, R. A., & Heringa, J. (2002). An analysis of protein domain linkers: Their classification and role in protein folding. *Protein Engineering, 15*, 871–879.

Glatter, O. (1980). Evaluation of small-angle scattering data from lamellar and cylindrical particles by the indirect Fourier transformation method. *Journal of Applied Crystallography, 13*, 577–584.

Glatter, O., & Kratky, O. (1982). *Small-angle X-ray scattering*. New York: Academic Press.

Golynskiy, M. V., Rurup, W. F., & Merkx, M. (2010). Antibody detection by using a FRET-based protein conformational switch. *Chembiochem, 11*, 2264–2267.

Guthe, S., Kapinos, L., Moglich, A., Meier, S., Grzesiek, S., & Kiefhaber, T. (2004). Very fast folding and association of a trimerization domain from bacteriophage T4 fibritin. *Journal of Molecular Biology, 337*, 905–915.

Hammersley, A. P., Svensson, S. O., Hanfland, M., Fitch, A. N., & Häusermann, D. (1996). Two-dimensional detector software: From real detector to idealised image or two-theta scan. *High Pressure Research, 14*, 235–248.

Huang, P. S., Boyken, S. E., & Baker, D. (2016). The coming of age of de novo protein design. *Nature, 537*, 320–327.

Igarashi, N., Watanabe, Y., Shinohara, Y., Inoko, Y., Matsuba, G., Okuda, H., et al. (2011). Upgrade of the small angle X-ray scattering beamlines at the photon factory. *Journal of Physics: Conference Series, 272*, 012026.

Kawakami, N., Kondo, H., Matsuzawa, Y., Hayasaka, K., Nasu, E., Sasahara, K., et al. (2018). Design of hollow protein nanoparticles with modifiable interior and exterior surfaces. *Angewandte Chemie International Edition, 57*, 12400–12404.

Kleffner, R., Flatten, J., Leaver-Fay, A., Baker, D., Siegel, J. B., Khatib, F., et al. (2017). Foldit standalone: A video game-derived protein structure manipulation interface using Rosetta. *Bioinformatics*, *33*, 2765–2767.

Kobayashi, N., & Arai, R. (2017). Design and construction of self-assembling supramolecular protein complexes using artificial and fusion proteins as nanoscale building blocks. *Current Opinion in Biotechnology*, *46*, 57–65.

Kobayashi, N., Inano, K., Sasahara, K., Sato, T., Miyazawa, K., Fukuma, T., et al. (2018). Self-assembling supramolecular nanostructures constructed from *de novo* extender protein nanobuilding blocks. *ACS Synthetic Biology*, *7*, 1381–1394.

Kobayashi, N., Yanase, K., Sato, T., Unzai, S., Hecht, M. H., & Arai, R. (2015). Self-assembling nano-architectures created from a protein nano-building block using an intermolecularly folded dimeric *de novo* protein. *Journal of the American Chemical Society*, *137*, 11285–11293.

Kozin, M. B., & Svergun, D. I. (2000). Automated matching of high- and low-resolution structural models. *Journal of Applied Crystallography*, *34*, 33–41.

Lai, Y. T., Cascio, D., & Yeates, T. O. (2012). Structure of a 16-nm cage designed by using protein oligomers. *Science*, *336*, 1129.

Lai, Y. T., Jiang, L., Chen, W., & Yeates, T. O. (2015). On the predictability of the orientation of protein domains joined by a spanning alpha-helical linker. *Protein Engineering, Design and Selection*, *28*, 491–499.

Lai, Y. T., Tsai, K. L., Sawaya, M. R., Asturias, F. J., & Yeates, T. O. (2013). Structure and flexibility of nanoscale protein cages designed by symmetric self-assembly. *Journal of the American Chemical Society*, *135*, 7738–7743.

Maeda, Y., Ueda, H., Kazami, J., Kawano, G., Suzuki, E., & Nagamune, T. (1997). Engineering of functional chimeric protein G-Vargula luciferase. *Analytical Biochemistry*, *249*, 147–152.

Marqusee, S., & Baldwin, R. L. (1987). Helix stabilization by Glu-..Lys+ salt bridges in short peptides of de novo design. *Proceedings of the National Academy of Sciences of the United States of America*, *84*, 8898–8902.

McCormick, A. L., Thomas, M. S., & Heath, A. W. (2001). Immunization with an interferon-gamma-gp120 fusion protein induces enhanced immune responses to human immunodeficiency virus gp120. *Journal of Infectious Diseases*, *184*, 1423–1430.

Padilla, J. E., Colovos, C., & Yeates, T. O. (2001). Nanohedra: Using symmetry to design self assembling protein cages, layers, crystals, and filaments. *Proceedings of the National Academy of Sciences of the United States of America*, *98*, 2217–2221.

Pieters, B. J., van Eldijk, M. B., Nolte, R. J., & Mecinovic, J. (2016). Natural supramolecular protein assemblies. *Chemical Society Reviews*, *45*, 24–39.

Robinson, C. R., & Sauer, R. T. (1998). Optimizing the stability of single-chain proteins by linker length and composition mutagenesis. *Proceedings of the National Academy of Sciences of the United States of America*, *95*, 5929–5934.

Schneidman-Duhovny, D., Hammel, M., Tainer, J. A., & Sali, A. (2016). FoXS, FoXSDock and MultiFoXS: Single-state and multi-state structural modeling of proteins and their complexes based on SAXS profiles. *Nucleic Acids Research*, *44*, W424–W429.

Sciore, A., Su, M., Koldewey, P., Eschweiler, J. D., Diffley, K. A., Linhares, B. M., et al. (2016). Flexible, symmetry-directed approach to assembling protein cages. *Proceedings of the National Academy of Sciences of the United States of America*, *113*, 8681–8686.

Shimizu, N., Yatabe, K., Nagatani, Y., Saijyo, S., Kosuge, T., & Igarashi, N. (2016). Software development for analysis of small-angle X-ray scattering data. *AIP Conference Proceedings*, *1741*, 050017.

Svergun, D. I. (1999). Restoring low resolution structure of biological macromolecules from solution scattering using simulated annealing. *Biophysical Journal*, *76*, 2879–2886.

Svergun, D., Barberato, C., & Koch, M. H. J. (1995). CRYSOL—A program to evaluate x-ray solution scattering of biological macromolecules from atomic coordinates. *Journal of Applied Crystallography*, *28*, 768–773.
Terpe, K. (2003). Overview of tag protein fusions: From molecular and biochemical fundamentals to commercial systems. *Applied Microbiology and Biotechnology*, *60*, 523–533.
Turner, S. L., Russell, G. C., Williamson, M. P., & Guest, J. R. (1993). Restructuring an interdomain linker in the dihydrolipoamide acetyltransferase component of the pyruvate dehydrogenase complex of Escherichia coli. *Protein Engineering*, *6*, 101–108.
Unger, T., Jacobovitch, Y., Dantes, A., Bernheim, R., & Peleg, Y. (2010). Applications of the restriction free (RF) cloning procedure for molecular manipulations and protein expression. *Journal of Structural Biology*, *172*, 34–44.
van den Ent, F., & Lowe, J. (2006). RF cloning: A restriction-free method for inserting target genes into plasmids. *Journal of Biochemical and Biophysical Methods*, *67*, 67–74.
Volkov, V. V., & Svergun, D. I. (2003). Uniqueness of ab initio shape determination in small-angle scattering. *Journal of Applied Crystallography*, *36*, 860–864.
Yeates, T. O., Liu, Y., & Laniado, J. (2016). The design of symmetric protein nanomaterials comes of age in theory and practice. *Current Opinion in Structural Biology*, *39*, 134–143.
Yu, K., Liu, C., Kim, B. G., & Lee, D. Y. (2015). Synthetic fusion protein design and applications. *Biotechnology Advances*, *33*, 155–164.
Zhao, H. L., Yao, X. Q., Xue, C., Wang, Y., Xiong, X. H., & Liu, Z. M. (2008). Increasing the homogeneity, stability and activity of human serum albumin and interferon-alpha2b fusion protein by linker engineering. *Protein Expression and Purification*, *61*, 73–77.

CHAPTER NINE

Synthetic protein switches: Combinatorial linker engineering with iFLinkC

Alexander Gräwe[a,b], Jan Ranglack[a,b], Anastasia Weyrich[a], and Viktor Stein[a,b,*]

[a]Fachbereich Biologie, Technische Universität Darmstadt, Darmstadt, Germany
[b]Centre for Synthetic Biology, Technische Universität Darmstadt, Darmstadt, Germany
[*]Corresponding author: e-mail address: stein@bio.tu-darmstadt.de

Contents

Methods in Enzymology, Volume 647
ISSN 0076-6879
https://doi.org/10.1016/bs.mie.2020.09.009

Abstract

Linker engineering constitutes a critical, yet frequently underestimated aspect in the construction of synthetic protein switches and sensors. Notably, systematic strategies to engineer linkers by predictive means remain largely elusive to date. This is primarily due to our insufficient understanding how the biophysical properties that underlie linker functions mediate the conformational transitions in artificially engineered protein switches and sensors. The construction of synthetic protein switches and sensors therefore heavily relies on experimental trial-and-error. Yet, methods for effectively generating linker diversity at the genetic level are scarce. Addressing this technical shortcoming, iterative functional linker cloning (iFLinkC) enables the combinatorial assembly of linker elements with functional domains from sequence verified repositories that are developed and stored in-house. The assembly process is highly scalable and given its recursive nature generates linker diversity in a combinatorial and exponential fashion based on a limited number of linker elements.

1. Introduction

1.1 Engineering artificial sensory systems

A central goal in synthetic biology concerns the development of systematic approaches to engineer artificial sensory functions (Bashor & Collins, 2018; Kiel, Yus, & Serrano, 2010). Applications of artificially engineered signaling functions are numerous, for instance, in industrial biotechnology regulating host cell metabolism (Gao, Xu, Ye, Chen, & Liu, 2019), in biomedicine co-ordinating a response against pathologic processes (Tolle, Stücheli, & Fussenegger, 2019) and in basic research exploring the design principles of naturally occurring signaling circuits (Lim, 2010).

The construction of artificial signaling functions has been primarily realized through transcriptional circuits where signals are processed through networks of mutually regulating transcription factors that receive, compute and actuate signals in response to distinct molecular queues (Brophy & Voigt, 2014). The construction of transcriptional circuits has been greatly facilitated by the modular organization of regulatory elements that control gene expression, especially in prokaryotes. This includes promoters, operators, ribosome binding sites and terminators that function sufficiently context independent so that, in practice, they can be exchanged with sufficient ease and comparatively predictive outcomes. Applications of transcriptional circuits range from simple sensory functions where transcription factors control the expression of a reporter gene such as GFP to report on the presence or absence of a metabolite (Lim, Jang, Jang, Seo, & Jung, 2018; Michener, Thodey, Liang, & Smolke, 2012) to increasingly complex circuits that are composed of mutually regulating transcription factors that are capable of computing Boolean logic gates and complex response functions such as bistability or adaptation (Brophy & Voigt, 2014).

In contrast, engineering signaling functions entirely based on proteins where molecular queues are entirely processed by post-translational means has proven substantially more challenging. Arguably, this is because proteins display less functional modularity compared to transcriptional units—yet principally promise to sense, propagate and actuate signals orders of magnitude faster while drawing on a large repertoire of naturally occurring receptor proteins and actuators (Golynskiy, Koay, Vinkenborg, & Merkx, 2011; Grünberg & Serrano, 2010; Stein & Alexandrov, 2015).

1.2 Engineering synthetic protein switches and sensors

At the molecular level, protein switches constitute the most elementary components in biological signal processing as they generate a tailored output in response to a specific input. Depending on the type of switch, an input queue can be actuated in a number of different ways. For instance, fluorescent and bioluminescent protein sensors frequently undergo changes in their optical properties rendering them able to report on the presence or absence of an analyte in a direct, non-invasive and stoichiometric fashion (Kaczmarski, Mitchell, Spence, Vongsouthi, & Jackson, 2019; Suzuki & Nagai, 2017). Alternatively, synthetic protein switches may form part of more complex signaling functions as they become embedded in a network of mutually regulating signaling proteins. In this case, molecular queues are relayed through post-translational modifications such as phosphorylation (Furukawa & Hohmann, 2013), proteolysis (Gräwe, Ranglack, Weber, & Stein, 2020) or non-covalent interactions that control the assembly of larger signaling complexes.

Synthetic protein switches with tailored response functions are typically engineered through recombination of an allosteric receptor with an actuator (Golynskiy et al., 2011; Stein & Alexandrov, 2015). Mechanistically, signals are generally triggered through structural rearrangements in the allosteric receptor upon binding its target analyte and subsequently transmitted via defined linker elements to the actuator to modulate its functional properties. Depending how an actuator is recombined with an allosteric receptor, structural rearrangements can either feature subtle distortions in the tertiary structure of an enzyme (Guntas, Mansell, Kim, & Ostermeier, 2005) or a fluorescent protein (Baird, Zacharias, & Tsien, 1999) or comprise the large relative movements of structurally distinct domains that result in an inhibitor being dislodged from the active site of an enzyme (Nirantar, Yeo, Chee, Lane, & Ghadessy, 2013; Stein & Alexandrov, 2014).

A key, yet poorly understood factor concerns the identity of linker elements that connect individual domains and transmit the conformational changes in the allosteric receptor to the actuator. In particular, linkers need to be sufficiently flexible to allow the relative movement of functional domains, but at the same time sufficiently stiff to effectively transmit an adequate amount of force and trigger the structural rearrangements in the actuator (Ma, Tsai, Haliloğlu, & Nussinov, 2011). In addition, linkers are increasingly shown to influence the expression and folding of adjacent protein domains which is particularly relevant in the context of fusion proteins that are composed of multiple structurally distinct domains (Arviv & Levy, 2012; Levy, 2017).

2. Combinatorial strategies for linker engineering

With an increasing repertoire of allosteric receptors and actuators, the key challenge is to functionally recombine the component parts into functional protein switches (Stein, 2017; Stein & Alexandrov, 2015). In practice, this critically depends on (i) the relative orientation of functional domains and (ii) the identity of the linkers connecting individual domains. The relative orientation of functional domains is largely determined by the tertiary structure of the allosteric receptor and actuator with their N- and C-terminal ends fixed in space. In a limited number of cases, these can be relocated by means of circular permutation (Yu & Lutz, 2011). Here, the original N- and C-termini are covalently joined through a linker while new N- and C-termini emerge in an alternative location. The second aspect concerns the functional properties of linkers that functionally couple an allosteric receptor with an actuator. These critically depend on length, flexibility and their relative combination especially in the context of fusion proteins that are composed of multiple domains. In addition, linkers affect the capacity of functional domains to fold and move relative to each other. This means linkers cannot be considered in isolation, but are highly context-dependent and therefore always need to be co-optimized, especially in the context of a protein switch.

To optimize the composition of linkers in the context of a synthetic protein switch or sensor, one can draw on a large number of different strategies to effectively diversify the functional properties of a protein at the genetic level and build libraries for subsequent screening or selection. This includes methods to randomize amino acids at either distinct or multiple sites by means of saturation mutagenesis (Cozens & Pinheiro, 2018; Kegler-ebo, Docktor, & Dimaio, 1994), randomly throughout the protein by means of error-prone PCR (Cadwell & Joyce, 1992) or randomly shuffle the genetic diversity underlying homologous protein sequences by DNA shuffling (Zhao & Arnold, 1997).

In contrast, methods to effectively diversify functional linker properties are very limited: For instance, the length of linker and sites for inserting domains and circular permutations can be targeted with single amino acid resolution either by limited nuclease digests (Ribeiro, Warren, & Ostermeier, 2017; Tullman, Guntas, Dumont, & Ostermeier, 2011) or by PCR based on staggered sets of oligonucleotide primers (Ohlendorf, Schumacher, Richter, & Möglich, 2016). Alternatively, linkers with a variable number of repeat elements can be subject to limited restriction digests in order to generate a linker library of defined repeats, but variable length (Evers, Van Dongen, Faesen, Meijer, & Merkx, 2006; Van Dongen et al., 2007; Van Rosmalen, Krom, & Merkx, 2017). Finally, linkers can be composed of smaller repeat elements with defined flexibility and length that are mixed and matched (Li et al., 2016; Norris & Hughes, 2018). While these approaches provide useful starting points for diversifying the biophysical properties of linker elements, they are limited to single linker elements. To address these technical shortcomings and provide a method for the simultaneous recombination of multiple linker elements, our laboratory recently developed a scalable DNA assembly method that was termed iterative functional linker cloning (iFLinkC) (Gräwe, Ranglack, Weyrich, & Stein, 2020). This DNA assembly method is highly scalable as it allows to assemble fusion proteins composed of structurally distinct domains in a modular and combinatorial fashion. In particular, iFLinkC is recursive and therefore allows to assemble fusion proteins in a hierarchical manner to provide combinatorial access to genetic linker diversity based on a limited number of linker elements.

3. Combinatorial linker engineering with iFLinkC

The majority of synthetic protein switches and sensors are engineered through recombination of naturally occurring allosteric receptor proteins with an actuator. With a relatively poor capacity to engineer the functional properties of a linker by predictive means, the identification of optimal linker compositions heavily relies on experimental trial-and-error. Yet, only a very limited number of methods are available for randomizing the diversity of linker elements at the genetic level in an effective manner. Addressing this technological shortcoming, iFLinkC comprises an iterative DNA assembly method (Fig. 1) which enables the combinatorial assembly of functional domains and defined linker elements independent of their length and separated by a single Gly residue (Gräwe, Ranglack, Weyrich, & Stein, 2020).

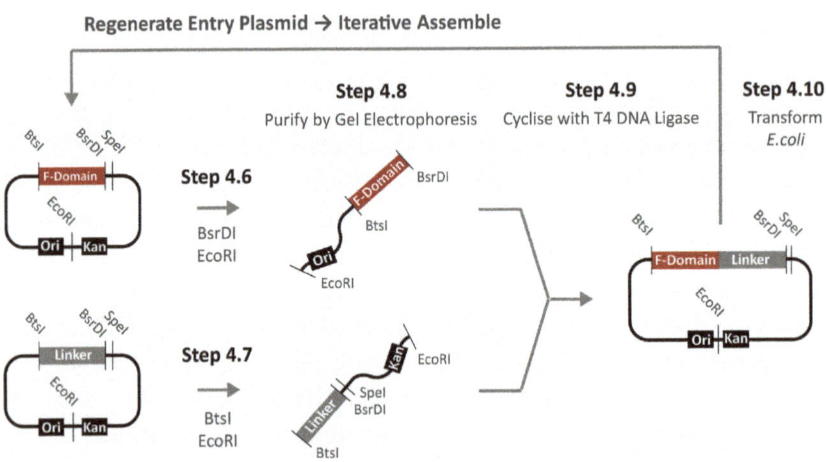

Fig. 1 Schematic overview of the elementary iFLinkC assembly process and the key experimental Steps 4.6–4.10. The iFLinkC assembly reaction relies on the combined action of type IIS restriction enzymes and T4 DNA ligase in order to fuse two protein coding regions in-frame bridged by a single Gly residue. Regenerating the entry plasmid in turn enable the iterative assembly of combinatorial linker libraries. *Illustration adapted with modifications from Gräwe, A., Ranglack, J., Weyrich, A., & Stein, V. (2020). iFLinkC: An iterative functional linker cloning strategy for the combinatorial assembly and recombination of linker peptides with functional domains.* Nucleic Acids Research, 48, *e24. https://doi.org/10.1093/nar/gkz1210.*

3.1 Vectors and DNA coding for linker elements and functional domains

1. Entry vectors pFD and pL2 for storing functional domains and linker elements, respectively. To this end, any pFD (Addgene ID #137107–#137113) or pL2 vector (Addgene ID #137091–137106) can serve as a starting point to generate plasmid backbones for inserting new functional domains or linker elements.
2. Empty expression vector pFLinkC-XE to express proteins in *E. coli* (Addgene ID #137114).
3. Selection of linker elements cloned in pL2 (Table 1, Addgene ID #137091–137106).
4. Selection of receptor and actuators cloned in pFD (Addgene ID #137107–#137113).
5. Custom linker elements and functional domains (commercial DNA synthesis).
6. Primers for linearizing pFD and pL2 backbones and sequencing DNA parts and libraries (Table 2).

3.2 Cloning strains

1. Competent DH10β (or alternative cloning strain such as Top10) for building a repository of linker elements, functional domains and combinatorial linker libraries. In general, chemically competent *E. coli* cells are transformed with single constructs. In this case, chemically competent *E. coli* cells may be prepared by $CaCl_2$ precipitation and transformed by heat-shock based on standard laboratory procedures (Sambrook & Russell, 2006a). For large libraries, transformation of *E. coli* is preferably done by electroporation (Sambrook & Russell, 2006b).

3.3 Media and reagents for bacterial cell culture

1. Lysogeny broth (LB) medium: 1% tryptone (w/v), 0.5% yeast extract (w/v) and 1% NaCl (w/v).
2. LB agar plates: LB medium supplemented with 1.5% agar (w/v).
3. Super optimal broth with catabolite repression (SOC): 2% tryptone (w/v), 0.5% yeast extract (w/v), 10 mM NaCl, 2.5 mM KCl, 10 mM $MgCl_2$, 20 mM $MgSO_4$ and 20 mM glucose.
4. 1000 × ampicillin stock solution (100 mg/mL).
5. 1000 × kanamycin stock solution (50 mg/mL).

Table 1 Linker elements available for iFLinkC assembly (with Addgene ID #137090–137106).

Linker name	Amino acid sequence	Length	Description
	Short		
G	G	1	Short linkers are operationally defined to comprise less than three amino acids, two of which constitute Gly-residues
G2	GG	2	
GPG	GPG	3	
G3	GGG	3	
	Flexible		
GGS4	GGSG	4	Flexible linkers are enriched in Gly and Ser residues of varying length
GGS7	GGSGGSG	7	
GGS10	GGSGGSGGSG	10	
GGS15	GGSGGSGGSGGSGSG	15	
	Rigid		
P5	GPPPG	5	Rigid linkers feature poly-Pro, poly-PT and poly-Ala motifs stabilized by E/K salt-bridges
P9	GPPPPPPPG	9	
TP11	GTPTPTPTPTG	11	
EAAAK16	GGAEAAAKEAAAKAGG	16	
	Semi-flexible		
PAS5	GSPAG	5	Semi-flexible linkers based on a so-called PASylation motif; Motif is based on a combination of Pro, Ala and Ser residues and enable linker expansion
PAS8	GGASPAGG	8	
PAS12	GGASPAAPAPAG	12	

3.4 Enzymes and reagents for manipulating DNA

1. Restriction Enzymes *Bts*I (R0667S), *Bsr*DI (R0574S), *Eco*RI-HF (R3101S), *Spe*I-HF (R3133S), *Nhe*I-HF (R3131S) and *Dpn*I (R0176S) all from New England Biolabs (NEB).
2. T4 DNA Ligase (M0202S, NEB).
3. Shrimp Alkaline Phosphatase (M0371S, NEB).

Table 2 Oligonucleotides for sequencing and amplifying entry vector backbones by PCR.

Name	DNA sequence	Comments
pFD amplify fwd	5′-GGCATTGCATAAAGCCCCGACAG-3′	Linearizing pFD for subsequent Gibson assembly (applicable to Step 4.3)
pFD amplify rev	5′-CCCACTGCTGGTGCGGAAAATTG-3′	Linearizing pFD for subsequent Gibson assembly (applicable to Step 4.3)
Template Sequence for cloning into pFD	5′-GAGTCACCCGCGAAATACAAGAAGACAATTTTCCC GCACCAGCAGTGGG**[Insert-FD]**GGCATTGCATAAAG CCCCGACAGTAAAAAGAGACCCTGACTAGTCTCCG −3′	Template DNA sequence for generating synthetic DNA fragments to insert a protein coding sequence into pFD. *Bts*I and *Bsr*DI are underlined
Template sequence for cloning into pL2	5′-GAGTCACCCGCGAAATACAAGAAGACAATTTTCCCGC ACCAGCAGTGGGT**[Insert-Linker]**GGGCATTGCATAAAG CCCCGACAGTAAAAAGAGACCCTGACTAGTCTCCG-3′	Template DNA sequence for generating synthetic DNA fragments to insert a protein coding sequence into pL2. BtsI and BsrDI are underlined
Gibson Overhang Fwd	5′-GAGTCACCCGCGAAATACAAG-3′	Optional: Forward primer for PCR amplification of dsDNA fragment for subsequent cloning by Gibson Assembly
Gibson Overhang Rev	5′-CGGAGACTAGTCAGGGTCTC-3′	Optional: Forward primer for PCR amplification of dsDNA fragment for subsequent cloning by Gibson Assembly
Control on bb fwd	5′-GCGGCCTTTTTACGGTTCCTGG-3′	Sequencing of DNA inserts in pFD or pL2 from the 5′ end (see Steps 4.4 and 4.10)
Control on bb rev	5′-CCTCTGACACATGCAGCTCCCGG-3′	Sequencing of DNA inserts in pFD or pL2 from 3′ end (see Steps 4.4 and 4.10)
Insert Fwd	5′-T**[Insert-Linker]**GGGCATTGCATAAAGCCCCGACAGT-3′	Template oligonucleotides for hybridizing into dsDNA cassettes into pL2
Insert Rev	3′-CCA**[Insert-Linker]**CCCGTAACGTATTTCGGGGCTGT CATTTT-5′	

4. Gibson Assembly Master Mix (E2611S, NEB).
5. Thermostable Phusion High-Fidelity DNA Polymerase (M0530S, NEB).
6. 10 mM dNTPs each.
7. Commercial plasmid DNA purification kit (e.g. NucleoSpin, Macherey & Nagel).
8. Commercial DNA gel extraction kit (e.g. NucleoSpin Gel and PCR Clean-Up, Macherey & Nagel).

3.5 Agarose DNA gel electrophoresis

1. 1 × Tris-Acetate EDTA (TAE) buffer: 40 mM Tris-Base, 20 mM AcOH and 1 mM EDTA.
2. DNA loading dye (e.g. 6 × Gel Loading Dye, Purple, NEB).
3. Agarose (e.g. Sigma-Aldrich).
4. Fluorescent dye for visualizing DNA in agarose gels (e.g. Midori Green, Nippon Genetics).
5. Safe Imager Blue-Light Transilluminator.
6. Agarose gel electrophoresis equipment.

3.6 Other

1. UV Spectrophotometer for quantifying DNA at 260 nm and purity relative to 230 and 280 nm.
2. A benchtop incubator for 1.5 mL tubes with cooling and heating function.
3. Shaking and static incubator(s) for growing bacterial cultures in flasks or on agar plates.
4. Thermocycler(s) for PCR as well as *Bsr*DI and *Bts*I restriction digests at 65 and 55 °C, respectively.

4. Protocol

4.1 Considerations for the design of new linker elements and functional domains

1. Linker elements and functional domains are stored in sequence verified entry vectors pL2 and pFD, respectively (Table 1). In this way, a repository of DNA parts can be continuously developed as an in-house resource and reused in subsequent construction efforts.

2. Alternatively, linker elements can also be cloned in pFD and functional domains in pL2 entry vectors. In this way, any two linker elements or functional domains can be directly fused in-frame bridged by a single Gly residue.
3. We recommend sourcing new parts through commercial DNA synthesis as linkers and functional domains need to be strictly devoid of *Bts*I, *Bsr*DI, *Spe*I and *Eco*RI restriction sites. In addition, any DNA parts should be devoid of repetitive DNA sequences and feature a balanced GC content to avoid any complications in the assembly or DNA sequencing due to secondary structures.
4. For operational flexibility, it is also preferable to avoid restriction sites for *Bsa*I, *Nhe*I and *Mfe*I: For instance, BsaI can be used as an alternative to treating the restriction digest with shrimp alkaline phosphatase (rSAP) in Step 4.6 in order to prevent the undesired circularization of the DNA fragment in the iFLinkC assembly reaction in Step 4.9. If the DNA insert coding for the N- or C-terminal portion is very small and the DNA fragments generated after the restriction digest in Steps 4.6 and 4.7 turn out difficult to separate by agarose gel electrophoresis, *Mfe*I can be used as an alternative to *Eco*RI. Furthermore, *Nhe*I can be used to transfer the fully assembled library into the expression vector pFLinkC-XE in Step 4.11.
5. Linker elements and functional domains can be cloned into the pL2 and pFD entry vectors by a range of conventional cloning methods: In our experience, DNA fragments greater than 40 bp that code for functional domains or large linker elements are preferably cloned by means of Gibson Assembly (Gibson et al., 2009) as exemplified in Steps 4.2–4.4. Alternatively, the ligase cycling reaction (De Kok et al., 2014) can also be employed to insert DNA fragments in the pL2 and pFD entry vectors as well as conventional restriction-digestion-ligation strategies.
6. DNA parts smaller than 50 bp coding for short linker elements and other short tags required for affinity purification or site-specific protein labeling can be effectively derived through hybridization of short single-stranded oligonucleotides. The resultant dsDNA cassettes featuring GG and AAAA overhangs can be readily inserted via the existing *Bts*I and *Bsa*I sites into pL2. In this case, plasmid backbones need to be restriction digested with *Bts*I and *Bsa*I in Step 4.2. Template sequences for designing short single-stranded oligonucleotides to insert dsDNA cassettes in pL2 are provided (Table 2). Note, optimal efficiency is achieved if dsDNA cassettes are phosphorylated at the 5′ end. Phosphorylation can

either be introduced through commercial DNA synthesis services or through treatment with T4 polynucleotide kinase.

7. Note, unless a functional domain, a linker element or any other type of tag is exclusively meant to define the N- or C-terminal end of a protein, the coding DNA should be devoid of any start or stop codon.

4.2 Backbone preparation for cloning new DNA parts by restriction digestion (option 1)

1. Pipette 2 μg pFD or pL2 vector into a 1.5 mL tube.
2. Fill up to 44 μL with MQ.
3. Add 5 μL 10 × NEBuffer CutSmart.
4. Add 1 μL BtsI (10 U/μL) in a total volume of 50 μL.
5. Incubate for 60–90 min at 55 °C.
6. Purify with a commercial PCR or DNA Gel Extraction Kit and elute with 44 μL with MQ.
7. Add 5 μL 10 × NEBuffer 2.1.
8. Add 1 μL *BsrDI* (5 U/μL) in a total volume of 50 μL.
9. Incubate for 60–90 min at 65 °C.
10. For optimal results, purify the restriction digest by means of agarose gel electrophoresis in combination with a commercial DNA gel extraction kit.

4.3 Backbone preparation for cloning new DNA parts by PCR (option 2)

1. Pipette 20 ng pFD or pL2 vector into 0.2 mL PCR tube.
2. Add 5 μL forward primer pFD amplify fwd (5 μM stock giving rise to 500 nM final).
3. Add 5 μL reverse primer pFD amplify rev (5 μM stock giving rise to 500 nM final).
4. Add 1 μL dNTPs (10 mM stock each giving rise to 200 μM final each).
5. Add 10 μL 5 × Phusion buffer.
6. Fill up to 49.5 μL with MQ.
7. Add 0.2 μL Phusion DNA polymerase.
8. Run a PCR according to NEB Phusion Protocol.
9. Optional: Add 1 μL *Dpn*I and incubate for another 1 h at 37 °C to digest the plasmid template and prevent it from being carried forward.
10. Purify the PCR with a commercial DNA clean-up kit.
11. Optional: Check the correct size of the amplified DNA fragment by means of analytical agarose gel electrophoresis.

4.4 Cloning of new DNA parts by Gibson assembly

1. Depending whether a DNA insert is cloned into pFD or pL2 backbones, mix a suitably prepared plasmid backbone (see Steps 4.2 and 4.3 for details) with a DNA fragment either coding for a functional domain or a linker element into a 0.2 mL PCR tube. Flanking arms for efficient Gibson Assembly into pFD2 and pL2 are summarized in Table 2, respectively.
2. The total amount of DNA should be approximately 0.005–0.125 pmol DNA and feature a 2–5-fold excess of DNA insert relative to plasmid backbone depending on its size according to manufacturer's instructions (NEB). Note, to save reagent, Gibson Assembly reactions can be set up in a quarter of the recommended volume.
3. Fill up to 2.5 μL with MQ.
4. Add 2.5 μL 2 × Gibson Assembly Mix.
5. Incubate for 60 min at 50 °C.
6. Add up to 5 μL the Gibson Assembly to 50 μL chemically competent *E. coli* and transform by heat-shock.
7. Recover and outgrow transformed cell suspension in 1 mL SOC medium for 1 h at 37 °C.
8. Plate 50 μL on LB agar plate supplemented with 50 μg/mL kanamycin and incubate overnight at 37 °C.
9. Pick 2–3 colonies and inoculate 5 mL LB supplemented with 50 μg/mL kanamycin each.
10. Incubate overnight at 37 °C in shaking incubator at 250 rpm.
11. Purify plasmid DNA with a commercial DNA purification kit.
12. Verify the insert by DNA sequencing (Table 2).
13. Strains harboring sequence verified clones of functional domains and linker elements in pFD and pL2 entry vectors can be stored long term at −80 °C in LB supplemented with 25% glycerol.

4.5 Planning an iFLinkC assembly process

1. Devise an iFLinkC assembly process deciding on the order of functional domains and different linker elements in pairwise assembly reactions (Fig. 2).
2. The only requirement is that DNA fragments originating from pFD and pL2 alternate in the fusion protein. This is to prevent frame shift as pL2 feature an additional G base at the immediate 5′ and 3′ of the coding reading frame.

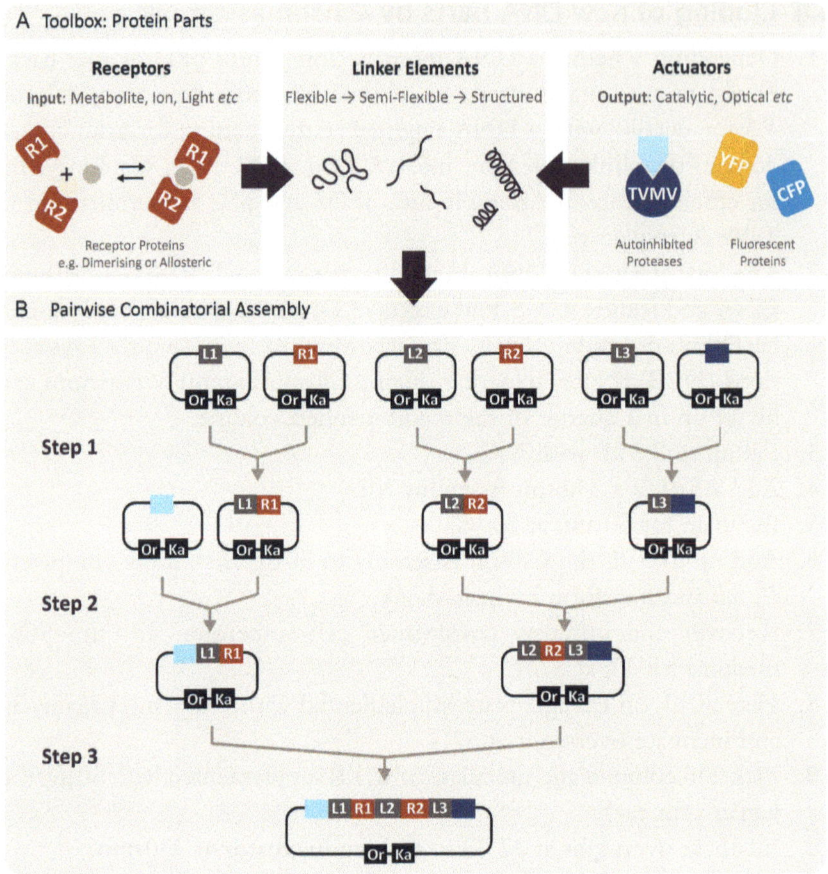

Fig. 2 Devising an iFLinkC assembly process: (A) A toolbox of linker elements and functional domains can be developed in-house and stored in sequence verified entry vectors. (B) Fusion protein libraries consisting of structurally distinct receptors and actuators are assembled through pairwise combinations of individual parts and, in subsequent steps, pairwise assembly reactions. In this way, synthetic protein switches can be assembled in a combinatorial fashion based on a limited number of sequence verified building blocks. *Illustration adapted with modifications from Gräwe, A., Ranglack, J., Weyrich, A., & Stein, V. (2020). iFLinkC: An iterative functional linker cloning strategy for the combinatorial assembly and recombination of linker peptides with functional domains.* Nucleic Acids Research, 48, *e24. https://doi.org/10.1093/nar/gkz1210.*

3. The DNA fragment coding for the N-terminal part is generally restriction digested with *Bsr*DI while the DNA fragment coding for the C-terminal part is restriction digested with *Bts*I irrespective of whether it constitutes a functional domain or a linker element.

4. Depending on the size of the insert, the DNA fragment coding for the C-terminal protein part can be restriction digested either with *Spe*I, *Eco*RI or *Mfe*I. This means, DNA fragments >400 bp are preferably restriction digested with *Spe*I. Shorter DNA fragments smaller than 400 bp are preferably restriction digested with *Eco*RI to maintain a sufficiently sized DNA fragment that can be purified by means of agarose gel electrophoresis. If the two restriction digested DNA fragments turn out very similar in size and therefore difficult to separate and purify by means of agarose gel electrophoresis, *Mfe*I can be used as an alternative to *Eco*RI.
5. To prevent circularization of the DNA fragments in the iFLinkC assembly reaction, we recommend that one of the two restriction digests is additionally treated with recombinant shrimp alkaline phosphatase (rSAP) dephosphorylated any 5′ DNA ends.
6. To monitor the efficiency of transformation, we generally recommend monitoring the number of transformants following every assembly reaction.
7. A no insert control ligation should generally be included in Step 4.9 to assess the efficiency of non-specific background transformants.
8. An elementary iFLinkC assembly cycle starts with Step 4.6 and is completed by Step 4.10.
9. Note, an elementary assembly reaction is exemplified with a linker (stored in pL2) and a functional domain (stored pFD) being fused in an N- to C-terminal orientation. Alternatively, the orientation can be readily reversed by including pFD in Step 4.6 and pL2 in Step 4.7 instead.
10. Depending on the duration of the ligation in Step 4.9, an elementary iFLinkC assembly cycle can be completed over the course of 1–2 days.

4.6 Restriction digest the DNA fragment coding for the N-terminal protein part with *Bsr*DI

1. Add equimolar amounts of purified pL2 coding for a desired combination of different linker elements. The total DNA mass should not exceed 1.2 μg. The choice of linker element generally constitutes a semi-rational estimate and ultimately depends on the structural organization of an underlying design.
2. Fill up to a volume of 43 μL with MQ water.
3. Add 5 μL 10 × NEBuffer 2.1.
4. Add 1 μL *Bsr*DI (5 U/μL) and incubate at 65 °C for 60 min.

5. Cool down to 37 °C and—depending on the assembly strategy—*either* add 1 μL *Spe*I-HF (20 U/μL) *or* *Eco*RI-HF (20 U/μL).
6. Incubate restriction digest for another 30 min at 37 °C.
7. Add 1 μL rSAP (1 U/μL) and continue incubation for 60 min at 37 °C.
8. Alternatively, the restriction digest can be treated with *Bsa*I instead of rSAP in order to minimize any non-specific re-ligation in the subsequent iFLinkC assembly reaction (Step 4.9).

4.7 Restriction digest the DNA fragment coding for the C-terminal protein part with *Bts*I

1. Add approx. 1.2 μg purified pFD
2. Fill up to a volume of 43 μL with MQ water.
3. Add 5 μL 10 × NEBuffer CutSmart.
4. Add 1 μL *Bts*I (10 U/μL).
5. Incubate restriction digest for 60 min at 55 °C.
6. Cool down to 37 °C and—depending on the assembly strategy—*either* add 1 μL *Spe*I-HF (20 U/μL) *or* *Eco*RI-HF (20 U/μL).
7. Incubate restriction digest for another 30 min at 37 °C.
8. Comment: BtsI has 100% activity at 55 °C and only 75% activity at 37 ° C. It is therefore recommended to conduct the restriction digest first at 55 °C and then lower the temperature to 37 °C for cleavage with *either* *Spe*I-HF *or* *Eco*RI-HF.

4.8 Purification of DNA fragments by agarose gel electrophoresis

1. The purification of DNA fragments by agarose gel electrophoresis applies to both restriction digests in Steps 4.6 and 4.7, respectively.
2. Stop restriction digest by adding 10 μL DNA loading dye (6 × stock) to 50 μL restriction digest.
3. Separate DNA fragments by means of gel electrophoresis on 1% agarose gels run in 1 × TAE buffer for approx. 40 min at 120 V.
4. Depending on the differences in size, the DNA fragments from individual restriction digests may also be separated for a longer period of time at reduced voltage.
5. Excise the desired DNA fragments using a clean scalpel.
6. To this end, we generally recommend visualizing DNA on a blue light transilluminator in association with suitable dye such as 1 × Midori Green to prevent any UV damage.

7. Purify DNA fragments from the agarose gel with a commercial gel extraction kit.

4.9 iFLinkC assembly reaction: Ligation of pFD and pL2 DNA fragments

1. Add equimolar amounts of gel purified pFD and pL2. Total amount of DNA should be between 5 and 100 ng.
2. Fill up to a volume of 10 μL with MQ.
3. Add 1.2 μL T4 DNA ligase buffer (NEB).
4. Add 0.8 μL T4 DNA ligase (400 U/μL).
5. Incubate ligation reaction at 16–18 °C either for 1 h or overnight.
6. Heat-inactivate ligation reaction at 65 °C for 20 min.

4.10 Transform iFLinkC assembly reaction

1. Add up to 5 μL of the iFLinkC assembly reaction from Step 4.9 to 50 μL chemically competent *E. coli* prepared by $CaCl_2$ precipitation and transform by heat-shock (Sambrook & Russell, 2006a).
2. Recover and outgrow transformed *E. coli* in 1 mL SOC medium for 1 h at 37 °C.
3. Assess the efficiency of transformation by plating approx. 40 μL of transformed cells on LB agar plates that have been supplemented with 50 μg/mL kanamycin.
4. If the efficiency of transformation turns out low, libraries may be alternatively transformed by electroporation (Sambrook & Russell, 2006b).
5. To assess the fidelity of the assembly reaction—i.e. whether all DNA fragments have been assembled in the correct order—we recommend verifying between two and three clones by DNA sequencing.
6. Use the remaining 960 μL of the transformed cell suspension to inoculate up to 10 mL LB that has been supplemented with 50 μg/mL kanamycin.
7. Incubate the cell culture overnight at 37 °C in a shaking incubator at 250 rpm.
8. Purify the plasmid library using a commercial plasmid purification kit.
9. The resultant plasmid is functionally equivalent to the iFLinkC entry vectors and can thus be recombined with additional linker elements, functional domains or thereof derived intermediate assembly products that have been assembled in parallel. Therefore, Steps 4.6–4.10 can be iteratively repeated to assemble linker libraries in a combinatorial

fashion. The only requirement is that DNA fragments originating from pFD and pL2 alternate.

4.11 Transfer iFLinkC library into an expression vector

1. For expression and screening of a library, the assembled library needs to be transferred into a suitable expression vector. To this end, we devised piFLinkC-XE to control the expression of a combinatorial linker library with T7 RNA polymerase under the control of an IPTG inducible promoter. The elementary pFLinkC-XE expression plasmid also features a copy of the orange fluorescent protein mKO_{κ} which is continuously expressed to monitor any undesired carryover of the source plasmid that may interfere with the quality of the assembled library.
2. Transfer of the assembled library into an expression vector is ideally realized through restriction digestion ligation via *Bts*I and *Nhe*I restriction sites that include the coding DNA of the fully assembled fusion protein. Alternatively, iFLinkC linker libraries can be amplified by PCR and transferred into alternative expression plasmid of choice.
3. Pipette 2 μg iFLinkC linker library and 2 μg pFLinkC-XE expression vector into two separate 1.5 mL tubes.
4. Fill up to 43 μL with MQ each.
5. Add 5 μL 10 × NEBuffer CutSmart each.
6. To the assembled iFLinkC linker library, add 1 μL *Bts*I (10 U/μL) and 1 μL *Spe*I-HF (20 U/μL) in a total volume of 50 μL.
7. To the FLinkC-XE expression vector, add 1 μL *Bts*I (10 U/μL) and 1 μL *Nhe*I-HF (20 U/μL) in a total volume of 50 μL.
8. Incubate for up to 90 min at 37 °C each.
9. Purify the insert of the iFLinkC linker library and the backbone of the pFLinkC-XE by means of DNA agarose gel electrophoresis (see Step 4.8 for details).
10. Add a 5-fold molar excess of the iFLinkC linker library relative to the pFLinkC-XE backbone. The total amount of DNA should be between 5 and 100 ng.
11. Fill up to a volume of 10 μL with MQ.
12. Add 1.2 μL T4 DNA ligase buffer.
13. Add 0.8 μL T4 DNA ligase.
14. Incubate ligation reaction at 16–18 °C overnight.
15. Heat-inactivate ligation reaction at 65 °C for 20 min.

16. Chemically-competent DH10β are transformed with the library by heat-shock (see Step 4.10 for details), but this time LB is supplemented with ampicillin at a concentration of 100 μg/mL (instead of kanamycin).
17. Following overnight growth and purification of the library using a commercial DNA purification kit, the plasmid prep is now ready for use in transformation of bacterial expression strains. The function of synthetic protein switches and sensors is preferably assayed in high-throughput in the presence and absence of a target analyte looking to identify mutants with the maximum differential response.
18. In the past, iFLinkC was successfully applied to engineer allosteric protease switches based on autoinhibited TVMV modules in combination with a conventional microwell plate screening protease assay (Gräwe, Ranglack, Weyrich, & Stein, 2020; Stein & Alexandrov, 2017) and should be readily applicable to any type of protein switch or sensor. Notably, the frequency of protein switches that display a strong differential response upon addition of their target analyte is sufficiently high for them to be identified in conventional microwell plate assays with a typical throughput of several 100 clones per round of screening.

4.12 Safety considerations and standard

The iFLinkC assembly process can be completed by exercising standard laboratory safety practices. Strains used for cloning of individual linker elements, functional domains and thereof composed libraries are laboratory safety strains. Care should be taken during the excision of dsDNA fragments from agarose gels with a scalpel to avoid cuts.

4.13 Analysis and statistics

To maintain a sufficiently large library size, the efficiency of transformation should be closely monitored following every assembly reaction counting the number of colony-forming units (CFUs) from a small aliquot of the transformed cell suspension (see Step 4.10). This is particularly critical in subsequent assembly steps given linker diversity grows exponentially and therefore requires to maintain a high transformation efficiency in order to continue covering the theoretical library diversity.

4.14 Pros and cons

Table 3 Pros and cons.

Pros	Cons
iFLinkC relies on type IIS restriction enzymes. This means, the assembly efficiency is independent of the underlying sequences of linker elements and functional domains that may otherwise hamper homology-dependent cloning methods such as a high GC content or repeat elements	Linker elements need to be strictly devoid of restriction sites *Bsr*DI, *Bts*I and *Spe*I/*Eco*RI and preferably lack *Bsa*I and *Nhe*I restriction sites, too
iFLinkC enables the simultaneous fusion of very short and very long linker elements	A minimal scar site of a single Gly residue remains upon connecting a linker element with a functional domain
Linker elements and functional domains can be developed and stored in-house in sequence verified repositories of plasmid DNA that can be reused in subsequent construction efforts	Establishing an in-house repository for iFLinkC assembly is time and cost-intensive in the initial stages as functional domains and linker elements are preferably sourced through commercial DNA synthesis
Intermediate assembly products of linker elements fused to functional domains can be stored in sub-libraries and readily reused in subsequent construction efforts	For high-fidelity assembly, the procedure currently relies on an agarose gel electrophoresis and extraction step that currently restricts its capacity for automation
Sub-libraries can be generated by alternative cloning methods and readily integrated into the iFLinkC assembly process affording a high operational flexibility	If any given protein switch depends on a defined C-terminus, a stop codon needs to be included in the DNA part coding for the most C-terminal protein element. Otherwise, a GIA-motif is appended
Linker diversity scales exponentially with the number of individual assembly steps and thus provides rapid access to a very large diversity in combinatorial linker space	Once assembled, individual parts cannot be easily exchanged
All reagents for iFLinkC assembly are commercially available or can be purchased through public repositories such as Addgene	

4.15 Troubleshooting and optimization

Table 4 Troubleshooting.

Problem	Solution
A high proportion of mutants originates from non-specific re-ligation events to yield the source/entry plasmids following an iFLinkC assembly reaction	The DNA fragments cut with either *Bts*I or *Bsr*DI in combination with *Eco*RI/*Spe*I are not efficiently separated by agarose gel electrophoresis. This can be overcome by separating individual DNA fragments by agarose gel electrophoresis for a longer period of time at reduced voltage. Alternatively, *Mfe*I can be used instead of *Eco*RI and *Spe*I
Following verification of individual assembly products by DNA sequence, deletions occur around the restriction sites	While the exact reason for this remains to be established, rSAP must only be included in one of the two restriction digests in Step 4.6
Low transformation efficiency	For higher transformation efficiencies prepare chemically competent cells by means of $RbCl_2$ precipitation or transform by electroporation

5. Summary

iFLinkC comprises a highly scalable DNA assembly method for recombining linker elements with functional domains following a rule-based protocol independent of tailored synthetic oligonucleotides. It relies on the combined action of the type IIS restriction enzymes *Bsr*DI, *Bts*I and T4 DNA ligase in order to fuse any two protein coding sequences in-frame bridged by a single Gly residue. Crucially, the vectors are effectively regenerated following every elementary iFLinkC assembly reaction cycle. In this way, the assembly process can be performed iteratively and generate linker diversity in a combinatorial fashion based on a limited number of assembly steps featuring sequence verified linker elements and functional domains. At the same time—as long as adjacent domains and linker elements are assembled in a pairwise fashion—the order of assembly is generally flexible. This means, fusion proteins can be assembled inside-out or outside-in. This carries the advantage that intermediate assembly products can be assembled first and readily reused in subsequent construction efforts. Such a scenario is conceivable in the construction of protein sensors that share a common actuator such as for example a fluorescent protein, but different types of allosteric receptors as input.

In this way, a set of highly potent rapamycin-inducible protease switches were generated through modular recombination of autoinhibited TVMV protease modules (Stein & Alexandrov, 2014; Stein, Nabi, & Alexandrov, 2017) and a single-chain version of the FKBP12-FRB receptor module. Crucially, the screening of combinatorial linker libraries highlighted rigid Pro-rich linkers as optimal solutions contradicting the conventional use of Gly-rich linkers. Given the availability of increasing number of modular-organized protease modules, iFLinkC should be readily applicable to the optimization of alternative protein switches, for instance, based on auto-inhibited β-lactamases (Adamson et al., 2019; Banala, Aper, Schalk, & Merkx, 2013; Nirantar et al., 2013) or the continuously developing toolbox of optical sensors that resolve molecular signals through distance-dependent FRET (Höfig et al., 2018) or BRET (Ni, Arts, & Merkx, 2019) between two fluorescent proteins or a luciferase and a fluorescent protein, respectively.

Acknowledgments

LOEWE iNAPO, Hessen State Ministry of Higher Education, Research and the Arts; Pioneer ACTIVATOR (Project No. 527-00-962), TU Darmstadt.

References

Adamson, H., Ajayi, M. O., Campbell, E., Brachi, E., Tiede, C., Tang, A. A., et al. (2019). Affimer-enzyme-inhibitor switch sensor for rapid wash-free assays of multimeric proteins. *ACS Sensors*, *4*, 3014–3022. https://doi.org/10.1021/acssensors.9b01574.

Arviv, O., & Levy, Y. (2012). Folding of multidomain proteins: Biophysical consequences of tethering even in apparently independent folding. *Proteins: Structure, Function and Bioinformatics*, *80*, 2780–2798. https://doi.org/10.1002/prot.24161.

Baird, G. S., Zacharias, D.a., & Tsien, R. Y. (1999). Circular permutation and receptor insertion within green fluorescent proteins. *Proceedings of the National Academy of Sciences of the United States of America*, *96*, 11241–11246. https://doi.org/10.1073/pnas.96.20.11241.

Banala, S., Aper, S. J. A., Schalk, W., & Merkx, M. (2013). Switchable reporter enzymes based on mutually exclusive domain interactions allow antibody detection directly in solution. *ACS Chemical Biology*, *8*, 2127–2132. https://doi.org/10.1021/cb400406x.

Bashor, C. J., & Collins, J. J. (2018). Understanding biological regulation through synthetic biology. *Annual Review of Biophysics*, *47*, 399–423. https://doi.org/10.1146/annurev-biophys-070816-033903.

Brophy, J. A. N., & Voigt, C. A. (2014). Principles of genetic circuit design. *Nature Methods*, *11*, 508–520. https://doi.org/10.1038/nmeth.2926.

Cadwell, R. C., & Joyce, G. F. (1992). Randomization of genes by PCR mutagenesis. *Genome Research*, *2*, 28–33. https://doi.org/10.1101/gr.2.1.28.

Cozens, C., & Pinheiro, V. B. (2018). Darwin assembly: Fast, efficient, multi-site bespoke mutagenesis. *Nucleic Acids Research*, *46*, e51. https://doi.org/10.1093/nar/gky067.

De Kok, S., Stanton, L. H., Slaby, T., Durot, M., Holmes, V. F., Patel, K. G., et al. (2014). Rapid and reliable DNA assembly via ligase cycling reaction. *ACS Synthetic Biology*, *3*, 97–106. https://doi.org/10.1021/sb4001992.

Evers, T. H., Van Dongen, E. M. W. M., Faesen, A. C., Meijer, E. W., & Merkx, M. (2006). Quantitative understanding of the energy transfer between fluorescent proteins connected via flexible peptide linkers. *Biochemistry*, *45*, 13183–13192. https://doi.org/10.1021/bi061288t.

Furukawa, K., & Hohmann, S. (2013). Synthetic biology: Lessons from engineering yeast MAPK signalling pathways. *Molecular Microbiology*, *88*, 5–19. https://doi.org/10.1111/mmi.12174.

Gao, C., Xu, P., Ye, C., Chen, X., & Liu, L. (2019). Genetic circuit-assisted smart microbial engineering. *Trends in Microbiology*, *27*, 1011–1024. https://doi.org/10.1016/j.tim.2019.07.005.

Gibson, D. G., Young, L., Chuang, R.-Y., Venter, J. C., Hutchison, C. A., Smith, H. O., et al. (2009). Enzymatic assembly of DNA molecules up to several hundred kilobases. *Nature Methods*, *6*, 343–345. https://doi.org/10.1038/nmeth.1318.

Golynskiy, M. V., Koay, M. S., Vinkenborg, J. L., & Merkx, M. (2011). Engineering protein switches: Sensors, regulators, and spare parts for biology and biotechnology. *ChemBioChem*, *12*, 353–361. https://doi.org/10.1002/cbic.201000642.

Gräwe, A., Ranglack, J., Weber, W., & Stein, V. (2020). Engineering artificial signalling functions with proteases. *Current Opinion in Biotechnology*, *63*, 1–7. https://doi.org/10.1016/j.copbio.2019.09.017.

Gräwe, A., Ranglack, J., Weyrich, A., & Stein, V. (2020). iFLinkC: An iterative functional linker cloning strategy for the combinatorial assembly and recombination of linker peptides with functional domains. *Nucleic Acids Research*, *48*, e24. https://doi.org/10.1093/nar/gkz1210.

Grünberg, R., & Serrano, L. (2010). Strategies for protein synthetic biology. *Nucleic Acids Research*, *38*, 2663–2675. https://doi.org/10.1093/nar/gkq139.

Guntas, G., Mansell, T. J., Kim, J. R., & Ostermeier, M. (2005). Directed evolution of protein switches and their application to the creation of ligand-binding proteins. *Proceedings of the National Academy of Sciences of the United States of America*, *102*, 11224–11229. https://doi.org/10.1073/pnas.0502673102.

Höfig, H., Otten, J., Steffen, V., Pohl, M., Boersma, A. J., & Fitter, J. (2018). Genetically encoded Förster resonance energy transfer-based biosensors studied on the single-molecule level. *ACS Sensors*, *3*, 1462–1470. https://doi.org/10.1021/acssensors.8b00143.

Kaczmarski, J. A., Mitchell, J. A., Spence, M. A., Vongsouthi, V., & Jackson, C. J. (2019). Structural and evolutionary approaches to the design and optimization of fluorescence-based small molecule biosensors. *Current Opinion in Structural Biology*, *57*, 31–38. https://doi.org/10.1016/j.sbi.2019.01.013.

Kegler-ebo, D. M., Docktor, C. M., & Dimaio, D. (1994). Codon cassette mutagenesis: A general method to insert or replace individual codons by using universal mutagenic cassettes. *Nucleic Acids Research*, *22*, 1593–1599. https://doi.org/10.1093/nar/22.9.1593.

Kiel, C., Yus, E., & Serrano, L. (2010). Engineering signal transduction pathways. *Cell*, *140*, 33–47. https://doi.org/10.1016/j.cell.2009.12.028.

Levy, Y. (2017). Protein assembly and building blocks: Beyond the limits of the LEGO brick metaphor. *Biochemistry*, *56*, 5040–5048. https://doi.org/10.1021/acs.biochem.7b00666.

Li, G., Huang, Z., Zhang, C., Dong, B. J., Guo, R. H., Yue, H. W., et al. (2016). Construction of a linker library with widely controllable flexibility for fusion protein design. *Applied Microbiology and Biotechnology*, *100*, 215–225. https://doi.org/10.1007/s00253-015-6985-3.

Lim, W. A. (2010). Designing customized cell signalling circuits. *Nature Reviews Molecular Cell Biology*, *11*, 393–403. https://doi.org/10.1038/nrm2904.

Lim, H. G., Jang, S., Jang, S., Seo, S. W., & Jung, G. Y. (2018). Design and optimization of genetically encoded biosensors for high-throughput screening of chemicals. *Current Opinion in Biotechnology*, *54*, 18–25. https://doi.org/10.1016/j.copbio.2018.01.011.

Ma, B., Tsai, C. J., Haliloğlu, T., & Nussinov, R. (2011). Dynamic allostery: Linkers are not merely flexible. *Structure*, *19*, 907–917. https://doi.org/10.1016/j.str.2011.06.002.

Michener, J. K., Thodey, K., Liang, J. C., & Smolke, C. D. (2012). Applications of genetically-encoded biosensors for the construction and control of biosynthetic pathways. *Metabolic Engineering*, *14*, 212–222. https://doi.org/10.1016/j.ymben.2011.09.004.

Ni, Y., Arts, R., & Merkx, M. (2019). Ratiometric bioluminescent sensor proteins based on intramolecular split luciferase complementation. *ACS Sensors*, *4*, 20–25. https://doi.org/10.1021/acssensors.8b01381.

Nirantar, S. R., Yeo, K. S., Chee, S., Lane, D. P., & Ghadessy, F. J. (2013). A generic scaffold for conversion of peptide ligands into homogenous biosensors. *Biosensors and Bioelectronics*, *47*, 421–428. https://doi.org/10.1016/j.bios.2013.03.049.

Norris, J. L., & Hughes, R. M. (2018). protaTETHER—A method for the incorporation of variable linkers in protein fusions reveals impacts of linker flexibility in a PKAc-GFP fusion protein. *FEBS Open Bio*, *8*, 1029–1042. https://doi.org/10.1002/2211-5463.12414.

Ohlendorf, R., Schumacher, C. H., Richter, F., & Möglich, A. (2016). Library-aided probing of linker determinants in hybrid photoreceptors. *ACS Synthetic Biology*, *5*, 1117–1126. https://doi.org/10.1021/acssynbio.6b00028.

Ribeiro, L. F., Warren, T. D., & Ostermeier, M. (2017). Construction of protein switches by domain insertion and directed evolution. *Methods in Molecular Biology*, *1596*, 43–55. https://doi.org/10.1007/978-1-4939-6940-1_3.

Sambrook, J., & Russell, D. W. (2006a). Preparation and transformation of competent E. coli using calcium chloride. *Cold Spring Harbor Protocols*. https://doi.org/10.1101/pdb.prot3932.

Sambrook, J., & Russell, D. W. (2006b). Transformation of E. coli by electroporation. *Cold Spring Harbor Protocols*. https://doi.org/10.1101/pdb.prot3933.

Stein, V. (2017). Synthetic protein switches: Theoretical and experimental considerations. *Methods in Molecular Biology*, *1596*, 3–25. https://doi.org/10.1007/978-1-4939-6940-1_1.

Stein, V., & Alexandrov, K. (2014). Protease-based synthetic sensing and signal amplification. *Proceedings of the National Academy of Sciences of the United States of America*, *111*, 15934–15939. https://doi.org/10.1073/pnas.1405220111.

Stein, V., & Alexandrov, K. (2015). Synthetic protein switches: Design principles and applications. *Trends in Biotechnology*, *33*, 101–110. https://doi.org/10.1016/j.tibtech.2014.11.010.

Stein, V., & Alexandrov, K. (2017). Engineering and characterizing synthetic protease sensors and switches. *Methods in Molecular Biology*, *1596*, 197–218. https://doi.org/10.1007/978-1-4939-6940-1_13.

Stein, V., Nabi, M., & Alexandrov, K. (2017). Ultrasensitive scaffold-dependent protease sensors with large dynamic range. *ACS Synthetic Biology*, *6*, 1337–1342. https://doi.org/10.1021/acssynbio.6b00370.

Suzuki, K., & Nagai, T. (2017). Recent progress in expanding the chemiluminescent toolbox for bioimaging. *Current Opinion in Biotechnology*, *48*, 135–141. https://doi.org/10.1016/j.copbio.2017.04.001.

Tolle, F., Stücheli, P., & Fussenegger, M. (2019). Genetic circuitry for personalized human cell therapy. *Current Opinion in Biotechnology*, *59*, 31–38. https://doi.org/10.1016/j.copbio.2019.02.003.

Tullman, J., Guntas, G., Dumont, M., & Ostermeier, M. (2011). Protein switches identified from diverse insertion libraries created using S1 nuclease digestion of supercoiled-form plasmid DNA. *Biotechnology and Bioengineering*, *108*, 2535–2543. https://doi.org/10.1002/bit.23224.

Van Dongen, E. M. W. M., Evers, T. H., Dekkers, L. M., Meijer, E. W., Klomp, L. W. J., & Merkx, M. (2007). Variation of linker length in ratiometric fluorescent sensor proteins allows rational tuning of Zn(II) affinity in the picomolar to femtomolar range. *Journal of the American Chemical Society*, *129*, 3494–3495. https://doi.org/10.1021/ja069105d.

Van Rosmalen, M., Krom, M., & Merkx, M. (2017). Tuning the flexibility of glycine-serine linkers to allow rational design of multidomain proteins. *Biochemistry*, *56*, 6565–6574. https://doi.org/10.1021/acs.biochem.7b00902.

Yu, Y., & Lutz, S. (2011). Circular permutation: A different way to engineer enzyme structure and function. *Trends in Biotechnology*, *29*, 18–25. https://doi.org/10.1016/j.tibtech.2010.10.004.

Zhao, H., & Arnold, F. H. (1997). Optimization of DNA shuffling for high fidelity recombination. *Nucleic Acids Research*, *25*, 1307–1308. https://doi.org/10.1093/nar/25.6.1307.

CHAPTER TEN

Enzyme-DNA chimeras: Construction, allostery, applications

Chiao-Yu Tseng[a,*], Yong Wang[b], and Giovanni Zocchi[c]
[a]Institute of Physics, Academia Sinica, Taipei, Taiwan
[b]Department of Physics, Cell and Molecular Biology Program, Materials Science and Engineering Program, University of Arkansas, Fayetteville, AR, United States
[c]Department of Physics and Astronomy, University of California Los Angeles, Los Angeles, CA, United States
*Corresponding author: e-mail address: tseng6499@gmail.com

Contents

Abstract

We describe the operational principle, synthesis, and applications of the enzyme-DNA chimeras. These are supramolecular constructions where a DNA spring is coupled to an enzyme and introduces artificial allosteric control of the enzyme. This method is universal and can be applied to various enzymes and proteins. In addition, this method is

Methods in Enzymology, Volume 647
ISSN 0076-6879
https://doi.org/10.1016/bs.mie.2020.09.010

versatile as the stresses applied by the DNA spring on the enzymes can be fine-tuned semi-continuously and thus their enzymatic activities can be modulated gradually. We give detailed protocols for the synthesis of these molecules. Summarizing our experience with different enzymes, we explain their use for fundamental studies of conformational plasticity, as well as the potential as molecular probes.

1. Introduction

The living cell is all about functional supra-molecular complexes; in contrast, our ability to imagine and produce such constructions artificially is at present quite limited. The purpose of this chapter is to introduce, in a practical way, a family of artificial enzyme-DNA complexes: the principle of their function, some applications, and most importantly the detailed protocols we use for their synthesis. Our objective is to explain what sort of interesting questions might be pursued through these constructions and to put these tools in the hands of whoever may want to use them or build upon them. Most of what we report here has appeared in print before, but the material is scattered through several publications, the more practical details often hidden in the "supplementary material" sections. Here we aim to present a more unified view of the subject matter.

In order to understand the working principle of the enzyme-DNA chimeras, it is useful to recapitulate the evolution of ideas concerning certain generic features of enzyme-substrate interactions, culminating in a generalized notion of allostery. Enzymes typically bind their substrates (reactants) through a large contact surface, i.e. through several weak bonds rather than one strong bond. The weak bonds may involve a combination of Van der Waals, electrostatic, and hydrophobic interactions, and hydrogen bonds. The large interaction surface is implicit in Fisher's lock and key theory, the first successful attempt to explain the specificity of enzymes on the basis of generic (not enzyme specific) features. However, the enzyme-substrate interaction surface is not dissimilar from other internal interaction surfaces of the folded polypeptide chain, such as the surface between two adjacent elements of the secondary structure. Therefore one might expect that substrate binding will in general significantly perturb the structure of the protein. Koshland's great discovery, the induced fit theory, was arrived at through rather more biochemical reasoning (Koshland, 1995), but it actually reconciles this state of affairs and the lock and key mechanism: substrate and enzyme do fit together snugly, in a deformed state of the enzyme stabilized by substrate binding. Here again, discoveries followed their own path, but conceptually induced fit must lead to allostery, as vividly demonstrated in the recent evolutionary model of allostery by Tlusty and collaborators (Tlusty, Libchaber, & Eckmann, 2017). If binding of a ligand can deform the enzyme structure, two separate binding sites on the same enzyme will in general "talk" through the structure, so that binding of one ligand (the regulator) will affect binding and/or catalysis at the other site.

The explosion in structure determination over the past 50 years has transformed lock-and-key, induced fit, and allostery from biochemical theories to structure-based mechanisms (Fig. 1A). The beautiful structures of an enzyme in different conformations invite, as the basis for further experiments, simplified models which regard the molecule as a clockwork designed to move from conformation A to conformation B under the driving force of ligand binding. We may say that the clockwork provides a route from conformation A to conformation B, and, somewhat tautologically, that ligand binding stabilizes conformation B. Experiments may then consist of perturbing the clockwork through mutations to establish its components (Johnston, Whitney, Volkman, Doe, & Prehoda, 2011). However, as far back as the 1970s Frauenfelder and collaborators established experimentally that a protein in equilibrium explores a large number of conformational substates (Frauenfelder, Parak, & Young, 1988), a conclusion arrived at also since the earliest "all atom" simulations by Karplus and collaborators (Karplus & McCammon, 2002). More recent experiments have shown that enzymes are viscoelastic and can "flow" from one conformation to another (Alavi & Zocchi, 2018; Wang & Zocchi, 2011). The existence of a myriad conformational substates suggests that by applying a field of force, the enzyme can be reversibly deformed into many different structures, besides the ones resulting from ligand binding. If that is the case, a sufficient mechanical stress applied to the molecule will in general modulate the enzymatic activity, in other words, any field of force can in principle act as an allosteric regulator! The method of the DNA springs is based on this concept.

2. Enzyme-DNA chimera

2.1 Basic principle of the method

The enzyme-DNA chimeras are in vitro synthesized supra-molecular constructions consisting of an enzyme and a synthetic DNA oligomer of 40 to 60 bases (Fig. 1B). The two ends of the DNA are covalently attached through crosslinkers to two sites on the surface of the enzyme, away from

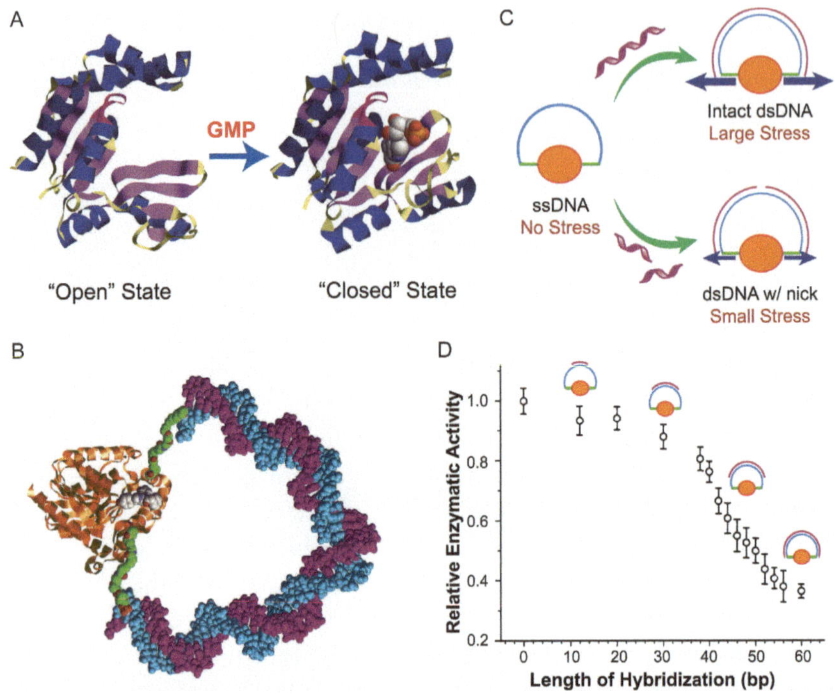

Fig. 1 (A) Example of structurally documented induced fit binding. The figure shows the structure of the enzyme Guanylate Kinase from yeast (PDB 1EX6 (apo enzyme) and 1EX7 (with GMP bound, shown in all atom representation)). The enzyme undergoes a large conformational change upon binding the substrate GMP. Namely, the two lobes of the protein swivel towards each other to form the "closed" state. The p-loop is colored in magenta. (B) Cartoon of an enzyme-DNA chimera. The enzyme is Renilla Luciferase (PDB 2PSJ), with the substrate coelenterazine-h shown in all atom representation. The DNA spring is cut out from the structure L8sa in the MD simulation of Lankas et al. (Lankaš, Lavery, & Maddocks, 2006), which shows a kink in the stressed DNA. The kink is at the origin of the softening transition mentioned in Section 5.2, so we believe this structure is a fair representation of the average conformation of the DNA in the enzyme-DNA chimeras. The DNA spring is shown attached at sites 161 and 188, through crosslinkers shown in green. The protein, DNA, and crosslinkers are drawn approximately to scale. (C) Schematic representation of how mechanical stress is controlled in the enzyme-DNA chimeras. With the DNA spring in the ss state there is essentially no stress on the enzyme. Adding the complementary DNA strand rigidifies the DNA, which acts as a leaf spring, pulling the attachment points on the enzyme apart (blue arrows). With a nick in the DNA spring, the stress on the enzyme is smaller. (D) The stress on the enzyme can be controlled semi-continuously by hybridizing the ss DNA of the chimera with complementary oligomers of different lengths. The plot shows the speed of the enzymatic reaction measured for a GK-DNA chimera with a 60mer DNA spring where the central portion is hybridized to oligomers of different lengths. The measurements are normalized to the activity of the ss chimera. *Panel (B) Reprinted with permission from Ariyaratne, A., Wu, C., Tseng, C.-Y., & Zocchi, G. (2014). Dissipative dynamics of enzymes.* Physical Review Letters, 113*(19), 198101. doi:10.1103/PhysRevLett.113.198101. Copyright (2014) by the American Physical Society. Panel (D) Reproduced with permission from Choi, B., & Zocchi, G. (2007). Guanylate kinase, induced fit, and the allosteric spring probe.* Biophysical Journal, 92*(5), 1651–1658. doi:10.1529/biophysj.106.092866 via Copyright Clearance Center. Copyright (2007) by Elsevier.*

the catalytic site so as not to impair function. In our constructions, the attachment sites on the enzyme are specific, defined by the reactive thiol groups of two Cysteines introduced by mutagenesis. The typical outcome (though there are exceptions) is that the enzyme-DNA chimera with the DNA in the single stranded (ss) form works just as well as the enzyme without DNA, in terms of catalytic activity. On the other hand, the ss DNA of the chimera can be hybridized with a complementary strand. This rigidifies the DNA, applying a stress on the enzyme. Namely, whereas a ss DNA molecule of 40–60 nucleotides (nt) is a very flexible polymer (the persistence length of ss DNA being around 3 bases or $\ell_{ss} \approx 1.5$ nm) (Smith, Cui, & Bustamante, 1996), the same length of ds DNA is a semi rigid rod (for ds DNA the persistence length is around 150 base pairs (bp) or $\ell_{ds} \approx 50$ nm) (Wang, Yin, Landick, Gelles, & Block, 1997). The ds DNA of the chimera therefore acts as a leaf spring (Fig. 1B), introducing a repulsive force between the attachment points, thus putting the enzyme under mechanical stress (Fig. 1C).

The virtue of the method is the external control of the force field: the experimenter can turn the stress on and off, on the same sample, by adding the complementary DNA oligomer (cDNA) to a solution of ss chimeras, or displacing the cDNA from the chimera by adding a competitor. Another, irreversible way to turn the stress off is to cut the DNA spring with a restriction enzyme. Moreover, different states of stress can be achieved with the same chimeric construct by partially hybridizing the DNA spring, using complementary oligomers of different lengths (Fig. 1D). The external control allows, on one hand, to perform quantitative experiments on the effect of mechanical stress on enzyme function, and, on the other hand, is the basis for turning the enzyme-DNA chimeras into molecular probes.

2.2 Examples of different enzymes modulated by the enzyme-DNA method

The enzyme-DNA method has been successfully utilized in modulating several enzymes. Our first successful target, maltose binding protein (MBP), is actually a transport protein, not an enzyme; we mention it because historically it was the first protein with which the method of the DNA springs was tried out (Choi et al., 2005). Under stress, a modest decrease (about a factor 2) in the binding affinity for maltose was achieved. This construction used a 60 bp DNA spring. MBP is a classic example of induced fit binding; its general structure consists of two lobes, with the maltose binding pocket located in between. Upon binding of maltose, the two lobes of the protein tilt towards each other, enclosing the maltose. The attachment points of the DNA spring were chosen so that the mechanical tension would work against the conformational motion driven by the binding of maltose (this consists of the two lobes of the molecule tilting closer to each other). On general thermodynamic grounds (Le Chatelier's principle), a perturbation which works against the conformational change induced by maltose binding must decrease the binding affinity for maltose. However, the method is far more general, as shown by the next example.

Guanylate Kinase (GK) is a 22kDa monomeric enzyme that phosphorylates guanosine monophosphate (GMP). Binding of GMP induces a large conformational change similar to the MBP, in that the two lobes of the molecule "close" around the substrate (Fig. 1A). We have studied three different chimeras of this molecule, with the DNA spring attached as shown in Fig. 2A. In the presence and absence of mechanical stress, we measured the enzymatic speed as a function of substrates' concentration, and extracted the Michaelis-Menten (MM) constants $K_{M,G}$ (for GMP) and $K_{M,A}$ (for ATP), and the catalytic rate k_{cat} (Tseng, Wang, & Zocchi, 2010). The important result is that the three different configurations of the DNA spring (i.e. the three different states of internal stress applied to the enzyme) modulate the parameters $K_{M,G}$, $K_{M,A}$, k_{cat} in different ways (as opposed to modulating the parameters in the same way but to different degrees). Specifically, tension along the 40/171 direction decreases k_{cat} to about 40% of the original value, while $K_{M,G}$ and $K_{M,A}$ remain essentially unaffected. In contrast, tension along the 75/171 direction affects mainly $K_{M,G}$ (Fig. 2B). We conclude that the effect of mechanical stress is not simply to bias the structure towards one or the other of the two conformations ("open" and "closed") seen in the X-ray structures, but is, in general, a stress specific deformation of the molecule. The DNA spring allows to access deformed states of the enzyme beyond the free and ligand bound states. Note that proteins are "soft" in the sense that thermal fluctuations around the average structure are relatively large. Fluctuations are also functionally important. In this context, a "deformed state" is a state where the average position of some atoms is different with respect to the reference state, or else a state where the fluctuations of some atoms around their average position is different compared to the reference state (Cooper & Dryden, 1984; Zocchi, 2018).

The next example is an enzyme that is not of the induced fit kind as it does not undergo apparent conformational change of the structure with and without substrates bound. Renilla Luciferase is a monomeric, 36kDa enzyme that catalyzes the oxidation reaction of the luciferin coelenterazine,

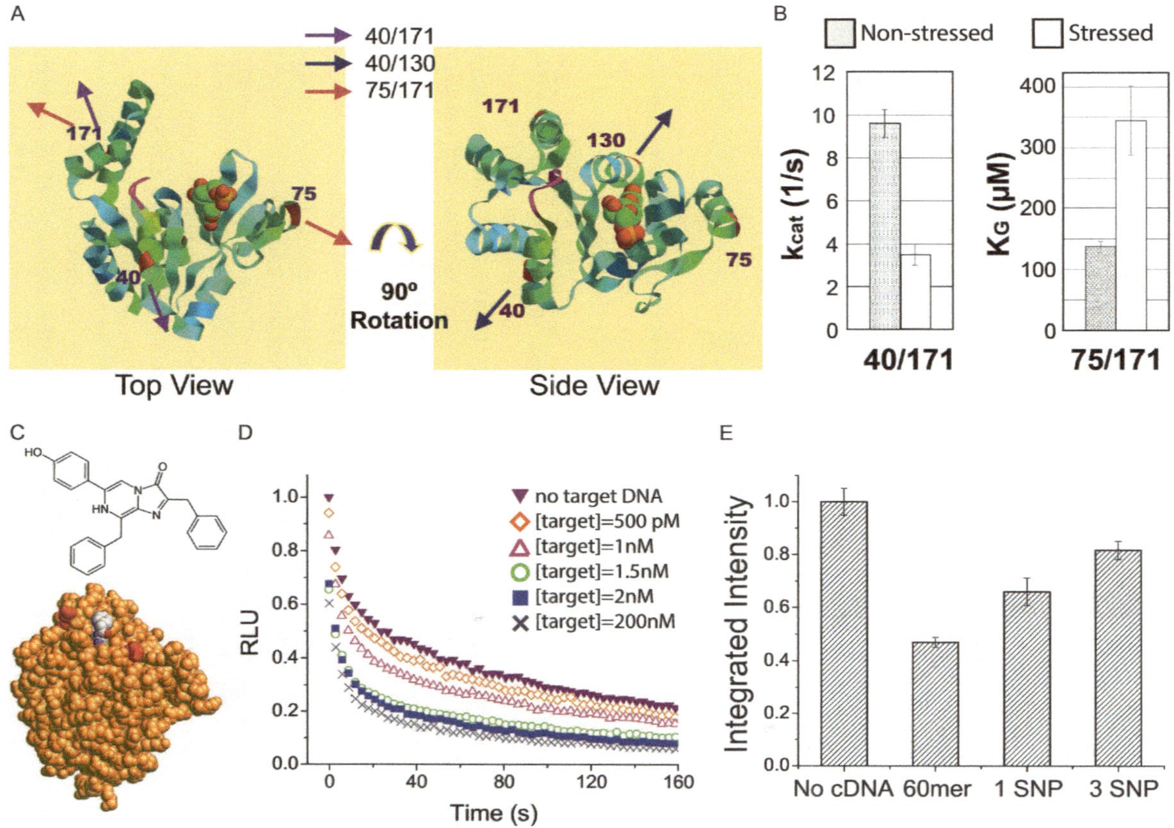

Fig. 2 See figure legend on next page.

using molecular oxygen dissolved in the medium. The reaction results in the emission of a blue photon (center wavelength is 470 nm, quantum efficiency is about 5%). Because of the light emission, this system is well suited as a platform to develop various molecular probes. Coelenterazine binds in a deep pocket in the enzyme (Figs. 1B and 2C), with no evident conformational change of the average structure with and without the ligand bound. Nonetheless, through a DNA spring attached as shown in Fig. 1B we could substantially modulate the enzymatic activity. With a 60 bp DNA spring we could dim the light emission to 36% of the original value (Tseng & Zocchi, 2013), and with a 40 bp spring it is reduced to 15% of the original value (Tseng & Zocchi, 2014). The applied stress does not simply unfold a fraction of the molecules. This is seen, for example, from the time course of the emitted light intensity under stress (Fig. 2D), which cannot be rescaled to coincide with the no-stress time course. We evince that the DNA spring induces a stress-specific deformation of the folded enzyme.

Fig. 2 (A) The figure shows three pairs of attachment points for the DNA spring on the enzyme Guanylate Kinase, which we used in the experiments. The GK structure (PDB: 1ZNX) is shown with bound GMP (space-filled). The three pairs of modified sites for DNA spring attachments (residue 40/171, 40/130, 75/171) are colored red. For each pair, the arrows illustrate the direction of the average mechanical force applied to the enzyme by the DNA spring in the "rigid" (ds) state. (B) The major effect of mechanical stress on the kinetic parameters measured for two of the GK-DNA chimeras described in (A). Shown is the catalytic rate k_{cat} for the 40/171 configuration, and the Michaelis-Menten constant relative to GMP for the 75/171 configuration, in the presence (white bars) and absence (gray bars) of mechanical stress. (C) Crystal structure of Renilla luciferase with bound substrate (coelenterazine-h, chemical formula shown) from Protein Data Bank (PDB) structure 2PSJ. Residues 161 and 188 (colored red) are mutated to cysteines in the experiments, serving as the attachment points for the DNA spring in the RLuc chimera. (D) The time course of luminescence intensity (arbitrary units) for the RLuc chimera from a titration experiment with varying amounts of complementary DNA (cDNA). The chimera DNA and the cDNA are both 60 bases long. From these data one can show that the fraction of hybridized chimeras is stoichiometric with the target cDNA down to the smallest concentration tested (500 pM) (Tseng & Zocchi, 2013), providing sensitivity in target detection. The chimera concentration was approximately 2 nM. (E) SNP detection with the RLuc chimera. By measuring the integrated intensity, the RLuc chimera we developed is able to detect one mismatch (1SNP) and three mismatches (3SNP) in the center of the DNA spring (60 bp). *Panels (C–E) are reprinted with permission from Tseng, C.-Y., & Zocchi, G. (2013). Mechanical control of Renilla luciferase.* Journal of the American Chemical Society, 135*(32), 11879–11886. doi:10.1021/ja4043565. Copyright (2013) by American Chemical Society.*

3. Principles of designing enzyme-DNA chimeras

3.1 Selection of the enzyme and DNA attachment sites

If we use the DNA spring as an experimental tool to study conformational plasticity in enzymes, the choice of enzyme is evidently dictated by the program of study. However, there may be other considerations and constraints. Our most successful constructions to date were with monomeric proteins in the 20 to 40kDa range. With multimeric enzymes, obtaining functional enzyme-DNA chimeras may be more difficult, and require new protocols. Possible concerns include whether the DNA will interfere with the multimerization, whether the DNA should be coupled to the individual monomers or to the multimeric complex, and the yield of different species. There are also opportunities: for a dimeric enzyme complex, chimeras could be constructed with the DNA spring acting on one of the monomers, or the other, or across the two. Our only experience of controlling a multimeric complex is with the Protein Kinase A (PKA), which we will discuss in detail in Section 5.1.

When planning a chimera construction, one of the first questions is of course where to attach the DNA spring on the surface of the enzyme in order to obtain control of the enzymatic activity. In our experience so far, virtually any configuration of the DNA spring likely will result in some effect on enzymatic activity, which is consistent with our view of the enzyme as a deformable viscoelastic solid. On the other hand, predicting where the DNA spring should be attached in order to obtain optimum control of the enzyme, is another matter, and we do not have an answer to this most interesting question. However, we may provide some suggestions. There have been attempts to identify "hot spots" for allosteric control on the surface of enzymes (Halabi, Rivoire, Leibler, & Ranganathan, 2009; Lockless & Ranganathan, 1999) and experiments where a light sensitive conformational switch was inserted at these locations obtained some control (Lee et al., 2008; Reynolds, McLaughlin, & Ranganathan, 2011). From our experience, using the DNA spring to mechanically control enzymes against their conformational changes during their catalytic cycles is a promising strategy, for example, in the cases of Maltose Binding Protein and Guanylate Kinase. Perturbing the substrate binding or the catalytic motif of enzymes is also demonstrated to work well, for example, in the cases of GK and Renilla Luciferase (see Sections 2.2 and 5.1 for more details).

To construct the enzyme-DNA chimeras, it is desired to have exactly two cysteines on the surface of the enzyme, defining the DNA attachment points. These cysteines are generally introduced by mutagenesis; in doing so, the objective is to obtain reactive (i.e. exposed) Cys, and to avoid significantly perturbing the enzyme. For the latter requirement, one must verify a posteriori that the enzymatic activity of the protein is not too much affected by the mutagenesis. Our strategy was to identify from the crystal structure well exposed surface residues chemically similar to Cys—ideally Ser, but we also successfully substituted Thr, Glu, Arg, Lys—and substitute with Cys. Removing internal cysteines might also help to improve the success and yield of the desired enzyme-DNA chimeras, and remove any ambiguity in identifying correct constructions by gel electrophoresis or HPLC. For example, for the study corresponding to Fig. 2A (Tseng et al., 2010; Wang, Wang, Qu, & Zocchi, 2009) we substituted the two internal Cys of the enzyme with Ser, and the DNA attachment points are chosen away from the active site of the enzyme, so as not to interfere with enzyme function. It is worthwhile to note that if a native cysteine is internal and inaccessible to the crosslinkers, it may be unnecessary to replace the Cys with Ser. For example, we have been successful to construct functional chimeras with RLuc without removing the internal cysteines in the protein. Our approach is to conduct a preliminary test by trying to react the wild-type protein with excess amount of crosslinker-conjugated DNA (Cx-DNA) and inspect the products; if Cx-DNA do not attach to the proteins, i.e. no protein-Cx-DNA (one-arm chimera) or protein with multi DNA arms are formed, then Cysteine substitution is not required. Namely, even if a protein contains a structurally important Cysteine or disulfide bond, the method may still be applicable.

3.2 Design of the sequence of the DNA arms

It is suggested that according to the size of target enzyme, the length of the DNA arms generally falls in the range of 40 to 60 bases, which, upon hybridization and bending, introduce significant mechanical stress on the enzyme of interest. The question of the optimal length of the DNA spring is delicate; from a theoretical perspective, it touches on several fundamental but still imperfectly known aspects of protein (and DNA) mechanics; experimentally, it could certainly be (laboriously) determined by construction for any given enzyme and attachment sites (and cross-linkers). For a quantitative understanding we need a model, which is briefly discussed in Section 5.2.

For the interested readers, we refer to these previous publications (Qu, Wang, Tseng, & Zocchi, 2011; Qu & Zocchi, 2011). The sequence of DNA arms should be carefully designed to avoid hairpin formation, self-dimerization, and self-hybridization, which can be achieved by using online web servers (e.g., Nupack) or the analyzer tools provided from DNA manufacturers (e.g., OligoAnalyzer from Integrated DNA Technologies). The DNA arms should also be modified to have amino groups, which will be cross-linked to the cysteines on the surface of the enzyme of interest via bifunctional cross-linkers. One of the DNA arms should also be phosphate-modified to enable the ligation step of the construction. We provide the sequences of our DNA arms in Section 4.2 as examples.

3.3 Two-step chimera construction

Enzyme-DNA chimeras are constructed in two sequential steps (see Fig. 3 for a flowchart). In the first step, the first DNA arm is attached to one surface cysteine of the protein, forming a one-arm (1R) chimera, through a crosslinker. The crosslinker we utilize here is $SM(PEG)_2$, consisting of N-hydroxysuccinimde (NHS) ester and maleimide reactive group linked by a 17.6 Å PEG spacer. The NHS esters react with the amino modifications at the end of the DNA arms while the maleimides react with the sulfhydryl groups of cysteines on the protein surface. Since there are two reactive cysteines on the protein, unavoidably some two-arm chimeras (with identical arms) are also obtained, as well as protein dimers (bound by one di-sulfide bond) with one or two DNA arms; therefore, reduction by tris (2-carboxyethyl)phosphine) (TCEP) and purification by HPLC with size-exclusion columns and ion-exchange columns are necessary after the first step. Size-exclusion column is not only used to filter out protein aggregation, but also to remove TCEP, which interferences DNA conjugation in 1R construction. We tested and found the maximum allowable concentration of TCEP is 10 μM (Wang et al., 2009) (Fig. 4D). For the second step in the synthesis, an excess amount of the second DNA arm reacts with the purified 1R, followed by Ni-NTA chromatography to remove uncoupled DNA. The purified two-arm (2R) chimeras are subsequently ligated with T4 DNA ligases, followed by Ni-NTA chromatography to obtain the final product, i.e., the ligated two-arm (L2R) chimeras. The two-step chimera construction is superior to one-step method, in which a single DNA strand is attached on the protein surface at both ends in one step, for its much higher yield (see Wang et al., 2009 for detailed discussions).

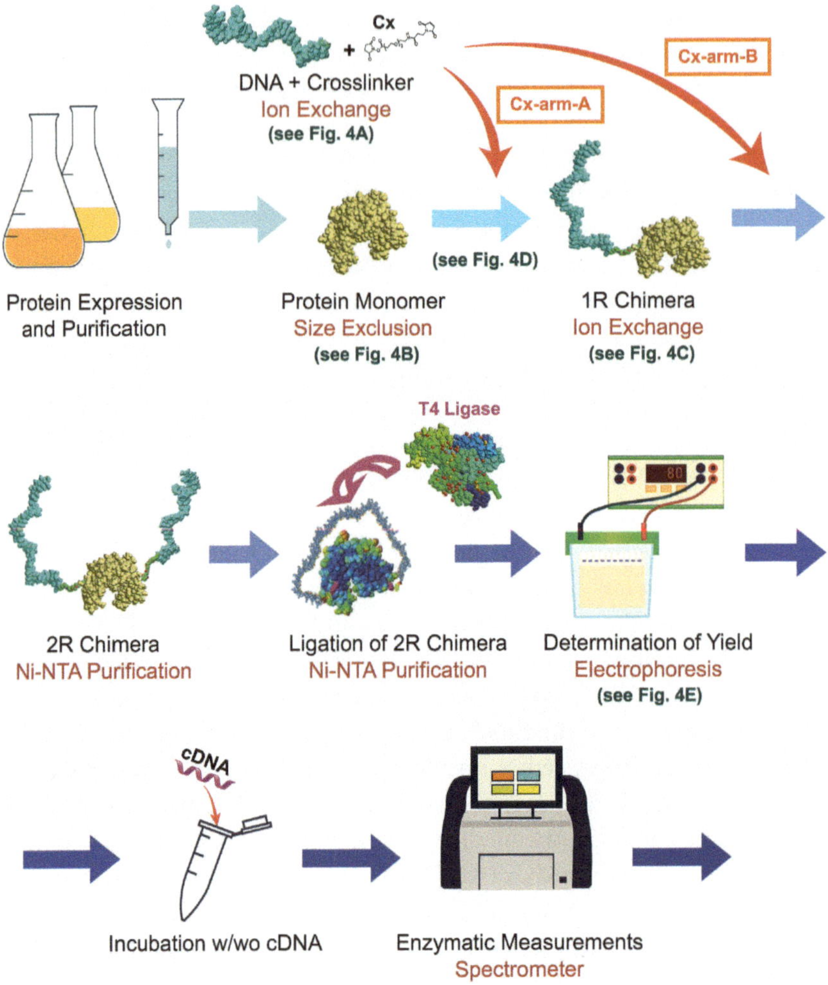

Fig. 3 Flowchart for synthesizing ligated two-arm chimeras. The step-by-step protocol is given in Section 4. The corresponding HPLC traces or gel images of the steps are noted and can be referred to Fig. 4.

4. Design and synthesize enzyme-DNA chimeras

4.1 Equipment

Protein expression system, high performance liquid chromatography (HPLC), centrifuge, gel electrophoresis system, spectrometer.

4.2 Chemicals and materials

Protein expression materials and reagents (plasmid, competent bacteria, bacteria growth media, isopropyl β-D-1-thiogalactopyranoside (IPTG), etc.), Ni-NTA purification reagents, cross-linker (e.g., $SM(PEG)_2$, an amine-to-sulfhydryl crosslinker with 17.6Å spacer arm, Thermo Scientific, Cat. No: 22102), Tris(2-carboxyethyl)phosphine (TCEP), T4 ligase.

- Synthesized DNA (with end modifications):
  - DNA arm-A: 5′-amino modification-sequence (20~30 nt)-3′
  - DNA arm-B: 5′-phosphate modification-sequence (20~30 nt)-amino modification-3′
  - DNA splint for ligation (12–18 nt)

Target cDNA: complementary DNA to arm-A plus arm-B

- Examples of DNA arms (from Integrated DNA Technologies):
  - arm-A (30 nt): 5′-/5AmMC6/GAGTGTGGAGCCTAGACCGT GAGTTGCTGG-3′
  - arm-B (30 nt): 5′-/Phos/CAGGAGTCCACGGTACCATCCAAG CAGCTG/3AmMO/-3’
  - Splint (18 nt): 5’-GGACTCCTG CCAGCAACT-3’
  - Target cDNA: 5’-CAGCTGCTTGGATGGTACCGTGGACTC CTGCCAGCAACTCACGGTCTAGGCTCCACACTC-3’

Note: 5AmMC6 and 3AmMO are amino-modifications positioned at the 5′-end and 3′-end, respectively, of synthetic DNA provided by Integrated DNA Technologies (IDT).

- Buffers for HPLC:
  - Low salt: 100 mM sodium phosphate, and 1 mM EDTA at pH 7.
  - High salt: 100 mM sodium phosphate, 1 M NaCl, and 1 mM EDTA at pH 7.

4.3 Protocols

4.3.1 Two-arm chimera (DNA-protein-DNA) synthesis

1. Protein expression and purification
   - Refer to other protocols for protein expression and purification.
   - The protein of interest should contain a His-tag or another affinity-based purification tag, which will facilitate the purification using affinity purification protocols.
2. DNA-crosslinker conjugation and purification
   - Dissolve DNA arm-A in phosphate buffer at pH 7.2–7.5 to prepare 1 mM stock and store at −20 °C.

- Dissolve crosslinker (Cx) $SM(PEG)_2$ in dimethyl sulfoxide (DMSO) to prepare 1 M stock solution, which is then divided into small aliquots and stored at −20 °C. Freshly prepared Cx solution (<2 months) is preferred.
- Mix Cx and DNA solution at the ratio of 50 (or more):1 in phosphate buffer at pH 7.2–7.5 and incubate for 1 h at RT.
- Remove excess crosslinkers by using ion-exchange chromatography (e.g. UNO Q1 Column, Bio-Rad, Cat. No: 7200001; Fig. 4A). The pH of the HPLC collection should be adjusted to ~7 to facilitate the following 1R conjugation. Measure the Cx-DNA concentration. The yield of Cx-DNA can be determined by using 15% TBE-Urea gel (Fig. 4D). The Cx-DNA conjugate should be used freshly (<1 week stored at 4 °C).
- Repeat the above steps for DNA arm-B.

3. Protein monomer purification

- If the target proteins form dimers or polymers through di-sulfide bond between cysteines exposed on the surfaces (e.g. GK), perform the following steps; otherwise, skip this part (e.g. RLuc).
- Add 1–5 mM TCEP to the protein solution at a final concentration of 1–5 mM and incubate for 30 min—1 h.
- Remove excess TCEP using size-exclusion chromatography (e.g. Bio-Sil SEC 125-5 Column, Bio-Rad, Cat. No: 125-0060; Fig. 4B). The pH of the HPLC collection should be adjusted to ~7 for later one-arm (1R) conjugation. Measure the protein concentration. Perform 1R conjugation immediately to minimize protein re-polymerization.

4. One-arm (DNA-protein) and two-arm (DNA-protein-DNA) chimera synthesis

- Add Cx-arm-A to protein monomer and incubate for 3–5 h at RT. The ratio of Cx-DNA to protein depends on the properties of the protein and should be optimized to achieve higher yield of 1R constructions compared to 2R constructions. The optimized ratios vary from protein to protein, or even to protein with different attachment sites. For example, for GK we used 1:2 while for RLuc we used 1:9.
- Purify the 1R chimera using ion-exchange chromatography (Fig. 4C) and measure the concentration.
- Add Cx-arm-B to the purified 1R chimera at the ratio higher than 2:1 and incubate at 4 °C overnight to maximize the yield.

5. Two-arm chimera purification
    - Purify 2R chimera by Ni-NTA affinity purification procedures. Note that, due to the excess use of Cx-arm-B, the amount of 1R chimeras after the conjugation is significantly low compared to the desired 2R chimeras. The main goal of this step is to remove the DNA; therefore, Ni-NTA affinity purification is simple yet sufficient to achieve the purpose, while purification based on chromatography is expected to work well. Also, note that as the conjugation buffer contains EDTA, which interfere with the Ni-NTA affinity, the 2R chimera should be exchanged into other buffers using centrifugal concentrators before purification. However, certain enzymes (e.g., RLuc) are unstable and vortexing/centrifuging should be avoided at all times. In these cases, dilute the conjugation reaction by adding PBS buffer (pH 7.4) instead.

4.3.2 Ligation of two-arm chimera

1. Ligation and purification of two-arm chimera
    - Mix 2 μM 18mer DNA splint, 2 μM two-arm chimera, 4 unit/μL T4 ligase and 0.4 mg/mL BSA in recommended reaction buffer and incubate at 16 °C overnight. If the concentration of 2R chimera is low, optimize the concentrations of the reagents and the incubation time (e.g., 1.25 μM DNA splint, 20 nM chimera, 1.2 unit/μL ligase and 0.1 mg/mL BSA were used for RLuc)
    - To purify the ligated 2R chimera, follow the steps for purifying 2R chimera in Protocol 4.3.1
2. Determination of yield
    - Perform SDS-PAGE electrophoresis to determine the yield of L2R chimera. Molecules with different molecular weight as well as different configurations (e.g. linear and circular shape) can be separated by using gel electrophoresis (Fig. 4E).

4.3.3 Enzymatic activity measurements

1. Measurements on effects by mechanical stress
    - For the sample: add target DNA (complementary DNA of the arm A+B, i.e. cDNA) to the chimera in the reaction buffer with ~0.4 mg/mL BSA and incubate for 3 h to overnight. Generally, 5:1 of cDNA:chimera is sufficient to saturate the enzymatic reactions

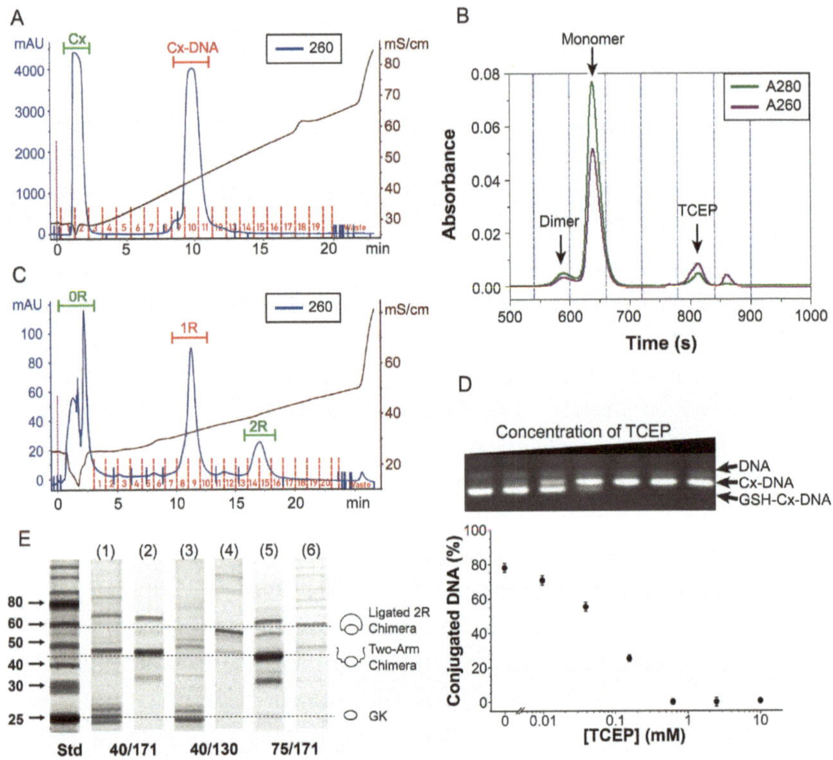

Fig. 4 Representative examples of preparative HPLC traces and gels at various stages during the synthesis of ligated two-arm chimeras. (A) High performance ion-exchange chromatography (IE-HPLC) profile for removing excess cross-linkers (SM(PEG)$_2$) from Cx-DNA. The blue curve is the absorption at 260 nm; the brown curve represents the conductivity. Red dashed lines and numbers indicate collection fractions. (B) Size-exclusion HPLC trace in terms of absorptions at 280 nm (green line) and 260 nm (purple line) of the purification step for removing TCEP from the reduced proteins (GK)prior to conjugation to DNA arms. Blue dash-dotted lines indicate different collection fractions. (C) IE-HPLC profile of 1R chimera purification (RLuc). The blue curve is the absorption at 260 nm; the brown curve represents the conductivity. Red dashed lines and numbers indicate collection fractions. (D) Verification of conjugation of crosslinkers (Cx) to DNA arms and conjugation of Cx-DNA to thiol groups in the presence of TCEP at different concentrations. Top: representative TBE-urea PAGE gel showing bands for unconjugated DNA (DNA), crosslinker-conjugated DNA (Cx-DNA), and glutathione conjugated Cx-DNA (GSH-Cx-DNA). Bottom: the dependence of the percentage of successfully conjugated GSH-Cx-DNA on the concentration of TCEP in the conjugation solution. These measurements give an estimate for the maximum residual concentration of TCEP which can be tolerated in the DNA-protein conjugation reaction. TBE-Urea gel can also be used to determine the crosslinking efficiency of the DNA-arms and the cross-linkers. (E) Gel electrophoresis of products after protein, 2R chimera and L2R chimera purification (GK). The quantities of the desired products and side products can be determined

(Continued)

- For the control: add nothing (if non-mechanical effects are absent, see Section 4.5.3) or two complementary DNA strands (one for arm-A and the other for arm-B) to the chimera in the reaction buffer with ~0.4 mg/mL BSA and incubate for 3 h to overnight. The cDNA:chimera ratio should be the same as the sample
- Perform enzymatic activity measurements for the sample and control side by side

4.4 Example of protocol—RLuc-DNA chimera synthesis

In addition to the general protocols (Section 4.3), here we provide the exact steps that we used to synthesize RLuc-DNA chimeras and the readers may follow directly.

1. Express Renilla Luciferase mutant (cysteine mutations at 161, 188) in *E. coli* BL21(DE3)pLysS and purify it using Ni-NTA chromatography. The final concentration is ~20–30 μM.
2. Dissolve 100 mg SM(PEG)$_2$ crosslinker (Thermo Scientific, 22102) (Cx) in 235 μL DMSO to a final concentration 1 M. Aliquot to 10 μL in 0.2 mL PCR tubes and store at −20 °C.
3. Dissolve customized DNA arm-A and arm-B (IDT) in 100 mM sodium phosphate buffer (pH 7.4) to a final concentration 1 mM. Store at −20 °C.
4. Mix 70 μL of arm-A, 10 μL of Cx with 100 μL phosphate buffer (pH 7.4) and incubate for 1 h. Inject the mixture onto HPLC ion-exchange column (Uno Q1 column, Bio-Rad, Cat. No: 7200001) and collect the Cx-arm-A fraction. The final concentration is ~50 μM.

Fig. 4—Cont'd by gel electrophoresis. After protein purification, the desired product is protein monomers and the side products include dimers and other multimers. After 2R chimera purification, the side products include 1R chimera while after L2R chimera purification, the side products include unligated 2R chimera. Lanes 1 and 2: 40/171 protein and 2R chimera purification; lanes 3 and 4: 40/130 protein and L2R purification; lanes 5 and 6: 75/171 2R and L2R purification. The small icons indicate the locations of protein monomers, 2R and L2R chimeras in the gel. Side products (e.g. dimers after protein purification and 1R chimeras after 2R chimera purification) are unavoidable. *Panel (D) Modified with permission of IOP Publishing, Ltd., from Wang, Y., Wang, A., Qu, H., & Zocchi, G. (2009). Protein–DNA chimeras: Synthesis of two-arm chimeras and non-mechanical effects of the DNA spring.* Journal of Physics: Condensed Matter, *21(33), 335103. doi:10.1088/0953-8984/21/33/335103); permission conveyed through Copyright Clearance Center. Panel (E) Modified with permission from Tseng, C.-Y., Wang, A., & Zocchi, G. (2010). Mechano-chemistry of the enzyme guanylate kinase.* Europhysics Letters Association, *91(1), 18005. doi:10.1209/0295-5075/91/18005). Copyright (2010) by European physical society.*

5. Mix 65 μL of arm-B, 10 μL of Cx with 100 μL phosphate buffer (pH 7.4) and incubate for 1 h. Inject the mixture onto HPLC ion-exchange column and collect the Cx-arm-B fraction. The final concentration is ~40 μM.
6. Mix 1000 μL RLuc mutant (~30 μM), 65 μL Cx-arm-A (~50 μM) with 900 μL phosphate buffer (pH 7) and incubate for 3 h. Inject the mixture onto HPLC ion-exchange column and collect the 1R chimera fraction.
7. Add 150 μL of Cx-arm-B into the collected fraction.
8. Repeat step 6 and 7 for three times to collect enough product.
9. Incubate the product at 4 °C overnight.
10. Mix 600 μL Ni-NTA beads, the product of step 9, with PBS buffer to a total volume ~50 mL. Incubate the mixture on shaker for 4 h at 4 °C.
11. Pass the mixture through gravity flow column. Discard the flow-through, wash the beads with 10 mL of PBS buffer, elute the product with 500 μL of elution buffer (containing 100 mM imidazole) for four times and collect each elution in individual tubes. Measure the luminescence intensity of each collection and keep the ones with high chimera concentration.
12. Determine the yield of successful construction (2R chimera) with 10% SDS-PAGE gel.
13. Add 100 μL of 10X ligation buffer, 10 μL of 100 μM DNA splint, 30 μL of BSA (20 mg/mL), 300 μL product of step 11, 20 μL of T4 ligase (1–3 unit/μL) with 540 μL of ultra pure water in a 1.5 mL tube and incubate at 16 °C overnight (or longer). We usually prepare three sets of samples to collect adequate quantity.
14. Mix 500 μL Ni-NTA beads, the product of step 13 with PBS buffer to a total volume ~50 mL. Incubate the mixture on shaker for 4 h at 4 °C.
15. Pass the mixture through gravity flow column. Discard the flow-through, wash the beads with 3 mL of PBS buffer, elute the product with 500 μL of elution buffer for four times and collect each elution in individual tubes. Measure the luminescence intensity of each collection and keep the ones with high chimera concentration.
16. Determine the yield of successful construction (Ligated 2R chimera) with 10% SDS-PAGE gel.
17. Incubate the L2R chimera with/without cDNA for 1 h (to overnight) and measure the luminescence activity.

4.5 Troubleshooting

4.5.1 Protein aggregation

Protein aggregation may be induced due to various reasons. Cysteine mutations introduced on the protein surface may lead to aggregation via the formation of disulfide bonds, for example, in the case of GK. We use TCEP to reduce the disulfide bonds and obtain protein monomers (Fig. 4B). The concentrations of TCEP and incubation time could be optimized. Reduced proteins will begin to polymerize again once TCEP is removed; therefore, it is strongly suggested that 1R conjugation should follow immediately the removal of TCEP. In addition, some proteins tend to aggregate (e.g. RLuc). Our strategy for this situation is to keep the protein concentration low (~25 μM for RLuc) and avoid vortexing at all times.

4.5.2 Low yield of 1R chimera

One-arm chimera is generally the most critical step and may require significant effort of optimization. Low yield of 1R chimera may be due to the failure in Cx/DNA or Cx-DNA/protein conjugation, or the formation of side products. Cross-linkers are highly sensitive to pH and storage time. If cross-linkers fail to conjugate to either DNA-arms or proteins, we suggest checking the pH at each step and using fresh cross-linkers (<2 months) and Cx-DNA (<1 week). The side products, including 2R chimeras (with identical arms) and dimers with one or two DNA arm, are unavoidable but can be reduced. If there is a large quantity of 2R chimera, we suggest increasing the protein/DNA ratio until the percentage of 1R chimera is satisfactory. If there are large quantities of dimer products, a higher concentration of TCEP or incubation time before 1R purification may be helpful.

4.5.3 Non-mechanical effect of DNA on enzymatic activity

Certain enzymes or proteins may suffer from non-mechanical effect due to the conjugation to DNA. Examples of such proteins include DNA-binding proteins (e.g., *lac* repressor) and proteins interacting with substrates containing the pyrimidine and purine bases (e.g., GK interacting with ATP and GMP). In these cases, an activation of the enzymatic activity could be resulted from the removal of DNA from the substrate binding domain rather than mechanical stress. To correctly evaluate the influence of the mechanical stress, special care should be taken by proper controls (Wang et al., 2009). For example, one can compare the enzymatic activity of the chimera hybridized with one cDNA, which forms an intact dsDNA spring and exerts mechanic stresses, with that of the chimera hybridized with two individual cDNA, which leaves nicks or gaps in the dsDNA and thus does not stress the enzyme.

5. Discussion

5.1 Selection of target enzyme: Possibilities and challenges

Our first chimera constructions were with Maltose Binding Protein and Guanylate Kinase; both proteins undergo large conformational changes induced by ligand binding, and the attachment of the DNA spring was initially designed to oppose part of that conformational change (e.g. the 75/171 attachments in Fig. 2A), and thus affect ligand binding. These designs were successful in mechanically controlling both proteins. In another case (the 40/171 attachments in Fig. 2A) we purposely tried to perturb elements of the catalytic motif (the so-called p-loop in GK), which also resulted in successful mechanical control of the enzyme. For the RLuc, we sought to perturb the substrate binding pocket (Figs. 1B and 2C), and obtained mechanical control even though substrate binding is not of the induced fit type.

So far, our only attempt to control a multimeric complex was with the Protein Kinase A (PKA). This allosterically controlled enzyme consists of a catalytic and a regulatory subunit; binding of cAMP to the regulatory subunit causes it to fall off the complex, exposing the catalytic site and thus activating the enzyme. We attached a DNA spring to the regulatory subunit, in a position such that the mechanical stress would favor part of the documented conformational change upon cAMP binding. We were able to activate the enzyme under stress, though the effect was about half of that obtained with cAMP (Choi & Zocchi, 2006). However, at the time we were using a different chimera synthesis method (coupling to the enzyme a single DNA oligomer modified at both ends), which resulted in a lower yield of correct constructions in the experiments. An interesting question is whether one can even improve on the natural, cAMP driven allosteric regulation (in terms of the ratio of enzymatic activity in the on and off states). There is a trade-off between the residual activity in the off state and the activity in the on state. Increasing the bond strength between the regulatory subunit (which acts as an inhibitor) and the catalytic subunit would decrease the residual activity in the off state, but also decrease the fraction of dissociated enzyme complexes (and thus the enzymatic activity) in the on state.

If one could engineer a stronger bond between the two subunits, and if the mechanical stress provided by the DNA spring were large enough to still drive almost complete dissociation of the subunits, one might increase the dynamic range of allosteric control.

5.2 Quantitative characterization of the mechanical perturbation

In Section 3.1 we noted that, in our experience, a mechanical perturbation at virtually any location on the enzyme will have some effect on the activity. However, obtaining a measurable effect must also depend on the strength of the perturbation. Therefore it would be desirable to have a measure to quantify this strength in mechanical terms, that is, in terms other than the effect on enzymatic activity. We now summarize our progress in this direction.

A complete microscopic characterization would be in terms of the stress field in the enzyme; at the other extreme, if one wished to characterize the perturbation with just one number, it would be the total elastic free energy injected into the enzyme, E_{enzyme} (in the following we will call it "elastic energy" for short). It is not easy to measure this quantity. We were able to measure a related quantity: the total elastic energy in the enzyme-DNA chimera under stress (i.e. the sum of the elastic energies in the enzyme and the DNA spring). These measurements are detailed here (Tseng, Wang, Zocchi, Rolih, & Levine, 2009), and an overview is also given here (Zocchi, 2018). The method we used exploits the elastic energy driven dimerization of certain enzyme-DNA chimeras (namely, those constructions with a nick in the DNA spring). With the same method, we measured the elastic energy of the DNA spring (Qu & Zocchi, 2011). For these cases we therefore know the elastic energy injected into the enzyme: $E_{enzyme} = E_{tot} - E_{DNA}$, where E_{tot} is the total elastic energy in the chimera, and E_{DNA} the elastic energy in the DNA spring. We note that it should also be possible to measure these elastic energies by calorimetry, though these would be quite delicate experiments. To give some actual numbers, for the GK-DNA chimera with a 60 bp DNA spring attached at the sites 75/171 (Fig. 2A), we find $E_{tot} \approx 9.1$ k_BT, $E_{DNA} \approx 6.6$ k_BT, and thus $E_{enzyme} \approx 2.5$ k_BT (Zocchi, 2018). Here k_BT is the thermal energy at room temperature, 1 $k_BT = 0.59$ kcal/mol.

To build an understanding of these numbers, and of the mechanical perturbation in general, one can build a model. The minimal such model considers the enzyme-DNA chimera as a system of two coupled springs: a DNA "leaf spring" which is compressed, and an "enzyme spring" which is stretched. We know the mechanical response of the DNA spring, from a combination of measurements and model building (Qu & Zocchi, 2011). It is a nonlinear spring, characterized by a softening transition that limits the force it can deliver to a few pN. It turns out that the enzyme spring is also nonlinear, with a similar softening transition (Tseng & Zocchi, 2014). With this model of two coupled nonlinear springs we can rationalize the measurements on the enzyme-DNA chimeras, and make some predictions. Given the geometry of the chimera construct (distance between surface attachment points in the undeformed enzyme, contour length of the cross-linkers, contour length of the DNA spring), and assuming a specific linear and nonlinear mechanical response of the enzyme, one can calculate the elastic energy injected in the enzyme (the energy in the "enzyme spring"), as well as other thermodynamic quantities. If we assume that enzymatic activity is a monotonically decreasing function of the elastic energy in the enzyme, we can then predict the effect of changing the length of the DNA spring (or other parameters) on the dynamic range of control obtained in the corresponding chimeras. In this way we understand, for example, why a 40 bp DNA spring elicits stronger control of the enzyme compared to a 60 bp spring, for the chimera of Fig. 1B. More details are provided in (Tseng & Zocchi, 2014).

5.3 Applications: Enzyme-DNA chimeras as molecular probes

Sensitive molecular detection systems often use an enzyme based amplification mechanism, as in enzyme-linked immunosorbent assay (ELISA) or polymerase chain reaction (PCR). An enzyme-DNA chimera does in principle constitute a detection system for the DNA (or RNA) sequence complementary to the chimera DNA, and more generally for any binding event or process which modifies the stiffness of the DNA spring. The RLuc-DNA chimera in particular—as well as future constructions utilizing any of the enzymes commonly used in ELISA—lends itself to develop in-solution (homogeneous) molecular probes. We have only taken the first steps in this direction, with a proof-of-concept of using the RLuc-DNA chimera for direct DNA sequence detection (Fig. 2D) and mismatch detection (Fig. 2E) in solution (Tseng & Zocchi, 2013). But there are wider possibilities. Namely, it turns out that, just as ligand binding to the enzyme is not impaired by the mere presence of the DNA spring, so is the DNA in the chimera still available for protein binding. For example, restriction enzymes can bind to the DNA spring and cut it, and the very construction of the chimeras uses ligation of the DNA spring. Our preliminary, unpublished results show that we can detect DNA binding proteins with the RLuc chimera. The mechanism is, to our present understanding, that binding of a protein to the ds DNA spring rigidifies the spring, causing a modulation of activity of the enzyme. In a different twist, we have used the RLuc-DNA chimera to measure the bending stiffness of exotic nucleic acids hybrids such as PNA/DNA and LNA/DNA (Zhan & Zocchi, 2017).

6. Summary

Various DNA-based techniques have been used to study and modulate the activity of enzymes (Engelen, Janssen, & Merkx, 2016; Zhang, Nangreave, Liu, & Yan, 2014). Among them, the method of enzyme-DNA chimera provides a distinctive approach; it works in solution and the stress applied on the enzyme can be fine-tuned semi-continuously. We have successfully constructed several enzyme-DNA chimeras using different enzymes, DNA springs, or attachment points, demonstrating that this is indeed a general method in artificial control of enzymes. With DNA springs, we can induce various states of deformation of a protein with tunable, rather than merely on/off, forces. For enzymes, deforming the molecule results in a modulation of enzymatic activity, in which the enzymatic parameters may be affected individually. For those interested in the general field of conformational changes and allostery, the method represents a relatively new experimental tool. For applications, the modulation of enzymatic activity in principle offers a basis to design various in solution and possibly in vivo molecular probes.

Acknowledgments

C-YT acknowledges support from Dr. Chin-Lin Guo and Academia Sinica of Taiwan (Grant No. AS-105-TP-A04, AS-TP-109-M04). YW acknowledges support from the National Science Foundation (Grant No. 1826642) and the Arkansas Biosciences Institute (Grant No. 0326). GZ acknowledges support from the National Science Foundation (Grant No. DMR-1809381).

We wish to acknowledge the fundamental contributions of, and the pleasure of working with, our colleagues Brian Choi, Andrew Wang, Hao Qu, Daniel Sanchez, Collin Joseph, and Yilin Zhan.

References

Alavi, Z., & Zocchi, G. (2018). Dissipation at the angstrom scale: Probing the surface and interior of an enzyme. *Physical Review E*, *97*(5), 052402. https://doi.org/10.1103/PhysRevE.97.052402.

Choi, B., & Zocchi, G. (2006). Mimicking cAMP-dependent allosteric control of protein kinase a through mechanical tension. *Journal of the American Chemical Society*, *128*(26), 8541–8548. https://doi.org/10.1021/ja060903d.

Choi, B., Zocchi, G., Canale, S., Wu, Y., Chan, S., & Perry, L. J. (2005). Artificial allosteric control of maltose binding protein. *Physical Review Letters*, *94*(3), 038103. https://doi.org/10.1103/PhysRevLett.94.038103.

Cooper, A., & Dryden, D. T. F. (1984). Allostery without conformational change. *European Biophysics Journal*, *11*(2), 103–109. https://doi.org/10.1007/BF00276625.

Engelen, W., Janssen, B. M. G., & Merkx, M. (2016). DNA-based control of protein activity. *Chemical Communications*, *52*(18), 3598–3610. https://doi.org/10.1039/C5CC09853J.

Frauenfelder, H., Parak, F., & Young, R. D. (1988). Conformational substates in proteins. *Annual Review of Biophysics and Biophysical Chemistry*, *17*(1), 451–479. https://doi.org/10.1146/annurev.bb.17.060188.002315.

Halabi, N., Rivoire, O., Leibler, S., & Ranganathan, R. (2009). Protein sectors: Evolutionary units of three-dimensional structure. *Cell*, *138*(4), 774–786. https://doi.org/10.1016/j.cell.2009.07.038.

Johnston, C. A., Whitney, D. S., Volkman, B. F., Doe, C. Q., & Prehoda, K. E. (2011). Conversion of the enzyme guanylate kinase into a mitotic-spindle orienting protein by a single mutation that inhibits GMP-induced closing. *Proceedings of the National Academy of Sciences of the United States of America*, *108*(44), 17863–17864.

Karplus, M., & McCammon, J. A. (2002). Molecular dynamics simulations of biomolecules. *Nature Structural Biology*, *9*(9), 646–652. https://doi.org/10.1038/nsb0902-646.

Koshland, D. E., Jr. (1995). The key–lock theory and the induced fit theory. *Angewandte Chemie International Edition in English*, *33*(23–24), 2375–2378. https://doi.org/10.1002/anie.199423751.

Lankaš, F., Lavery, R., & Maddocks, J. H. (2006). Kinking occurs during molecular dynamics simulations of small DNA Minicircles. *Structure*, *14*(10), 1527–1534. https://doi.org/10.1016/j.str.2006.08.004.

Lee, J., Natarajan, M., Nashine, V. C., Socolich, M., Vo, T., Russ, W. P., et al. (2008). Surface sites for engineering allosteric control in proteins. *Science*, *322*(5900), 438. https://doi.org/10.1126/science.1159052.

Lockless, S. W., & Ranganathan, R. (1999). Evolutionarily conserved pathways of energetic connectivity in protein families. *Science*, *286*(5438), 295–299. https://doi.org/10.1126/science.286.5438.295.

Qu, H., Wang, Y., Tseng, C.-Y., & Zocchi, G. (2011). Critical torque for kink formation in double-stranded DNA. *Physical Review X*, *1*, 21008. https://doi.org/10.1103/PhysRevX.1.021008.

Qu, H., & Zocchi, G. (2011). The complete bending energy function for nicked DNA. *Europhysics Letters Association*, *94*(1), 18003.

Reynolds, K. A., McLaughlin, R. N., & Ranganathan, R. (2011). Hot spots for allosteric regulation on protein surfaces. *Cell*, *147*(7), 1564–1575. https://doi.org/10.1016/j.cell.2011.10.049.

Smith, S. B., Cui, Y., & Bustamante, C. (1996). Overstretching B-DNA: The elastic response of individual double-stranded and single-stranded DNA molecules. *Science*, *271*(5250), 795–799. https://doi.org/10.1126/science.271.5250.795.

Tlusty, T., Libchaber, A., & Eckmann, J.-P. (2017). Physical model of the genotype-to-phenotype map of proteins. *Physical Review X*, 7(2), 021037. https://doi.org/10.1103/PhysRevX.7.021037.

Tseng, C.-Y., Wang, A., & Zocchi, G. (2010). Mechano-chemistry of the enzyme guanylate kinase. *Europhysics Letters Association*, *91*(1), 18005. https://doi.org/10.1209/0295-5075/91/18005.

Tseng, C.-Y., Wang, A., Zocchi, G., Rolih, B., & Levine, A. J. (2009). Elastic energy of protein-DNA chimeras. *Physical Review E*, *80*(6), 061912. https://doi.org/10.1103/PhysRevE.80.061912.

Tseng, C.-Y., & Zocchi, G. (2013). Mechanical control of Renilla luciferase. *Journal of the American Chemical Society*, *135*(32), 11879–11886. https://doi.org/10.1021/ja4043565.
Tseng, C.-Y., & Zocchi, G. (2014). Equilibrium softening of an enzyme explored with the DNA spring. *Applied Physics Letters*, *104*, 153702. https://doi.org/10.1063/1.4871872.
Wang, Y., Wang, A., Qu, H., & Zocchi, G. (2009). Protein–DNA chimeras: Synthesis of two-arm chimeras and non-mechanical effects of the DNA spring. *Journal of Physics: Condensed Matter*, *21*(33), 335103. https://doi.org/10.1088/0953-8984/21/33/335103.
Wang, M. D., Yin, H., Landick, R., Gelles, J., & Block, S. M. (1997). Stretching DNA with optical tweezers. *Biophysical Journal*, *72*(3), 1335–1346. https://doi.org/10.1016/S0006-3495(97)78780-0.
Wang, Y., & Zocchi, G. (2011). The folded protein as a viscoelastic solid. *Europhysics Letters Association*, *96*(1), 18003. https://doi.org/10.1209/0295-5075/96/18003.
Zhan, Y., & Zocchi, G. (2017). Flexibility of DNA/PNA, DNA/LNA, DNA/RNA hybrids measured with a nanoscale transducer. *Europhysics Letters Association*, *119*(4), 48005. https://doi.org/10.1209/0295-5075/119/48005.
Zhang, F., Nangreave, J., Liu, Y., & Yan, H. (2014). Structural DNA nanotechnology: State of the art and future perspective. *Journal of the American Chemical Society*, *136*(32), 11198–11211. https://doi.org/10.1021/ja505101a.
Zocchi, G. (2018). *Molecular machines: A materials science approach*. Princeton University Press.

CHAPTER ELEVEN

The Spy that links: Creation of nonlinear protein architectures and materials using SpyTag/SpyCatcher chemistry

Ri Wang and Fei Sun*
Department of Chemical and Biological Engineering, The Hong Kong University of Science and Technology, Hong Kong SAR, People's Republic of China
*Corresponding author: e-mail address: kefsun@ust.hk

Contents

Abstract

The peptide/protein pair, SpyTag/SpyCatcher, which is derived from split immunoglobulin-like collagen adhesin domain (CnaB2) from *Streptococcus pyogenes*, can spontaneously form a stable Lys-Asp isopeptide bond under physiological conditions. This enabling technology— also known as genetically encoded click chemistry owing to its marked efficiency and specificity—has led to a variety of applications in protein engineering, materials science and synthetic biology in recent years. In this chapter, we discuss the use of SpyTag/SpyCatcher chemistry to create nonlinear protein architectures and materials, with emphasis on its role in shaping up topology

Methods in Enzymology, Volume 647
ISSN 0076-6879
https://doi.org/10.1016/bs.mie.2020.09.011

engineering as an emerging branch of protein engineering. The synthesis of entirely protein-based molecular networks, Spy networks, is highlighted. The protocols for preparing Spy networks and applications thereof are also illustrated.

1. Introduction

1.1 Protein topology engineering

Chemical topology refers to the structural feature of a molecule that remains unchanged through arbitrary and continuous distortion and deformation without resorting to breaking a covalent bond or passing through one another (Wang & Zhang, 2018). Synthetic macromolecules are marked by their sophistication and diversity in chemical topologies, from branched to mechanically interlocked molecules (Forgan, Sauvage, & Stoddart, 2011; Mitchell & Chambron, 1995). On the other hand, naturally occurring biomacromolecules like DNA, RNA and proteins are predominantly linear polymers. Proteins or peptides with unusual chemical topologies, such as rings, knots, tadpoles, lassos and so on, do exist sporadically in biological realms though, some of which are associated with functional benefits like increased protein stability (Cascales & Craik, 2010; Giegerich, Voss, & Rehmsmeier, 2004). The contrast between synthetic polymers and natural biomacromolecules in chemical topology prompts an intriguing question for synthetic biologists: is it feasible to use cellular machinery to create biomacromolecules with diverse chemical topologies, rivaling the way chemists make synthetic polymers? It is conceivable that the creation of these biomacromolecules with unconventional topologies could open the door to new materials with properties that have never been explored before.

Initially attempts to create nonlinear biomacromolecules focused primarily on cyclic proteins, drawing on classical protein ligation strategies, including split-intein, sortase, butelase, etc. (Schmidt, Toplak, Quaedflieg, & Nuijens, 2017). Among these ligation methods, sortase has received significant attention due to ease of use; the enzyme ligates two simple peptide motifs, the C-terminal LPXT-G and N-terminal oligo-glycine, via transpeptidation (Mao, Hart, Schink, & Pollok, 2004). However, these ligation reactions based on sortase and alike can only occur in the termini of proteins, which is directional and invariably leads to the formation of cyclic products after intramolecular reactions. Besides, these methods often suffer from poor efficiency and slow reaction kinetics (Chen, Dorr, & Liu, 2011). These drawbacks have thus constrained their use in creating more diverse protein topologies.

1.2 SpyTag/SpyCatcher chemistry

Structural studies of the second immunoglobulin-like collagen adhesin domain (CnaB2) from the *S. pyogenes* fibronectin binding protein FbaB revealed an intramolecular isopeptide bond formed between side chains of Lys and Asp (or Asn), which contributes to increased protein stability and resilience (Hagan et al., 2010; Kang & Baker, 2011). Dissection of a CnaB2 domain by Howarth and coworkers yielded a peptide/protein pair, SpyTag (13 a.a.)/SpyCatcher (83 a.a.), that retains the ability to spontaneously form the Lys-Asp isopeptide bond, while in an intermolecular manner. A Glu residue in SpyCatcher activates the carbonyl group of the Asp residue nearby via hydrogen bonding, followed by nucleophilic addition by the Lys residue from SpyTag, thus leading to the formation of the isopeptide bond (Fig. 1) (Li, Fierer, Rapoport, & Howarth, 2014; Zakeri et al., 2012). This reaction has been marked by its efficiency and robustness, under diverse buffer conditions (pH 4–9; temperature 4°C–37°C), with a second-order rate constant $\sim 1.7 \times 10^3 \, M^{-1} s^{-1}$ (Zakeri et al., 2012). Moreover, unlike those traditional protein ligation methods such as sortase, intein, etc. that can only engage the protein termini (Popp, Antos, Grotenbreg, Spooner, & Ploegh, 2007; Shah & Muir, 2014), SpyTag and SpyCatcher can react efficiently almost everywhere in a protein molecule, as long as they are spatially accessible.

SpyTag/SpyCatcher chemistry, while reminiscent of the powerful click chemistry like Cu (I)-catalyzed azide-alkyne cycloaddition in terms of efficiency and selectivity, is unique in that both reactants are made of natural

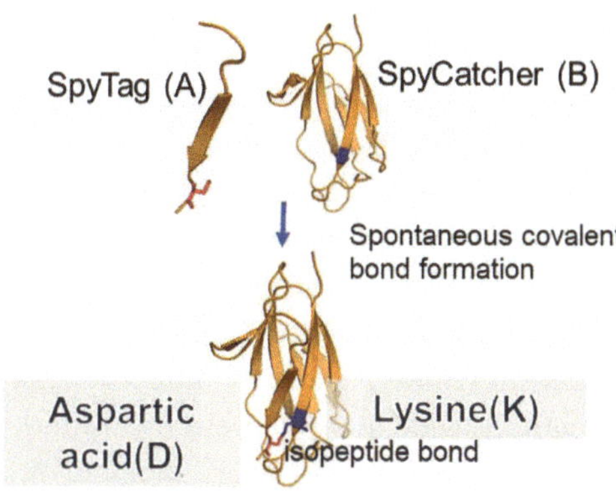

Fig. 1 SpyTag/SpyCatcher chemistry. Reactive aspartic acid and lysine are colored in red and blue, respectively. PDB code: 4MLI.

amino acids and thus are fully genetically encodable. For this reason, the Spy chemistry has been regarded as a prototype of genetically encoded click chemistry (GECC) (Fei Sun, 2017; Sun & Zhang, 2020). Because of such genetic encodability and evolvability, this chemistry has been optimized through the combined use of rational design and directed evolution, leading to further improved reaction kinetics. Using phage display, Howarth and coworkers developed the SpyTag002/SpyCatcher002 pair with a second-order rate constant, $\sim 2 \times 10^4 M^{-1} s^{-1}$, increased by >12 fold, making it applicable for reactions with a wider range of protein expression levels inside living cells (Keeble et al., 2017). A further engineered pair, SpyTag003/SpyCatcher003, has been shown to be able to react at a second-order rate constant of $\sim 5.5 \times 10^5 M^{-1} s^{-1}$, approaching the diffusion limit of similarly sized macromolecules (Keeble et al., 2019). This new pair is marked by its reaction efficiency at low nanomolar concentration and is therefore well-suited for in vivo applications. Using directed evolution, Zhang and coworkers also developed several SpyTag/SpyCatcher variant pairs that exhibited nearly orthogonal reactivities, which further demonstrated the feasibility of using protein engineering to expand this GECC toolbox (Liu et al., 2017). However, in some scenarios, the large size of SpyCatcher (83 a.a.) may impart certain limitations such as perturbation with protein function, increased chance of immunogenicity in vivo and so on. To overcome these limitations, two independent studies showed the creation of two types of protein ligases, SpyLigase and SpyStapler, by splitting SpyCatcher, which enabled the less-trace conjugation of SpyTag with the cognate small peptides, KTag (10 a.a.) and BDTag (28 a.a.), respectively (Fierer, Veggiani, & Howarth, 2014; Wu, Liu, Liu, Sun, & Zhang, 2018).

In this chapter, we focus on the earliest version of SpyTag/SpyCatcher chemistry, because of its wide use in a variety of scenarios–protein topology engineering in particular. One common use of SpyTag/SpyCatcher chemistry is to create cyclic enzymes for enhanced stability. For example, cyclization of β-lactamase and dihydrofolate reductase by placing SpyTag and SpyCatcher at the N- and C-terminals of these enzymes has led to significantly improved thermostability, a better outcome than the cyclization enabled by other methods such as split intein or disulfide bond formation (Iwai & Pluckthun, 1999; Schoene, Fierer, Bennett, & Howarth, 2014; Takahashi et al., 2007; Wang & Zhang, 2017). Similar functional benefits have also been observed from the catenation of enzymes (Wang & Zhang, 2017). Since these cases of enzyme cyclization enabled by Spy chemistry have already been well documented(Schoene, Bennett, & Howarth, 2016), the pertinent topics such as the selection of biocatalysts, the design of constructs and the characterization of cyclic molecules are not covered here. Instead of focusing on enzyme engineering, in this chapter, we illustrate the ways to create diverse nonlinear protein (non-enzyme) architectures and materials through the flexible use of diverse protein assembly strategies, including—but not limited to—SpyTag/SpyCatcher chemistry (Veggiani et al., 2016; Yang et al., 2020). Three recent studies serve as the basis for this chapter. The first study showed the creation of branched proteins (star- and H-shaped proteins) in vitro and cyclic proteins in vivo (Zhang, Sun, Tirrell, & Arnold, 2013). The second study demonstrated the synthesis of protein catenanes in vitro through the combined use of Spy chemistry and p53 dimerization domain (Wang & Zhang, 2016). The third study focused on the design and synthesis of information-rich, bioactive protein networks, also knowns as Spy networks, with no need for chemical modification (Sun, Zhang, Mahdavi, Arnold, & Tirrell, 2014). It is noteworthy that the building blocks in these three studies share great similarity in their biochemical composition, mainly comprising elatin-like polypeptides (ELPs), while differing in the order and numbers of motifs such as SpyTag and SpyCatcher needed for covalent assembly, as well as in the bioactive motifs such as RGD cell-binding ligands, matrix-metalloproteinase cleavage sites and leukemia inhibitory factor (LIF). Such modularity in design and genetic programmability renders protein topology engineering a powerful approach for designing bioactive materials.

2. Construct design

To create nonlinear protein architectures and materials, we would recommend careful consideration of a suitable parent protein to begin with. Several considerations need to be taken into account, including structural flexibility, expression yield and solubility in aqueous buffers. Sufficient structural flexibility ensures the accessibility of SpyTag and SpyCatcher for reactions; in the case of a compact, globally folded protein with a rigid 3D structure, a long, flexible spacer such as an ELP domain may be introduced to link the SpyTag or SpyCatcher domain with the protein of interest. Good expression yield and solubility makes it possible to prepare protein materials in a sufficient quantity for subsequent tests or applications. While being largely described as intrinsic properties to the protein of interest, poor expression yield and solubility could sometimes be overcome by genetically fusing the protein of interest with a highly solution fusion tag such as thioredoxin (Trx), glutathione S-transferase (GST), maltose-binding protein (MBP) or ELP.

Here we choose an engineered ELP as a model system to illustrate the principles of protein topology engineering. The ELP domain, originally derived from connective tissue proteins, often consists of a repetitive pentapeptide, VPGXG, where X can be either Val or Glu. One unique feature of ELP is its temperature-dependent phase transition behavior; the protein is structurally disordered below the lower critical solution temperature (LCST) and transitions into a coacervate phase in aqueous solution as temperature surpasses LCST. Thanks to great efforts by biochemists and protein engineers, a number of ELP compositions have been developed for variable heterologous expression yield, solubility in aqueous solutions and predicable physical properties, which, together with marked biocompatibility, have enabled their wide use as building blocks in biomaterials science and engineering. The particular ELP composition we chose consists of triple repeats of $(VPGVG)_2VPGEG(VPGVG)_2$, a composition marked by its excellent expression yield and solubility, as well as its thermally dependent phase transition behavior that is compatible with biological assays (Sun et al., 2014; Zhang et al., 2013).

To create cyclic proteins, we can design a gene construct, *SpyTag-ELP-ELP-SpyCatcher* (*AB*), with two consecutive *ELP* genes in the middle, flanked by the *SpyTag* and *SpyCatcher* genes (Fig. 2). Moving the *SpyTag* gene to the middle of ELP would lead to a gene construct, *ELP-SpyTag-ELP-SpyCatcher* (*EAEB*), encoding a tadpole-shaped protein. The *Escherichia coli* expression system transformed with these genes would be able to produce the corresponding protein constructs that would undergo intramolecular cyclization inside cells (Fig. 3A). Creation of protein catenanes requires the introduction of an additional motif, p53 (X) dimerization domain, that enables intertwined physical interactions between protein polymers. A simple gene construct, *SpyTag-ELP-p53-ELP-SpyCatcher* (*AEXEB*), can be used to produce recombinant proteins that undergo catenation inside cells—a process that integrates p53-mediated homodimerization and Spy-mediated intramolecular cyclization (Wang & Zhang, 2016; Yan & Dawson, 2001).

In addition to the *in cellulo* synthesis of the proteins with cyclic features, more diverse topological architectures such as branched protein polymers and protein networks can be created through in vitro assembly of purified protein building blocks. We can synthesize 3-arm star proteins by mixing

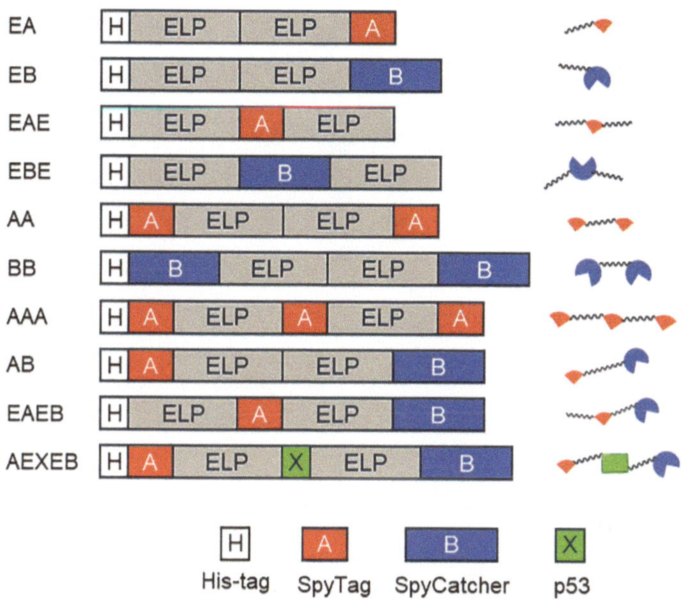

Fig. 2 Genetic constructs used in this protocol. Note: In the original study (Zhang et al., 2013), these constructs, EA, EB, AA and BB, were designed to contain an RGD cell-binding ligand in the middle of ELP. *Reproduced with permission from Wang, X. W., & Zhang, W. B. (2016). Cellular synthesis of protein catenanes.* Angewandte Chemie (International Ed. in English), 55*(10), 3442–3446. doi:10.1002/anie.201511640; Zhang, W. B., Sun, F., Tirrell, D. A., & Arnold, F. H. (2013). Controlling macromolecular topology with genetically encoded SpyTag-SpyCatcher chemistry.* Journal of American Chemical Society, 135*(37), 13988–13997. doi:10.1021/ja4076452.*

the purified recombinant proteins, ELP-SpyTag (EA) and ELP-SpyCatcher-ELP (EBE), or alternatively, ELP-SpyCatcher (EB) and ELP-SpyTag-ELP (EAE), at an equimolar ratio. Similarly, mixing EAE and EBE at a 1:1 molar ratio would give rise to a 4-arm star protein, while the reaction of AA with EBE (2 *eq*), or alternatively SpyCatcher-ELP-ELP-SpyCatcher (BB) with EAE (2 *eq*), would lead to the formation of H-shaped molecules (Fig. 3B).

To synthesize macroscopic molecular networks, we can use telechelic protein polymers such as SpyTag-ELP-SpyTag (AA), SpyTag-ELP-SpyTag-ELP-SpyTag (AAA), SpyCatcher-ELP-SpyCatcher (BB) and/or SpyCatcher-ELP-SpyCatcher-ELP-SpyCatcher (BBB). While mixing AA and BB at an equimolar ratio would only generate linear polymers, the reaction of the following combinations, AAA and BB at a 2:3 molar ratio, AA

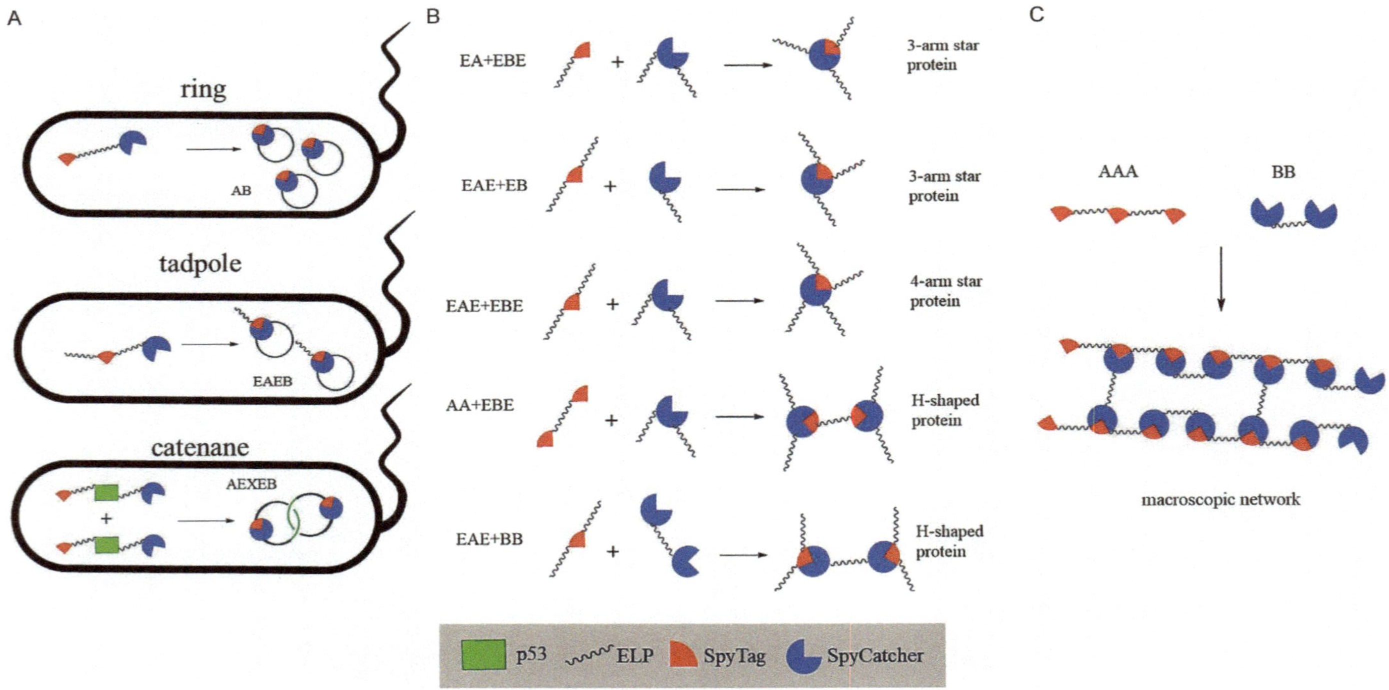

Fig. 3 (A) Cellular synthesis of protein rings, tadpoles, and catenanes. (B) In vitro synthesis of branched protein polymers. (C) In vitro synthesis of covalently crosslinked hydrogel networks. *Reproduced with permission from Sun, F., Zhang, W. B., Mahdavi, A., Arnold, F. H., & Tirrell, D. A. (2014). Synthesis of bioactive protein hydrogels by genetically encoded SpyTag-SpyCatcher chemistry.* Proceedings of the National Academy Sciences of the United States of America, 111*(31), 11269–11274. doi:10.1073/pnas.1401291111; Zhang, W. B., Sun, F., Tirrell, D. A., & Arnold, F. H. (2013). Controlling macromolecular topology with genetically encoded SpyTag-SpyCatcher chemistry.* Journal of American Chemical Society, 135*(37), 13988–13997. doi:10.1021/ja4076452.*

and BBB at a 3:2 molar ratio, and AAA and BBB at a 1:1 molar ratio, would not only lead to linear polymerization, but also interchain crosslinking between these polymers, because of the presence of the middle SpyTag in AAA or SpyCatcher in BBB, which ultimately results in the formation of viscoelastic hydrogel materials (Fig. 3C).

3. Materials and equipment

3.1 Strains and vectors

1. Bacterial strains: *E. coli* DH5α and BL21(DE3).
2. Vector: pQE80L (QIAGEN; Amp^R; an N-terminal His-tag).

3.2 Culture media

1. LB Agar plates with ampicillin
   LB agar consists of 10 g/L NaCl, 5 g/L yeast extract, 10 g/L tryptone, 15 g/L agar and 100 mg/L ampicillin.
2. 2 × YT medium
   The medium was prepared by dissolving 16 g/L tryptone, 10 g/L yeast extract and 5 g/L NaCl in deionized water, and then was autoclaved at 121°C for 20 min.

3.3 Reagents

1. Lysis buffer (50 mM NaH_2PO_4, 300 mM NaCl, 10 mM imidazole, pH 8.0, filtered by 0.22-μm filter before use).
2. Denaturing lysis buffer (100 mM NaH_2PO_4, 10 mM Tris–HCl, 8 M urea, pH 8.0; filtered by 0.22-μm filter before use).
3. Wash buffer (50 mM NaH_2PO_4, 300 mM NaCl, 25 mM imidazole, pH 8.0; filtered by 0.22 μm filter before use).
4. Denaturing wash buffer (100 mM NaH_2PO_4, 10 mM Tris–HCl, 8 M urea, pH 6.3; filtered by 0.22 μm filter before use).
5. Elution buffer (50 mM NaH_2PO_4, 300 mM NaCl, 250 mM imidazole, pH 8.0; filtered by 0.22 μm filter before use).
6. Denaturing elution buffer (100 mM NaH_2PO_4, 10 mM Tris–HCl, 8 M urea, 10 mM imidazole, pH 4.5; filtered by 0.22 μm filter before use).
7. Phosphate-buffered saline (PBS: 137 mM NaCl, 2.7 mM KCl, 10 mM Na_2HPO_4, 2 mM KH_2PO_4, pH 7.5; filtered by 0.22 μm filter before use).

8. Isopropyl-β-1-thiogalactopyranoside (IPTG) solution (1 M in distilled water).
9. Phenylmethylsulfonyl fluoride (PMSF) solution (100 mM in ethanol).
10. Sodium dodecyl sulfate-polyacrylamide gel electrophoresis (SDS-PAGE).
11. Silicone oil.
12. Liquid nitrogen.
13. Deionized water.

3.4 Equipment

1. An ATKA pure FPLC system (GE Healthcare).
2. Ni-NTA columns (GE Healthcare).
3. A Superdex 200 10/300GL column (GE Healthcare).
4. Gel electrophoresis systems (Bio-Rad).
5. A gel imaging/documentation system (Bio-Rad).
6. A probe ultrasonic homogenizer (MSL).
7. Freeze dryers.
8. A TA Instruments ARES-RFS Rheometer (with TA software).
9. A centrifuge with refrigeration (5000 × g, 18,000 × g).
10. A 5-digit analytical balance (Mettler Toledo).
11. Vortex mixers.
12. Autoclaving oven.
13. Incubators.
14. Shakers.
15. Refrigerators (4°C, −20°C and −80°C).
16. A lyophilizer (Labconco).
17. Bruker Ultraflex MALDI TOF/TOF mass spectrometer.

4. Methods

4.1 In vitro synthesis of branched proteins

4.1.1 Construct pQE80L plasmids that contain the synthetic genes encoding EA, EB, EAE, EBE, AA, and BB via standard molecular cloning. The sequence information of these genes have been described in the previously published studies (Zhang et al., 2013).

Note: This step can be performed by researchers themselves, or by outsourcing to contract research organizations such as GENEWIZ, IDT, etc.

4.1.2 Transform the recombinant plasmids into competent *E. coli* BL21 (DE3) strains via heat shock or electroporation.

4.1.3 Plate the culture onto LB agar plates with 100 μg/mL ampicillin. Incubate the plates at 37°C overnight until single colonies can be seen.

4.1.4 Pick a single colony and inoculate in 5 mL of 2 × YT medium containing 100 μg/mL ampicillin in a shaker at 220 rpm, 37°C overnight.

4.1.5 Inoculate the overnight culture (5 mL) into 1 L of 2 × YT medium containing 100 μg/mL ampicillin at 220 rpm, 37°C for 3–4 h until the mid-log phase (OD_{600}, ~0.6–0.8).

4.1.6 Add IPTG (1 mM) to the culture medium to induce protein expression at 37°C for 4 h.

4.1.7 Harvest the cells via centrifugation (5000 g, 15 min) and discard the supernatant. Cell pellets can be stored at −80°C before use.

4.1.8 Resuspend the cells in the lysis buffer supplemented with the serine protease inhibitor PMSF (1 mM), or sometimes in the denaturing lysis buffer containing 8 M urea if the protein possesses no globularly folded region (e.g., EA, EAE, AA or AAA).

4.1.9 Lyse cells using the ultrasonic homogenizer for 20 min at 20% amplitude, 5-s intervals, <10°C.

Note: The exact parameter setting may vary depending on the specifications of the homogenizer.

4.1.10 Spin down cell lysates (18,000 g, 30 min, 4°C) to obtain a clear supernatant.

4.1.11 Load the supernatant to a 5-mL Ni-NTA column at a flow rate below 5 mL/min to ensure sufficient protein binding.

4.1.12 Wash the resin with 5 column-volumes (CVs) of wash buffer (or denaturing wash buffer).

4.1.13 Elute proteins with 3 CVs of elution buffer (or denaturing elution buffer). Collect the flow-through. Assess the protein purity using SDS-PAGE (> 90% purity is expected).

4.1.14 Protein solutions can be dialyzed against distilled water (4 L × 5, 4 h each round) at 4°C to remove salts.

4.1.15 Centrifuge the solution at 5000 g for 30 min to pellet and remove precipitants. Transfer the supernatant to a clean falcon tube. Assess the protein purity using SDS-PAGE again (>95% purity is expected). Lyophilize the resulting solutions into protein powders.

4.1.16 To synthesize branched proteins in vitro, dissolve desirable amounts of protein powders in PBS to make 20 μM solutions. (Note: it is

Table 1 Molar equivalent of each component used for the synthesis of branched protein polymers.

	EA	EB	EAE	EBE	AA	BB
Block protein	1 *eq*	1 *eq*				
3-arm star	1 *eq*			1 *eq*		
3-arm star		1 *eq*	1 *eq*			
4-arm star			1 *eq*	1 *eq*		
H-shaped				2 *eq*	1 *eq*	
H-shaped			2 *eq*			1 *eq*

recommended to use a 5-digit analytical balance to weigh the protein powers for accuracy). Branched proteins can be synthesized by mixing the corresponding protein solutions according to the volumes indicated in Table 1.

4.1.17 After 2-h reaction at room temperature, analyze the resulting mixtures using SDS-PAGE and/or MALDI-TOF (Fig. 4 and Table 2).

4.2 In vivo synthesis of protein rings, tadpoles and catenanes

4.2.1 Construct pQE80L plasmids carrying the synthetic genes encoding AB, EAEB and AEXEB via molecular cloning. These genes have been designed to contain the sequence encoding TEV protease cleavage site (ENLYFQG/S), which can help determine the cyclic protein topology later

Note: This step can be performed by researchers themselves, or by outsourcing to contract research organizations such as GENEWIZ, IDT, etc.

4.2.2 Transform the recombinant plasmids into competent *E. coli* BL21 (DE3) strains via heat shock or electroporation.

4.2.3 Plate the culture onto LB agar plates with 100 μg/mL ampicillin. Incubate the plates at 37°C overnight until single colonies can be seen.

4.2.4 Pick a single colony and inoculate in 5 mL of 2 × YT medium containing 100 μg/mL ampicillin in a shaker at 220 rpm, 37°C overnight.

4.2.5 Cool the culture to a desirable temperature (16°C, 25°C, 30°C or 37°C), followed by addition of a proper amount of IPTG (0–1 mM)

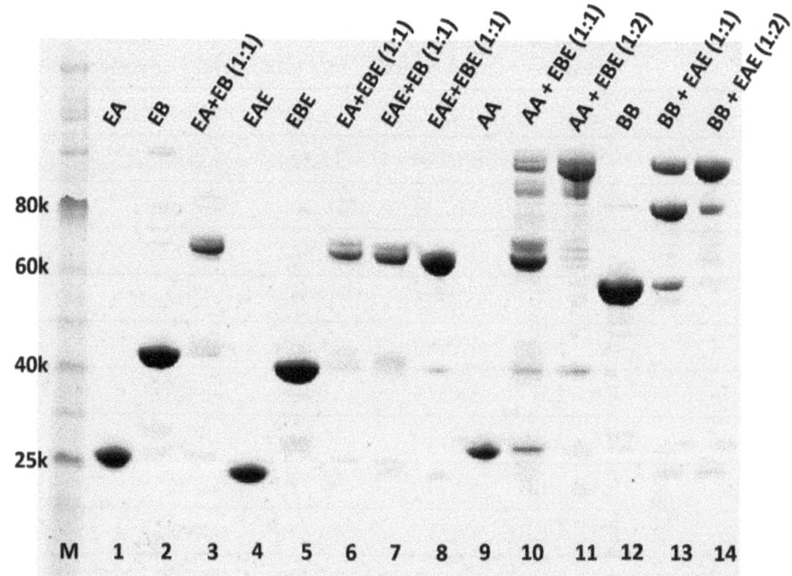

Fig. 4 SDS-PAGE analysis of branched proteins. The reaction between SpyTag and SpyCatcher is highly efficient as evidenced by the fact that scarce reactants are observed on SDS-PAGE. *Reproduced with permission from Zhang, W. B., Sun, F., Tirrell, D. A., & Arnold, F. H. (2013). Controlling macromolecular topology with genetically encoded SpyTag-SpyCatcher chemistry.* Journal of American Chemical Society, 135*(37), 13988–13997. doi:10.1021/ja4076452.*

to induce protein expression. (Note: intramolecular cyclization and intermolecular chain extension are two competing processes, largely depending on protein expression rates (i.e., IPTG concentration and induction temperature) inside cells. The rule of thumb is that a slow protein expression rate at low IPTG concentration and/or low temperature favors intramolecular cyclization).

4.2.6 Harvest the cells via centrifugation (5000 g, 15 min) and discard the supernatant. Cell pellets can be stored at −80°C before use.

4.2.7 Resuspend the cells in the denaturing lysis buffer containing 8 M urea.

4.2.8 Lyse cells using the ultrasonic homogenizer for 20 min at 20% amplitude, 5-s intervals, <10°C.

Note: The exact parameter setting may vary depending on the specifications of the homogenizer.

Table 2 Summary of observed MW (from MALDI-TOF), theoretical MW, and typical yield of linear proteins and branched proteins.

Sample	Construct	MW (calcd.)	*m/z* (obs)	Yield or conversion ratio
EA	–	19,242	19,259	~30 mg/L[a]
EB	–	32,319	32,324	~35 mg/L[b]
EAE	–	18,409	18,405	~25 mg/L[a]
EBE	–	31,486	31,463	~38 mg/L[b]
AA	–	20,925	20,926	~30 mg/L[a]
BB	–	47,080	47,065	~24 mg/L[b]
block	EA + EB	51,543	51,508	~84%[c]
3-arm star	EA + EBE	50,710	50,731	~87%[c]
3-arm star	EAE + EB	50,710	50,719	~84%[c]
4-arm star	EAE + EBE	49,877	49,870	~96%[c]
H-shaped protein	AA + 2EBE	83,862	83,903	~85%[c]
H-shaped protein	BB + 2EAE	83,862	83,865	~81%[c]

[a]Expression yield at 37°C in 2 × YT with 1 mM IPTG induction for 4 h.
[b]Expression yield at 30°C in 2 × YT with 1 mM IPTG induction for 4 h.
[c]Yields are based on densitometry analysis of SDS-PAGE gels.
Reproduced with permission from Zhang, W. B., Sun, F., Tirrell, D. A., & Arnold, F. H. (2013). Controlling macromolecular topology with genetically encoded SpyTag-SpyCatcher chemistry. *Journal of American Chemical Society*, *135*(37), 13988–13997. doi:10.1021/ja4076452, Copyright 2013, American Chemical Society.

4.2.9 Spin down cell lysates (18,000 g, 30 min, 4°C) to obtain a clear supernatant.

4.2.10 Load the supernatant to a 5-mL Ni-NTA column at a flow rate below 5 mL/min to ensure sufficient protein binding.

4.2.11 Wash the resin with 5 column-volumes (CVs) of denaturing wash buffer.

4.2.12 Elute proteins with 3 CVs of denaturing elution buffer. Collect the flow-through, followed by analysis with SDS-PAGE and size-exclusion chromatography (SEC) (Fig. 5).

4.2.13 The resulting samples may contain considerable amounts of by-products. The Superdex 200 increase 10/300GL column, equilibrated with 1 CV of PBS, can be used to further purify the desirable cyclic products.

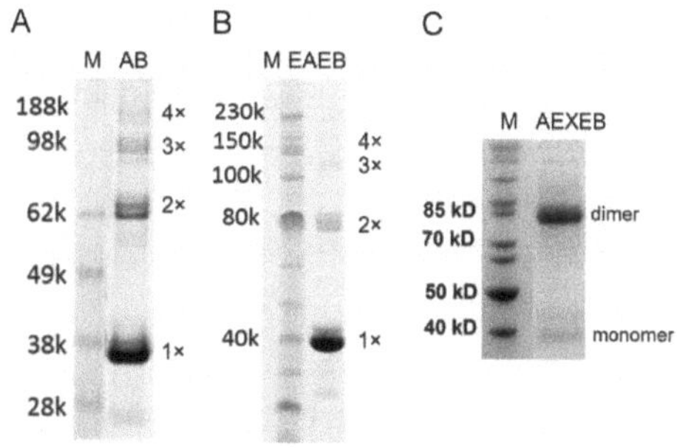

Fig. 5 SDS-PAGE analysis of cyclic (A), tadpole (B) and catenane (C) proteins. *Reproduced with permission from Wang, X. W., & Zhang, W. B. (2016). Cellular synthesis of protein catenanes.* Angewandte Chemie (International Ed. in English), 55*(10), 3442–3446. doi:10.1002/anie.201511640; Zhang, W. B., Sun, F., Tirrell, D. A., & Arnold, F. H. (2013). Controlling macromolecular topology with genetically encoded SpyTag-SpyCatcher chemistry.* Journal of American Chemical Society, 135*(37), 13988–13997. doi:10.1021/ja4076452.*

Table. 3 Summary of observed MW (from MALDI-TOF), theoretical MW, and typical yield of topological proteins.

Sample	MW (calcd.)	m/z (obs)	Yield
AB (ring)	33,984	33,973	~17 mg/L[a]
EAEB (tadpole)	32,154	32,161	~12 mg/L[a]
AEXEB (catenane)	73,442	73,412	~25 mg/L[b]

[a]Expression yield at 16°C in 2 × YT with 0.1 mM IPTG induction for 12 h.
[b]Expression yield at 16°C in 2 × YT with 1 mM IPTG induction for 12 h.
Reproduced with permission from Zhang, W. B., Sun, F., Tirrell, D. A., & Arnold, F. H. (2013). Controlling macromolecular topology with genetically encoded SpyTag-SpyCatcher chemistry. *Journal of American Chemical Society, 135*(37), 13988–13997. doi:10.1021/ja4076452, Copyright 2013, American Chemical Society.

4.2.14 After loading 1 mL of the sample into the column, elute with 1.5 CV of PBS at a flow rate of 0.5 mL/min. Collect the major protein fractions for analysis (Table 3).

Note: Multiple peaks may be observed in this experiment, corresponding to oligomers, dimers, monomers, etc. The exact composition can be determined by SDS-PAGE or MALDI-TOF mass spectrometry, coupled with TEV proteolysis (Wang & Zhang, 2016). For example, cleavage of monomeric cyclic products by TEV only linearizes the molecules and does not alter their molecular weight, while their linear counterparts, after TEV cleavage, break into two fragments.

4.3 In vitro synthesis of macroscopic protein networks

4.3.1 Refer to Steps 4.1.1–4.1.15 with regards to cloning, protein expression and purification of the gelation precursor such as AA, AAA, BB and BBB.

4.3.2 Dissolve protein powders in PBS to yield 10 wt% solutions at 4°C.

Note: It may take 8 h or longer to completely dissolve protein powders at 4°C. Pipetting and vortexing should be avoided, as fierce mechanical force may destabilize the protein molecules.

4.3.3 To prepare a macroscopic hydrogel network comprising AAA + BB (10 wt% polymers), mix 10 μL of AAA (10 wt%) and 33 μL of BB (10 wt%)—the molar ratio of AAA to BB is 2:3—at room temperature using gentle pipetting or stirring, while avoiding creating bubbles. The mixture should gradually become viscous and then turn into a solid hydrogel within several minutes (Fig. 6B).

4.3.4 Assess the gelation kinetics and the viscoelastic properties of the resulting material using dynamic shear rheology at room temperature; a set of 8-mm standard steel parallel plates can be used.

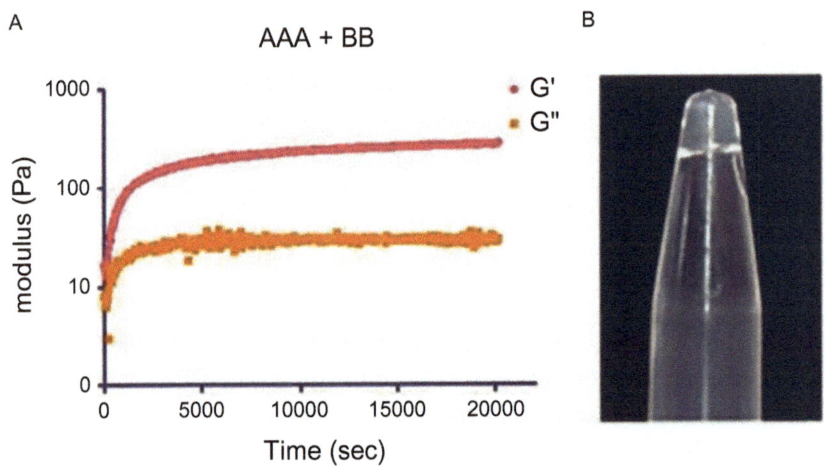

Fig. 6 A Spy network hydrogel comprising the molecular network AAA+BB. (A) Dynamic time-sweep test on the formation of this hydrogel at a fixed strain of 5% and frequency of 1 rad/s. (B) Representative image of a protein hydrogel. *Reproduced with permission from Sun, F., Zhang, W. B., Mahdavi, A., Arnold, F. H., & Tirrell, D. A. (2014). Synthesis of bioactive protein hydrogels by genetically encoded SpyTag-SpyCatcher chemistry.* Proceedings of the National Academy Sciences of the United States of America, 111*(31), 11269–11274. doi:10.1073/pnas.1401291111.*

4.3.5 Mount the sample on the fixed bottom plate and adjust the top plate to cover the sample.

4.3.6 Seal the sample with silicone oil to prevent water evaporation.

Note: Do not adjust parallel plates after addition of silicone oil, because silicone oil may seep through the gap, thus tampering the accuracy of the measurement.

4.3.7 Conduct the rheological measurements in a time-, frequency- or strain-sweep mode. Fix the shear stress and frequency at a proper value or range and then start the measurement.

5. Summary

In this chapter, the methods for creating topological protein structures using SpyTag/SpyCatcher chemistry have been described. These methods serve as examples to illustrate the considerations that may need to be taken when creating unusual protein architectures and materials in vitro or in vivo. These design principles are not limited to SpyTag/SpyCatcher chemistry and can readily be extended to other GECC-based assembly strategies such as SnoopTag/SnoopCatcher chemistry and those variants derived from Spy and Snoop chemistries. One obvious concern over the use of Spy chemistry is the large size of the Catcher domains, which may perturb protein function and may also increase the chance of immunogenicity for in vivo applications. By splitting these Catcher domains, the researchers have successfully developed protein ligases, such as SpyLigase, SpyStapler and SnoopLigase, capable of covalent assembly of small peptides in a less-trace manner while maintaining high efficiency and specificity. These emerging protein chemistries will open up interesting opportunities for protein topology engineering and materials biology.

References

Cascales, L., & Craik, D. J. (2010). Naturally occurring circular proteins: Distribution, biosynthesis and evolution. *Organic & Biomolecular Chemistry*, *8*(22), 5035–5047. https://doi.org/10.1039/c0ob00139b.

Chen, I., Dorr, B. M., & Liu, D. R. (2011). A general strategy for the evolution of bond-forming enzymes using yeast display. *Proceedings of the National Academy of Sciences of the United States of America*, *108*(28), 11399–11404. https://doi.org/10.1073/pnas.1101046108.

Fei Sun, W.-B. Z. (2017). Unleashing chemical power from protein sequence space toward genetically encoded "click" chemistry. *Chinese Chemical Letters*, *28*(11), 2078–2084. https://doi.org/10.1016/j.cclet.2017.08.052.

Fierer, J. O., Veggiani, G., & Howarth, M. (2014). SpyLigase peptide-peptide ligation polymerizes affibodies to enhance magnetic cancer cell capture. *Proceedings of the National Academy of Sciences of the United States of America*, *111*(13), E1176–E1181. https://doi.org/10.1073/pnas.1315776111.

Forgan, R. S., Sauvage, J. P., & Stoddart, J. F. (2011). Chemical topology: Complex molecular knots, links, and entanglements. *Chemical Reviews*, *111*(9), 5434–5464. https://doi.org/10.1021/cr200034u.

Giegerich, R., Voss, B., & Rehmsmeier, M. (2004). Abstract shapes of RNA. *Nucleic Acids Research*, *32*(16), 4843–4851. https://doi.org/10.1093/nar/gkh779.

Hagan, R. M., Bjornsson, R., McMahon, S. A., Schomburg, B., Braithwaite, V., Buhl, M., et al. (2010). NMR spectroscopic and theoretical analysis of a spontaneously formed Lys-Asp isopeptide bond. *Angewandte Chemie (International Ed. in English)*, *49*(45), 8421–8425. https://doi.org/10.1002/anie.201004340.

Iwai, H., & Pluckthun, A. (1999). Circular beta-lactamase: Stability enhancement by cyclizing the backbone. *FEBS Letters*, *459*(2), 166–172. https://doi.org/10.1016/S0014-5793(99)01220-X.

Kang, H. J., & Baker, E. N. (2011). Intramolecular isopeptide bonds: Protein crosslinks built for stress? *Trends in Biochemical Sciences*, *36*(4), 229–237. https://doi.org/10.1016/j.tibs.2010.09.007.

Keeble, A. H., Banerjee, A., Ferla, M. P., Reddington, S. C., Anuar, I., & Howarth, M. (2017). Evolving accelerated amidation by SpyTag/SpyCatcher to analyze membrane dynamics. *Angewandte Chemie (International Ed. in English)*, *56*(52), 16521–16525. https://doi.org/10.1002/anie.201707623.

Keeble, A. H., Turkki, P., Stokes, S., Anuar, I. N. K., Rahikainen, R., Hytönen, V. P., et al. (2019). Approaching infinite affinity through engineering of peptide–protein interaction. *Proceedings of the National Academy of Sciences of the United States of America*, *116*(52), 26523–26533.

Li, L., Fierer, J. O., Rapoport, T. A., & Howarth, M. (2014). Structural analysis and optimization of the covalent association between SpyCatcher and a peptide tag. *Journal of Molecular Biology*, *426*(2), 309–317. https://doi.org/10.1016/j.jmb.2013.10.021.

Liu, Y. J., Liu, D., Yang, W., Wu, X. L., Lai, L. H., & Zhang, W. B. (2017). Tuning SpyTag-SpyCatcher mutant pairs toward orthogonal reactivity encryption. *Chemical Science*, *8*(9), 6577–6582. https://doi.org/10.1039/c7sc02686b.

Mao, H. Y., Hart, S. A., Schink, A., & Pollok, B. A. (2004). Sortase-mediated protein ligation: A new method for protein engineering. *Journal of the American Chemical Society*, *126*(9), 2670–2671. https://doi.org/10.1021/ja039915e.

Mitchell, D. K., & Chambron, J. C. (1995). Chemical topology—The ins and outs of molecular structure. *Journal of Chemical Education*, 72(12), 1059–1064. Retrieved from <Go to ISI>://WOS:A1995TQ49700006.

Popp, M. W., Antos, J. M., Grotenbreg, G. M., Spooner, E., & Ploegh, H. L. (2007). Sortagging: A versatile method for protein labeling. *Nature Chemical Biology*, *3*(11), 707–708. https://doi.org/10.1038/nchembio.2007.31.

Schmidt, M., Toplak, A., Quaedflieg, P. J. L. M., & Nuijens, T. (2017). Enzyme-mediated ligation technologies for peptides and proteins. *Current Opinion in Chemical Biology*, *38*, 1–7. https://doi.org/10.1016/j.cbpa.2017.01.017.

Schoene, C., Bennett, S. P., & Howarth, M. (2016). SpyRings declassified: A blueprint for using Isopeptide-mediated cyclization to enhance enzyme thermal resilience. In *Vol. 580. Methods in Enzymology: Peptide, protein and enzyme design* (pp. 149–167). Academic Press.

Schoene, C., Fierer, J. O., Bennett, S. P., & Howarth, M. (2014). SpyTag/SpyCatcher cyclization confers resilience to boiling on a mesophilic enzyme. *Angewandte Chemie, International Edition*, *53*(24), 6101–6104. https://doi.org/10.1002/anie.201402519.

Shah, N. H., & Muir, T. W. (2014). Inteins: Nature's gift to protein chemists. *Chemical Science*, *5*(2), 446–461. https://doi.org/10.1039/c3sc52951g.

Sun, F., & Zhang, W.-B. (2020). Genetically encoded click chemistry†. *Chinese Journal of Chemistry*, *38*(8), 894–896. https://doi.org/10.1002/cjoc.202000101.

Sun, F., Zhang, W. B., Mahdavi, A., Arnold, F. H., & Tirrell, D. A. (2014). Synthesis of bioactive protein hydrogels by genetically encoded SpyTag-SpyCatcher chemistry. *Proceedings of the National Academy of Sciences of the United States of America*, *111*(31), 11269–11274. https://doi.org/10.1073/pnas.1401291111.

Takahashi, H., Arai, M., Takenawa, T., Sota, H., Xie, Q. H., & Wakura, M. (2007). Stabilization of hyperactive dihydrofolate reductase by cyanocysteine-mediated backbone cyclization. *Journal of Biological Chemistry*, *282*(13), 9420–9429. https://doi.org/10.1074/jbc.M610983200.

Veggiani, G., Nakamura, T., Brenner, M. D., Gayet, R. V., Yan, J., Robinson, C. V., et al. (2016). Programmable polyproteams built using twin peptide superglues. *Proceedings of the National Academy of Sciences of the United States of America*, *113*(5), 1202–1207. https://doi.org/10.1073/pnas.1519214113.

Wang, X. W., & Zhang, W. B. (2016). Cellular synthesis of protein catenanes. *Angewandte Chemie (International Ed. in English)*, *55*(10), 3442–3446. https://doi.org/10.1002/anie.201511640.

Wang, X. W., & Zhang, W. B. (2017). Protein catenation enhances both the stability and activity of folded structural domains. *Angewandte Chemie (International Ed. in English)*, *56*(45), 13985–13989. https://doi.org/10.1002/anie.201705194.

Wang, X. W., & Zhang, W. B. (2018). Chemical topology and complexity of protein architectures. *Trends in Biochemical Sciences*, *43*(10), 806–817. https://doi.org/10.1016/j.tibs.2018.07.001.

Wu, X.-L., Liu, Y., Liu, D., Sun, F., & Zhang, W.-B. (2018). An intrinsically disordered peptide-peptide stapler for highly efficient protein ligation both in vivo and in vitro. *Journal of the American Chemical Society*, *140*(50), 17474–17483.

Yan, L. Z., & Dawson, P. E. (2001). Design and synthesis of a protein catenane. *Angewandte Chemie, International Edition*, *40*(19), 3625–3627. https://doi.org/10.1002/1521-3773(20011001)40:19<3625::Aid-Anie3625>3.0.Co;2-Q.

Yang, Z., Yang, Y., Wang, M., Wang, T., Fok, H. K. F., Jiang, B., et al. (2020). Dynamically tunable, macroscopic molecular networks enabled by cellular synthesis of 4-arm star-like proteins. *Matter*, *2*(1), 233–249.

Zakeri, B., Fierer, J. O., Celik, E., Chittock, E. C., Schwarz-Linek, U., Moy, V. T., et al. (2012). Peptide tag forming a rapid covalent bond to a protein, through engineering a bacterial adhesin. *Proceedings of the National Academy of Sciences of the United States of America*, *109*(12), E690–E697. https://doi.org/10.1073/pnas.1115485109.

Zhang, W. B., Sun, F., Tirrell, D. A., & Arnold, F. H. (2013). Controlling macromolecular topology with genetically encoded SpyTag-SpyCatcher chemistry. *Journal of the American Chemical Society*, *135*(37), 13988–13997. https://doi.org/10.1021/ja4076452.